ADVANCES IN

Applied
Microbiology

VOLUME 31

CONTRIBUTORS TO THIS VOLUME

H. Babich

Daniel K. Brannan

Douglas E. Caldwell

Douglas Gunnison

John E. Herrmann

Carol A. Justice

B. Kristiansen

B. McNeil

Betty H. Olson

Joseph O'Sullivan

William L. Parker

S. Reuveny

Palmer Rogers

G. Stotzky

Richard B. Sykes

R. W. Thoma

N. Robert Ward

Roy L. Wolfe

ADVANCES IN

Applied
Microbiology

Edited by ALLEN I. LASKIN

Somerset, New Jersey

VOLUME 31

 1986

ACADEMIC PRESS, INC.

Harcourt Brace Jovanovich, Publishers

Orlando   San Diego   New York   Austin
London   Montreal   Sydney   Tokyo   Toronto

ACADEMIC PRESS, INC.
Orlando, Florida 32887

United Kingdom Edition published by
ACADEMIC PRESS INC. (LONDON) LTD.
24–28 Oval Road, London NW1 7DX

LIBRARY OF CONGRESS CATALOG CARD NUMBER: 59-13823

ISBN 0–12–002631–7

PRINTED IN THE UNITED STATES OF AMERICA

86 87 88 89   9 8 7 6 5 4 3 2 1

CONTENTS

Apparatus and Methodology for Microcarrier Cell Culture

S. Reuveny and R. W. Thoma

Naturally Occurring Monobactams

William L. Parker, Joseph O'Sullivan, and Richard B. Sykes

New Frontiers in Applied Sediment Microbiology

Douglas Gunnison

Ecology and Metabolism of *Thermothrix thiopara*

Daniel K. Brannan and Douglas E. Caldwell

Enzyme-Linked Immunoassays for the Detection of Microbial Antigens and Their Antibodies

John E. Herrmann

The Identification of Gram-Negative, Nonfermentative Bacteria from Water: Problems and Alternative Approaches to Identification

N. Robert Ward, Roy L. Wolfe, Carol A. Justice, and Betty H. Olson

CONTRIBUTORS

Numbers in parentheses indicate the pages on which the authors' contributions begin.

H. Babich, *Laboratory Animal Research Center, Rockefeller University, New York, New York 10021* (93)

Daniel K. Brannan, *The Procter and Gamble Co., Cincinnati, Ohio 45224* (233)

Douglas E. Caldwell, *Department of Applied Microbiology and Food Science, University of Saskatchewan, Saskatoon, Saskatchewan, Canada S7N 0W0* (233)

Douglas Gunnison, *Environmental Laboratory (WESES-A), USAE Waterways Experiment Station, Vicksburg, Mississippi 39180* (207)

John E. Herrmann, *Division of Infectious Diseases, University of Massachusetts Medical School, Worcester, Massachusetts 01605* (271)

Carol A. Justice, *Program in Social Ecology, University of California, Irvine, Irvine, California 92717* (293)

B. Kristiansen, *Center for Industrial Research, N-0314 Oslo 3, Norway* (61)

B. McNeil, *Department of Bioscience and Biotechnology, Applied Microbiology Division, University of Strathclyde, Glasgow G1 1XW, Scotland* (61)

Betty H. Olson, *Program in Social Ecology, University of California, Irvine, Irvine, California 92717* (293)

Joseph O'Sullivan, *The Squibb Institute for Medical Research, Princeton, New Jersey 08540* (181)

William L. Parker, *The Squibb Institute for Medical Research, Princeton, New Jersey 08540* (181)

S. REUVENY, *Israel Institute for Biological Research, Ness-Ziona 70450, Israel* (139)

PALMER ROGERS, *Department of Microbiology, Medical School, University of Minnesota, Minneapolis, Minnesota 55455* (1)

G. STOTZKY, *Laboratory of Microbial Ecology, Department of Biology, New York University, New York, New York 10003* (93)

RICHARD B. SYKES, *The Squibb Institute for Medical Research, Princeton, New Jersey 08540* (181)

R. W. THOMA, *The New Brunswick Scientific Co., Edison, New Jersey 08818* (139)

N. ROBERT WARD, *BioControl Systems, Kent, Washington 98032* (293)

ROY L. WOLFE, *Metropolitan Water District of Southern California, La Verne, California 91750* (293)

Genetics and Biochemistry of *Clostridium* Relevant to Development of Fermentation Processes

PALMER ROGERS

Department of Microbiology, Medical School
University of Minnesota
Minneapolis, Minnesota

I. Introduction

Over the past 5 years there has been an increase in research and development efforts involving clostridial fermentations, and, concomitantly, a renewal of interest in the biochemistry, physiology, and genetics of these organisms. The reasons for the renewed interest in using the clostridia as biocatalysts for the production of chemical feedstocks from biomass are rather straightforward. First, although we are not now facing an immediate threat of using up the world's petroleum reserves, sometime in the twenty-first century there is the assurance of dwindling nonrenewable resources now serving as major raw materials for fuels and chemicals (Office of Technology Assessment, 1984). As a number of reviewers have pointed out, it is technologically possible to produce essentially all commodity chemicals from renewable biomass feedstocks such as starch or cellulose (Lipinsky, 1981; Ng *et al.*, 1983). Also, from about 1915 until 1952 fermentations employing *Clostridium acetobutylicum* developed an excellent track record, typically producing about 1 ton of the solvents acetone, butanol, and ethanol in a

1

ADVANCES IN APPLIED MICROBIOLOGY, VOLUME 31

90,000-liter vessel with a cane-molasses feedstock (Prescott and Dunn, 1959; Spivey, 1978). These early fermentations carried out in North America (Beesch, 1953) and until 1984 in South Africa demonstrate the feasibility of large-scale commercial chemical production employing the strictly anaerobic clostridia (Ng *et al.*, 1983). In addition, the physiologic and biochemical characterization of a number of thermophilic and now truly cellulolytic species of *Clostridium* (Duong *et al.*, 1983) lend a clear hope that not only starch and sugars but the large potential of cellulosic biomass might be converted to chemicals. In the United States alone about 550 million dry tons of lignocellulose are easily collected and available for fermentation to chemicals each year (Office of Technology Assessment, 1984). In addition, the higher temperature of fermentation increases the efficiency of extraction of the products or of other separation methods used on the fermentation beers (Phillips and Humphrey, 1983). Finally, recent advances in molecular genetics of the clostridia promise that strain development will yield new chimeric organisms and engineered organisms capable of carrying out more specific and higher yielding chemical conversions than are typical of the natural species. Such an optimistic statement must be modified by pointing out that the more we know concerning the biochemistry and physiology of the clostridia the more effective will our genetic engineering of these organisms become (Wood, 1981). A usable genetic system in the clostridia will be a powerful force in nailing down the molecular basis of regulatory devices, morphogenesis, membrane function, and other systems important in the biology of these bacteria (Rabson and Rogers, 1981).

It is the purpose of this review to draw together the recent research adding to our understanding of the molecular nature of this fascinating array of anaerobic spore-forming bacteria, that we classify, often hesitantly, as the clostridia. The thrust here will be to emphasize those research advances in biochemistry and genetics that bear upon future development of fermentation processes employing the clostridia.

A number of excellent reviews have appeared recently that deal with specific aspects of the detailed biochemistry of the clostridia (Ljungdahl and Wood, 1982; Ljungdahl, 1983; Duong *et al.*, 1983; Zeikus, 1980, 1983), and with new developments and considerations in fermentation processes utilizing the clostridia (Linden and Moreira, 1983; Linden *et al.*, 1986; Moreira, 1983; Zeikus *et al.*, 1981). Only sparse but hopeful treatment has been given by reviewers to *Clostridium* genetics (Snedecor and Gomez, 1983; Walker, 1983), perhaps because there wasn't much to write about. In fact, an entire symposium appeared in 1982 about the "Genetic Engineering of Microorganisms for Chemicals" without a single word on the clostridia (Hollaender, 1982).

II. Biochemistry of the Fermentations and Their Regulation

A. THE CLOSTRIDIA AND THEIR GENERAL FERMENTATION STRATEGY

Most of the *Clostridium* species that are currently subjects of research and that are known to carry out fermentations of potential interest for the production of organic solvents and acids are listed in Table I. For clarity they have been divided into six groups named for the major chemical product formed. A number of these interesting species such as *C. thermoautotrophicum* (Wiegel *et al.*, 1981), *C. (strain H10)* (Giallo *et al.*, 1983), *C. cellulovorans* (Sleat *et al.*, 1984), *C. thermohydrosulfuricum* (Wiegel *et al.*, 1979), and *C. tetanomorphum* (Gottwald *et al.*, 1984) were described within the past 5 years. Thus, there is most likely still an untapped pool of uncharacterized organisms with interesting metabolic capabilities in nature worthy of the attention of industrial microbiologists searching for bacteria with specific unique properties. Searching first for natural organisms with new pathways, increased efficiency, or increased tolerance to physical or chemical stress, using new or established enrichment techniques, should perhaps precede a program of genetic engineering. This is particularly true of the clostridia, since our knowledge of the details of their fermentation metabolism and general biochemistry is somewhat limited and derived from only a few studies. Thauer *et al.* (1977) outlined the general mechanisms of energy metabolism by the chemotrophic anaerobic bacteria which include the clostridia. All the clostridia use almost exclusively the fructose biphosphate pathway (Embden–Meyerhof pathway) for conversion of one hexose to two pyruvates with the net production of 2 ATPs and 2 NADHs [Fig. 1, Reaction (1)]. When pentoses are fermented, after formation of pentose 5-phosphate, fructose 6-phosphate and glyceraldehyde 3-phosphate are formed using a combination of the enzymes transaldolase and transketolase [Fig. 1, Reaction (6); Zeikus, 1980]. The phosphorolytic 3–2 cleavage of xylose 5-phosphate to yield acetyl-phosphate and glyceraldehyde 3-phosphate catalyzed by phosphoketolase is not found in the clostridia or in most other obligate anaerobes (Thauer *et al.*, 1977). The intermediates from 3 mol of pentose 5-phosphate are one glyceradehyde 3-phosphate and two fructose 6-phosphates. These enter the fructose-biphosphate pathway and have the capability of producing five ATPs and five NADHs per three pentoses fermented. Most clostridia do not carry the enzymes required for oxidation of glucose 6-phosphate producing NADPH and pentose 5-phosphate (Jungermann *et al.*, 1973). However, there is a report of high glucose-6-phosphate

TABLE I

Substrates and Products of Important Clostridial Fermentations

| | | Major products (mmol/100 mmol hexose) | | | | | | | | | |
| | | Acids | | | | Solvents | | | Gases | | |
Organism	Substrates utilized	Acetate	Butyrate	Other acid	Ethanol	Butanol	Isopropanol	Acetone	CO₂	H₂	References
Acetate fermentations											
C. aceticum	Fructose, ribose	300[a]									Andreesen et al. (1970)
C. formicoaceticum	Hexose, xylose, H_2/CO_2, CO	300									Andreesen et al. (1970)
C. thermoautotrophicum	Hexose, pentose, lactic, H_2/CO_2	280		40[b]							Weigel et al. (1981)
C. thermoaceticum	Hexose, cellobiose, cellulose	172		17[c]	38						Fontaine et al. (1942)
C. (strain H10)									126	396	Giallo et al. (1983)
Propionate fermentation											
C. propionicum	Lactate, alanine	66		132[d]					73		Wood (1961)
Butyrate fermentations											
C. butyricum	Starch, hexose, pentose	42	76						188	235	Wood (1961)
C. thermosaccharolyticum	Hexose, pentose, cellobiose	49	60	26[c]					176	230	Sjolander (1937)
C. pasteurianum	Hexoses	60	70						200	260	Wood (1961)
C. perfringens	Starch, hexoses	34	60	33[c]	26				176	214	Wood (1961)
C. cellulovorans	Cellulose, cellobiose, pectin sucrose, hexose	26	88	67[e]					156	148	Sleat et al. (1984)
Caproic–butyric fermentation											
C. kluyveri	Ethanol–acetate[f]		33	13[g]						22	Thauer et al. (1968)

	Substrates									Reference
Ethanol fermentations										
C. *thermocellum* (strain AS39)	Cellulose, cellobiose, glucose	50	5[c]	104				146	55	Lamed and Zeikus (1980)
C. *thermohydrosulfuricum* (strain 39E)	Starch, hexose, pentose, cellobiose	11	2[c]	194				207	11	Zeikus (1980)
C. *saccharolyticum*	Cellobiose, hexose, pentose	19	4[c]	180				176	56	Murray and Khan (1983b)
Butanol fermentations										
C. *acetobutylicum*	Starch, hexose, pentose, cellobiose	14	6[h]	9	56		22	221	139	Papoutsakis (1984)
C. *beijerinckii*	Starch, hexose, pentose, cellobiose	3	1	12	68	26		220	81	Papoutsakis (1984)
C. *tetanomorphum* (syn. C. *butylicum*) (strain MG1)	Cellobiose, hexose, pentose	23	5	43	47			i	i	Gottwald et al. (1984)
C. *aurantibutyricum*	Starch, hexose pentose, sucrose, cellobiose	i	i	i	42	10	15	i	i	George et al. (1983)
VPI No. 10789		57	49	34	12	5	9	i	i	Prevot et al. (1967)

[a] Batch fermentation after 15 hours incubation.

[b] Formic acid formed after batch fermentation for 30 hours incubation.

[c] Lactic acid formed. Also some formic and/or succinic acid formed in this fermentation.

[d] Propionic acid formed. Also 8 mmol formic acid formed in this fermentation.

[e] Formic acid (51 mmol) and lactic acid (15 mmol) formed.

[f] Ethanol (66.8 mmol) and acetate (35.7 mmol) utilized. Products listed are per 100 mmol C_2 equivalents utilized.

[g] Caproic acid formed.

[h] Acetoin formed.

[i] Not reported.

5

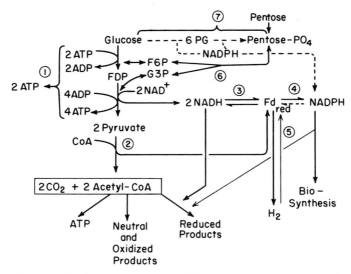

FIG. 1. The clostridial fermentation strategy. The numbers indicate either pathways or enzymes as follows: (1) Fructose biphosphate (Embden–Meyerhof) pathway; (2) pyruvate-ferredoxin oxidoreductase; (3) NADH-ferredoxin oxidoreductase; (4) NADPH-ferredoxin oxidoreductase; (5) hydrogenase; (6) transketolase + transaldolase; and (7) oxidative pentose phosphate pathway.

dehydrogenase and 6-phospho-gluconate dehydrogenase activities which are induced in sporulating cultures of *C. thermosaccharolyticum* (Hsu and Ordal, 1970). The possibility that this pathway [Fig. 1, Reaction (7)] is induced only during sporulation to provide either pentose or perhaps NADPH may have been overlooked in other species. Certainly, on the other hand, the NADPH-ferredoxin oxidoreductases, present in all clostridia investigated (Jungermann *et al.*, 1973; Petitdemange *et al.*, 1976), should provide sufficient NADPH for biosynthesis during both vegetative growth and sporulation. Most of the pyruvate generated from sugars or lactate in the clostridia is cleaved by pyruvate-ferredoxin oxidoreductase yielding CO_2, acetyl-CoA, and reduced ferredoxin (Fd_{red}) as shown in Fig. 1, Reaction (2). Acetyl-CoA is the central intermediate in all clostridial fermentations. Acetyl-CoA is an important source of ATP in most clostridial fermentations and the only source of substrate level phosphorylation in the acetogenic clostridia during unicarbonotrophic growth (Ljungdahl, 1983) and in *C. kluyveri* (Thauer *et al.*, 1968). It is the most important precursor for all solvents and fatty acids synthesized by clostridia, where these pathways serve as a major sink for NADH, which must be reoxidized and recycled in order that the fermentations continue to function (Fig. 1). A second vital element in the clostridial fermentation strategy is the electron distribution system which is shown in its simplest stripped-down form in Fig. 1, Reactions (3–5). Basically, it consists of one or more iron–sulfur proteins, such as ferredoxin, that can accept or donate electrons at a very low potential ($E^{\circ\prime} = -0.41$) near that of the hydrogen electrode; a NADH-ferredoxin oxidoreductase for equilibration of electrons between NADH and Fd; a NADPH-ferredoxin oxidoreductase for controlled production of NADPH required in biosynthesis of protoplasm (Jungermann *et al.*, 1973); and finally, one or more hydrogenases permitting the use of protons as an ultimate electron acceptor or, in some acetogenic clostridia, for using H_2 gas as a source of electrons and perhaps generation of ATP (Ljungdahl, 1983; LeGall *et al.*, 1982).

This electron distribution system operates in apparent close coordination with the branched fermentation pathways of the clostridia which usually yield from two to five major acids and/or solvents (see Table I). The amount of reduced versus neutral and oxidized products is always balanced with the amount of H_2 and also ATP produced, and has the potential of a great deal of natural variation. For example, *C. pasteurianum* produces both butyrate and acetate from acetyl-CoA (Fig. 2). If only acetate were produced, the fermentation would yield an efficient four ATPs per glucose since two ATPs are produced from glucose to pyruvate (Fig. 1) and two more come from

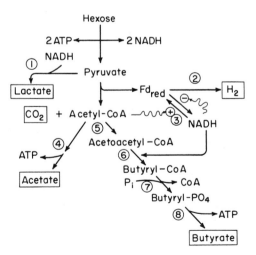

FIG. 2. The butyrate fermentations. The numbers indicate enzymes as follows: (1) Lactate dehydrogenase; (2) hydrogenase; (3) NADH:ferredoxin oxidoreductase; (4) phosphotransacetylase + acetate kinase; (5) acetyl-CoA acetyl transferase; (6) three enzymes + 2 NADH forming butyryl-CoA; (7) phosphotransbutyrylase; (8) butyrate kinase. The wavy arrows indicate allosteric effectors that activate (+) or inhibit (−) enzyme activity.

acetyl-phosphate when it is converted to acetate (Fig. 2). The overall reaction is

$$\text{Glucose} + 4H_2O \rightarrow 2 \text{ acetate} + 2HCO_3^- + 4H^+ + 4H_2$$
$$(\Delta G^\circ = -49.3 \text{ kcal/mol})$$

Since the value for the hydrolysis of ATP is about $\Delta G^\circ = -10.5$ kcal/mol the efficiency of this fermentation would be -42 (output) (kcal/mol)/-49.3 (input) (kcal/mol) = 0.85. If only butyrate were produced then only one additional ATP is produced from butyryl-phosphate [Fig. 2, Reaction (8)] and with the 2 ATPs produced from glucose to pyruvate as for acetate there are three ATPs per glucose.

The overall reaction is

$$\text{Glucose} + 2H_2O \rightarrow 1 \text{ butyrate}^- + 2HCO_3^- + 3H^+ + 2H_2$$
$$(\Delta G^\circ = -60.9 \text{ kcal/mol})$$

and here the efficiency of the fermentation would be

$$-31.5 \text{ (output) (kcal/mol)}/-60.9 \text{ (input) (kcal/mol)} = 0.52.$$

Natural butyrate fermentations vary around 0.6 acetate and 0.7 butyrate per glucose (see *C. pasteurianum* and *C. butyricum*, Table I) with about 3.3 ATPs and an efficiency of about 0.62. In other butyrate fermentations only 0.3 mmol of acetate are produced per mole glucose, with a third or even fourth pathway branch added yielding combinations of lactate, ethanol, and formate (Table I), thus lowering the ATP yield to 3.0 and the efficiency to 0.59 to 0.6. Thauer *et al.* (1977) propose a thermodynamic explanation for the choice made of how to balance flow through these branches, by these and other clostridia such as *C. kluyveri*. The view is that the synthesis of fatty acids provides the necessary entropy (TdS) to "drive the entire catabolic system"; thus the regeneration of ATP must be adjusted to some optimal ratio, $d(\text{ATP})/TdS$.

The proposed mechanism of coupling of the electron distribution with the double branched pathway in the butyrate fermentations was derived from direct biochemical experiments and is depicted in Fig. 2. It was shown earlier for a number of clostridia that NADH-ferredoxin reductase activity can react in both directions, and that the reaction direction, NADH $\rightarrow$ ferredoxin, requires acetyl-CoA as an obligate activator, and the reaction direction, ferredoxin $\rightarrow$ NAD$^+$, is inhibited by NADH (Jungermann *et al.*, 1973). Thus, adjustments in the electron traffic are probably monitored by the CoA/acetyl-CoA ratio and the NAD$^+$/NADH ratio (Petitdemange *et al.*, 1976). The above mechanism as suggested is certainly a reasonable working model for regulation and coupling of the two systems leading to proton reduction and ATP synthesis or just organic product reduction. However, we are left to demonstrate that the "driving force" theory for retaining such "energy-wasting" pathways as lactate or butyrate formation plays a role. Mutants blocked in one or more branches of the fermentation pathways or mutants with altered regulation of the NADH-ferredoxin oxidoreductase are not yet available for study. In addition, the effects of limiting nutrients or excess products or physical factors on the growth or energy efficiency of these organisms has not been studied. Rather, the emphasis of these physiologic studies, particularly in the case of *C. acetobutylicum*, discussed below, has been directed toward discovering how to maximize solvent production (e.g., Linden *et al.*, 1984).

B. THE HOMOACETOGENIC FERMENTATIONS

Three reviews with details on the physiology and biochemistry of the acetogenic bacteria have appeared recently (Ljungdhal, 1983; Ljungdahl and Wood, 1982; Zeikus, 1983). Here, the aspects of recent research that bear on understanding of this fermentation and the capabilities of this group of organisms for producing chemicals will be reviewed.

The first four organisms listed in Table I, *C. aceticum*, *C. formicoaceticum*, *C. thermoautotrophicum*, and *C. thermoaceticum*, are classified as homoacetogenic bacteria. The attractive feature of these organisms is that quantitative conversion of sugars to 100% acetic acid is theoretically possible, so that 1 glucose $\rightarrow$ 3 acetic acid or 1 xylose $\rightarrow$ 2.5 acetic acid. This quantitative conversion is a consequence of the ability of these organisms to fix carbon dioxide. In fact, these bacteria are among the few species that use carbon dioxide as an electron acceptor forming a third acetic acid. A picture of the mechanism of formation of acetic acid from pyruvate or CO_2 is shown in Fig. 3 and is derived from the extensive work on the enzymology of these organisms. The reduction of CO_2 to the methyl group of acetate apparently occurs via formate through a series of one carbon tetrahydrofolate intermediates to methyl tetrahydrofolate (Fig. 3, Paths 2 and 3). The methyl group, probably first transferred to form a methyl-corrinoid protein, is combined with the carboxyl group of pyruvate to yield acetate (Fig. 3, Reaction 5). This final reaction probably requires a five-component enzyme system (Drake *et al.*, 1981) and the products are two acetic acids. Typical of all the clostridia, a second pyruvate is cleaved to CO_2, Fd_{red}, and acetyl-CoA, which is subsequently converted by phosphotransacetylase and acetate kinase yielding the third acetic acid. In contrast to the butyrate fermentation, here the entire process appears tightly coupled and beginning with hexose or xylose there is little room for variation in the flow-through from this "branched pathway" since each branch of the pathway from pyruvate depends stoichiometrically upon the other for both carbon and electron balance. The ability to utilize

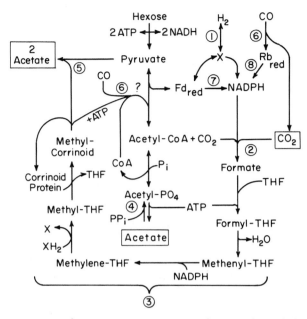

Fig. 3. Homoacetogenic fermentations. A summary of proposed metabolism of hexoses, hydrogen, carbon dioxide, and carbon monoxide to form acetate. The numbers indicate enzymes: (1) Hydrogenase; (2) $NADP^+$–formate dehydrogenase; (3) sequence of reactions reducing formate plus $4e$ to methyl-corrinoid protein requiring five enzymes, one ATP, and tetrehydrofolic acid (THF); (4) acetate kinase–PP_i; acetate kinase; (5) carboxy transferase (pyruvate donates the carboxyl group); (6) carbon monoxide dehydrogenase, rubredoxin (Rb_{red}) is a natural electron acceptor; (7) NADPH:ferrodoxin oxidoreductase; (8) unknown.

sugars is rather limited in the homoacetogenic clostridia. All of them can ferment fructose but only some can, in addition, utilize galactose and glucose. Only *C. thermoaceticum* and *C. thermoautotrophicum* can ferment xylose, and ribose is fermented by *C. aceticum* and *C. formicoaceticum*. *C. formicoaceticum* can also utilize uronic and aldonic acids via a modified Entner–Doudoroff pathway (Ljungdahl, 1983). Usually, these bacteria use the fructose-biphosphate pathway to pyruvate. The oxidation of sugars to pyruvate results in both ATP and NADH. The regeneration of NAD^+ from NADH is carried out indirectly by the reduction of formyltetrahydrofolate to methyltetrahydrofolate (Fig. 3).

The first step in the formation of acetate from the CO_2 released from pyruvate involves a NADP-dependent formate dehydrogenase catalyzing Reaction (2) of Fig. 3. Extensive studies of the nutrition of *C. thermoaceticum* and *C. formicoaceticum* as well as with *Acetobacterium woodii* showed that formation of formate dehydrogenase required either tungstate or molybdate, and selenite, as well as iron salts in the growth medium (see Ljungdahl, 1983). Recently, this extremely O_2-sensitive enzyme has been purified to homogeneity and contains 2 W, 2 Se, and 36 Fe as g-atoms per mole enzyme (Yamamoto *et al.*, 1983). Thus, production of formate, from CO_2 which is essential for acetic acid generation, and growth on hexoses or pentoses, requires the specific nutrients W, Se, and Fe present in the fermentation medium.

When the homoacetogens grow on hexoses, then, the electrons required for reduction of CO_2 come from NADH generated from the oxidation reactions, glyceraldehyde-3-phosphate dehydrogenase, and from reduced ferredoxin to NADPH via ferredoxin:NADP$^+$ oxidoreductase. Wiegel *et al.* (1981) showed that *C. thermoautotrophicum* grows on carbon dioxide and hydrogen and forms acetate. Clark *et al.* (1982) demonstrated hydrogenase in this bacteria as well as all the enzymes (Fig. 3) necessary to convert CO_2 to acetate. It is now clear, however, that perhaps all of the homoacetogens may use molecular hydrogen or carbon monoxide as electron sources for CO_2 reduction to acetate (Braun *et al.*, 1981; Braun and Gottschalk, 1981; Hu *et al.*, 1982). Hydrogenase has been demonstrated in both *C. aceticum* (Braun and Gottschalk, 1981) and *C. thermoaceticum* (Drake, 1982a) using artificial electron carriers such as methyviologen and benzylviologen. It is not clear as to what are the natural electron acceptors, but recently a large variety of electron transfer proteins have been identified in the acetogens, *C. thermoaceticum* (Ljungdahl and Wood, 1982) and *C. formicoaceticum* (Ragsdale and Ljungdahl, 1984). These include ferredoxins, rubredoxins, a flavodoxin, cytochrome *b*, and menaquinone. Although the exact pathway of electrons is unknown, the overall Reaction (1) does occur:

$$2CO_2 + 4H_2 \rightarrow CH_3COOH + 2H_2O \tag{1}$$

Acetogenic bacteria can also synthesize acetate from one carbon compounds such as carbon monoxide, methanol, and formate (Zeikus, 1983). For example, a reaction with carbon monoxide may be written

$$4CO + 2H_2O \rightarrow CH_3COOH + 2CO_2 \tag{2}$$

This reaction has been observed to occur in both *C. thermoaceticum* (Kerby and Zeikus, 1983) and *C. thermoautotrophicum* (Wiegel, 1982). Also, Martin *et al.* (1983) have shown a carbon monoxide-dependent evolution of hydrogen by *C. thermoaceticum* suggesting the reaction

$$CO + H_2O \rightarrow CO_2 + H_2 \tag{3}$$

Combining Eqs. (2) and (3), an inorganic synthesis of acetate can be visualized as occurring from carbon monoxide and molecular hydrogen according to Reaction (4):

$$2CO + 2H_2 \rightarrow CH_3\,COOH \tag{4}$$

As an example, Ljungdahl suggested that synthetic gas (syngas) produced from coal and water might be converted to acetate using the acetogenic bacteria as biocatalysts. In the future, perhaps, reactors containing immobilized homoacetogenic bacteria may be used to convert a number of one-carbon compounds to acetate.

The enzyme responsible for the first step in carbon monoxide metabolism in acetogenic bacteria is carbon monoxide dehydrogenase, which catalyzes the reaction $CO + H_2O \rightarrow CO_2 + 2e + 2H^+$. This enzyme is also found in methanogenic bacteria, sulfate-reducing bacteria, other members of the genus *Clostridium,* and in some photosynthetic bacteria. Early work showed that formation of this enzyme activity in *C. thermoaceticum* and *C. pasteurianum* required nickel during growth. The enzyme has been purified to homogeneity from both of these bacteria and has been shown to contain nickel (Drake *et al.,* 1980; Drake, 1982b). The purified enzyme from *C. thermoaceticum* is a hexamer ($\alpha3\beta3$) of M_r 440,000 containing per mole 6 Ni, 3 Zn, 33 Fe, and 42 acid-labile S atoms (Ragsdale and Ljungdahl, 1983). Again, as in the case of formate dehydrogenase, the importance of attention to the balance of required metal ions in the growth medium is vital to the fermentations requiring these enzymes. The carbon monoxide dehydrogenase activity is detected using methyl viologen as an electron acceptor. But recent studies predict that the iron protein rubredoxin is the most likely candidate for the natural electron acceptor for the enzyme from *C. thermoaceticum, C. formicoaceticum,* and *A. woodii* (Ragsdale *et al.,* 1983). Reactions (6) and (8) of Fig. 3 depict a possible route for electrons to NADPH and thence to reduction of CO_2 to acetate. Although the NADPH-ferredoxin oxidoreductase is widespread in all species of *Clostridium,* there is still some question as to how electrons from rubredoxin are distributed for reduction of CO_2 or CO to acetate (Ragsdale *et al.,* 1983). In these bacteria, the CO dehydrogenase may be directly involved in the formation of an intermediate in the synthesis of either the methyl or carboxyl group of acetate. Earlier, Hu *et al.* (1982) proposed a C_1 intermediate [HCOOH] produced from carbon monoxide according to Reaction (5) that substitutes for both the formate for methyl group synthesis and for the carboxyl group of acetate.

$$CO + H_2O \xrightarrow[\text{dehydrogenase}]{\text{CO}} [HCOOH] \rightarrow CO_2 + 2H^+ + 2e \tag{5}$$

It was shown experimentally that two protein fractions from *C. thermoaceticum* carried out Reaction (6).

$$\overset{\text{ATP}}{CO + CH_3-THF + CoA \rightarrow CH_3CO\text{-}CoA + THF} \tag{6}$$

Other experiments using extracts showed that pyruvate and CO were interchangeable (Hu *et al.*, 1982). Exactly how CO supplies the carboxyl group to acetate is not known. A Ni–C radical was detected with EPR after formation with carbon monoxide and carbon monoxide dehydrogenase (Ragsdale *et al.*, 1983), which may be the proposed C_1 intermediate.

The proposed production of the CH_3–THF directly from a C_1 intermediate derived from CO probably does not occur. It was not possible to demonstrate any formation of [^{14}C]formate using purified carbon monoxide dehydrogenase coupled with 10-formyl-tetrahydrofolate synthetase and ^{14}C-labeled CO as substrate (Ragsdale *et al.*, 1983). Thus, the total synthesis of acetate from carbon monoxide in these bacteria probably involves production of CO_2, reduction to formate, and the normal pathway to CH_3–THF [see Fig. 3, Reactions (6), (8), (2), and (3)]. The carboxyl group is then added probably by combination of the methyl-corrinoid protein, ATP, CoA and the "C_1 intermediate" drived from CO and carbon monoxide dehydrogenase to form acetyl-CoA [see Fig. 3, Reaction (6)]. Recently, synthesis of acetyl-CoA in extracts of *C. thermoaceticum* from H_2, CO_2, CoA, and methyltetrahydrofolate has been reported and suggest the C_1 intermediate may be formed from CO_2 and Fd_{red} (Pezacka and Wood, 1983, 1984).

The recent discoveries that *C. thermoautotrophicum* (Wiegel *et al.*, 1981) and *C. thermoaceticum* (Kerby and Zeikus, 1983) grow on $CO_2 + H_2$ or on carbon monoxide alone as the sole carbon and energy source shows that at least some of these homoacetogenic clostridia are very versatile organisms. *C. thermoautotrophicum* also grows rapidly on methanol and, in fact, will grow slowly even at 6.5% methanol indicating that it is quite tolerant to this solvent (Wiegel, 1982). Unicarbonotrophic growth by species of acetogenic clostridia yields acetate and CO_2. This indicates that ATP must be generated during the course of reduction of CO_2 by molecular hydrogen or utilization of carbon monoxide to synthesize acetate. Exactly where and how the ATP is generated to provide for unicarbonotrophic growth by these anaerobic bacteria is still a puzzle. However, the experiments with growth on hexoses provided the first clue to the existence of extra energy. From Fig. 3 one would predict a net production of only two ATPs per hexose from the level glyceraldehyde-3-phosphate to pyruvate. The one ATP that is formed from the one acetyl-CoA must be used to generate formyl tetrahydrofolate during the homoacetic fermentation. However, growth yields of *C. thermoaceticum* on fructose, glucose, or xylose indicate that as many as five ATPs are produced per hexose during the fermentation (Andreesen and Ljungdahl, 1973). A second clue is that these bacteria contain a large variety of electron

transport proteins mentioned above that must be essential in shuffling electrons from and to NAD, NADP, flavoproteins, and even molecular hydrogen. For example, cytochrome *b* from *C. thermoaceticum* is reduced by CO
using carbon monoxide dehydrogenase and may be involved in electron
transport phosphorylation (Drake, 1980). In *C. formicoaceticum* it may be
that during cytochrome *b* reduction of fumarate to succinate an ATP is
formed (Dorn *et al.*, 1978). More interesting, however, is the observation
that *C. aceticum* evolves hydrogen during fermentations of fructose,
glucose, and lactate (Braun and Gottschalk, 1981). Also the striking presence
of hydrogenase in all of these bacteria, even when grown under heterotrophic conditions (Kellum and Drake, 1984), together with the generation
of H_2 during hexose fermentation suggests an important role for this enzyme
in energy generation. Peck and collaborators have proposed a H_2 cycling
mechanism for energy generation in *Desulfovibrio* (Odom and Peck, 1984).
This idea applied to acetogens growing with H_2/CO_2 or carbon monoxide or
hexoses is shown in Fig. 4 (Ljungdahl, 1983). Basically, a cytoplasmic hydrogenase can generate H_2 from reduced ferredoxin (Fd_{red}) which when passed
to a membrane-bound hydrogenase is recycled through a set of electron
carriers. The protons produced outside the membrane may then generate
ATP via a membrane bound ATPase. The ATP, CO_2, and reduced carriers
form acetyl-CoA and other materials essential for growth. When growing on
added molecular hydrogen and CO_2 the membrane-bound hydrogenase of

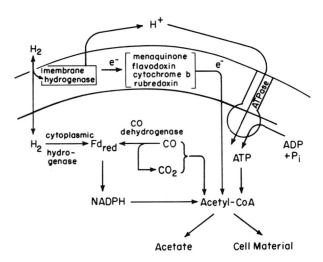

Fɪɢ. 4. Scheme for hydrogen cycling for the generation of ATP by acetogenic bacteria growing with carbon dioxide and molecular hydrogen or with carbon monoxide as sole energy and
carbon sources, as suggested by Ljungdahl (1983).

Clostridium would capture some or all of the hydrogen for production of ATP and reduced electron carriers. This mechanism involving two hydrogenases is reminiscent of that proposed for some of the hydrogen bacteria (Gottschalk, 1979).

A second mechanism for production of ATP is suggested by the discovery by Peck *et al.* (1983) that a pyrophosphate-dependent acetate kinase activity is found in a number of species of *Clostridium* including all the homoacetogenic bacteria tested, species of *Desulfotomaculum*, and a number of other bacteria [see Fig. 3, Reaction (4)]. The reactions yielding energy are

$$O_3POPO_3 + \text{acetate} \rightarrow \text{acetyl} - PO_4 + PO_4 \tag{7}$$

$$\text{Acetyl-}PO_4 + ADP \rightarrow ATP + \text{acetate} \tag{8}$$

It was, in fact, found that the obligate anaerobe *Desulfotomaculum orientis* can grow in the presence of acetate and sulfate with pyrophosphate as a sole source of energy. Although this capacity has not been investigated for all clostridia, the presence of significant pyrophosphate acetate kinase enzyme activity in the acetogenic bacteria may partially explain the high growth yields on hexoses as well as growth on C_1 compounds. The reason for this is that, for example, during biosynthesis of proteins and nucleic acids, there is a significant formation of pyrophosphate (PP_i) by the reaction, $XTP \rightarrow XMP + PP_i$. Assuming that this PP_i can be recycled efficiently, about 0.3–0.5 mol of additional ATP would be available per original mole of ATP produced.

In order to effectively use these organisms in acetate production, it is important that the basic energetics supporting growth and metabolism be understood. The recent development of a defined medium for growth of *C. thermoaceticum* will aid in future research (Lundie and Drake, 1984).

C. THE BUTYRATE AND OTHER ACID FERMENTATIONS

The remaining important acid-forming species of *Clostridium* are listed in Table I. In contrast to the homoacetogenic clostridia and the solvent-producing clostridia, the fermentation pathway of the acid producers have not been subject to detailed biochemical study over the past few years. However, earlier investigations of *C. butyricum*, *C. pasteuranium*, and *C. kluyveri* suggested important regulatory mechanisms controlling the energy production, and electron flow utilizing their branched fermentation pathways. This work has been reviewed and was described above (Thauer *et al.*, 1977). There is a recently described cellulolytic mesophilic *Clostridium* sp. strain H10 that produces mostly acetate and some ethanol and lactate when fermenting low levels (0–1 g/liter) of glucose or cellobiose (Giallo *et al.*, 1983). In contrast to the homoacetogenic bacteria, these organisms produce an enormous amount of hydrogen (See Table I). This strain shows a shift from high acetate levels to high lactate levels when the substrate concentration is increased from 1 to 4 g of cellobiose, with an attendant reduction in growth yield (Giallo *et al.*, 1983). The fermentation pathways of the acetate-lactate-hydrogen strains of the *Clostridium* are illustrated in Fig. 2, omitting the branch of the fermentation leading to butyrate. Thus, in contrast to the butyrate fermentors discussed above where there is apparently a balance between acetate and butyrate production controlled at the acetyl-CoA branch point; in these bacteria the branch point is at the level of pyruvate. Again, the strategy of channeling the fermentation to produce high acetate when only low concentrations of cellobiose are available appears to have adaptive value. In this circumstance the extra ATP produced from acetyl-phosphate increases the energy efficiency of the fermentation. Availability of high levels of cellobiose yields a fermentation shifted toward lactate (0.6 mol/mol hexose), lowering H_2 production and energy efficiency but ensuring a high rate of substrate flow-through. It is interesting that the level of ethanol production (0.13–0.21 mol/mol hexose) does not change much in these different conditions, indicating noninvolvement of this minor branch in the regulatory matrix. The biochemical mechanisms of these regulatory events is unknown, but these sorts of variations in the fermentation are typical of many other clostridia and should be understood prior to using these strains or developing them for industrial fermentations. This strain has the potential of serving as a biocatalyst for conversion of cellulosic biomass to mostly acetic acid and a large amount of hydrogen gas.

The fermentation of α-alanine, β-alanine, or lactate by *C. propionicum* yields primarily propionate and acetate (Table I). The mechanism of propionate production in this organism is quite different from that observed in the *Propionibacteria*. The overall pathway is shown in Fig. 5. Evidence for a direct reductive pathway is that added acrylate is reduced to propionate, [3-14C]lactate is directly incorporated into [3-14C]propionate and is not randomized, and finally, 14CO₂ is not fixed into the carboxyl carbon of propionate (Leaver *et al.*, 1955). Since no H_2 is formed in the fermentation, this branch of the fermentation pathway must account for all electrons produced during oxidation of a substrate (such as lactate) to pyruvate and then to acetyl-CoA. The energy for growth is provided solely through phosphotransacetylase and acetate kinase [Fig. 5, Reaction (6)]. Recently, it was shown that a small amount of acrylate accumulates transiently in the fermentation broth; also, extracts of *C. propionicum* convert acrylyl-pantetheine (Pa) to propionyl-pantetheine in the presence of a reduced dye (Sinskey *et al.*, 1981). When resting cells of *C. propionicum* are exposed to oxygen or to electron acceptors such as methylene blue while fermenting β-alanine, acrylic acid (1.5 mM) accumulates after 2 hours. Thus, the normal pathway for the reduction of acrylate is subverted due to oxidation of the electron

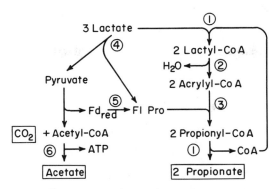

FIG. 5. The propionate fermentation. The numbers indicate enzymes as follows: (1) CoA-transferase; (2) unknown (Lactoyl–CoA-dehydrase); (3) flavoprotein (Fl Pro) dehydrogenase; (4) D-lactate dehydrogenase; (5) flavoprotein–ferredoxin oxidoreductase; (6) phosphotransacetylase + acetate kinase.

carriers, and then acrylate accumulates in the medium (Sinskey *et al.*, 1981). Finally, the same resting cells were found to convert lactate to acrylate in the presence of 3-butynoic acid, an analog of acrylate, and conversely to accumulate lactate from acrylate in the presence of 3-fluoro-pyruvate, an analog of lactate (Akedo *et al.*, 1983). Although the exact enzyme reactions from lactate to propionate have not been clarified, the proposed reaction is Eq. (9):

$$CH_3\text{—}\underset{\underset{OH}{|}}{CH}\text{—}CO\text{—}SCoA \underset{+H_2O}{\overset{-H_2O}{\rightleftharpoons}} CH_2\text{=}CH\text{—}CO\text{—}SCoA \qquad (9)$$

Lactyl-CoA   Acrylyl-CoA

This direction addition or removal of the α-hydroxyl group of lactate by a lactoyl-CoA dehydrase seems thermodynamically unlikely. However, the addition of a β-hydroxyl group has been shown to occur using extracts of *C. kluyveri* (Sinskey *et al.*, 1981) which catalyze Reaction (10):

$$CH_2\text{—}CH_2\text{—}CO\text{—}SCoA \underset{+H_2O}{\overset{-H_2O}{\rightleftharpoons}} CH_2\text{=}CH\text{—}CO\text{—}SCoA \qquad (10)$$
$$\underset{OH}{|}$$

β-OH-propionyl-CoA   Acrylyl-CoA

Recently, [^{3}H]OH elimination from D-[3-^{3}H]lactate was observed using *C. propionicum* extracts (Schweiger and Buckel, 1984). This oxygen-sensitive enzyme activity required coenzyme-A and acetyl-PO$_4$ but no acrylate or acrylyl-CoA intermediate to propionate formation could be isolated. Thus the reaction proposed for propionate formation from acrylate remains to be elucidated. Anaerobic fermentation of resting cells of *C. propionicum* on acrylic acid yields not only propionate but an equal amount of lactate and also acetate (Sinskey *et al.*, 1981). These preliminary studies are important since they illustrate how chemical subversion of a natural pathway (in this case by oxygen) can yield important organic chemicals otherwise not available. Development of this sort of approach using immobilized resting cells of acid or solvent-producing clostridia may yield a number of important chemicals. As yet the acrylic acid fermentation has not yielded sufficient product to make it the basis of a process for production of acrylate from lactate. Acrylate is of interest industrially since the U.S. produced about 350,000 tons/year in 1982 as a commodity chemical, valued at about $0.58/lb (Office of Technology Assessment, 1984).

The group of organisms listed in Table I under butyrate fermentations have the property of producing butyric acid as a major fermentation product, usually combined with acetic and lactic acid and often some ethanol. There is little new biochemical data on the details of the regulation of the fermentation pathways of the six species listed. *Clostridium thermosaccharolyticum* is of interest for future biochemical studies for two reasons, both involving morphogenetic shifts from the vegetative phase to the presporulation phase. First, this bacterium has been reported to either activate or induce a glucose oxidation pathway including glucose-6-phosphate dehydrogenase and 6-phosphogluconate dehydrogenase during the morphogenetic shift (Hsu and Ordal, 1970). Second, extreme changes from an acidogenic fermentation to a production of ethanol occurs when cell division is interrupted (Landuyt *et al.*, 1983). These peculiar findings deserve further investigation. The relationship of nitrogen fixation to energy generation in *Clostridium pasteurianum* is of interest because of the enormous amount of energy required to convert 1 mol of molecular nitrogen to 2 mol of ammonia. During nitrogen reduction, nitrogenase converts some of the reduced ferredoxin directly to H_2 instead of to ammonia. Also *C. pasteurianum* produces H_2 via hydrogenase (bidirectional hydrogenase) during fermentation (see Fig. 2). But this organism contains a second H_2-oxidizing hydrogenase (uptake hydrogenase) that converts H_2 back to protons and electrons, which may serve to reduce the net flow of electrons from ferredoxin to H_2 (Chen and Blauchard, 1984). The presence of two hydrogenases in *C. pasteurianum* suggests, that, in addition to conserving electrons, this organism may produce ATP from a proton gradient in a recycling process similar to that suggested for the homoacetogens (Fig. 4). Formate dehydrogenase is found 6 to 10 times higher in *C. pasteurianum* during N_2 fixation. This enzyme was isolated and purified and found to contain 2 mol of Mo, 24 mol of Fe, and 28 mol of acid-labile sulfur per mole of enzyme, (Liu and Mortenson, 1983). The regulation of hydrogen evolution and recycling for energy as connected with molecular nitrogen reduction in the clostridia has not been carefully investigated.

The inclusion of *C. perfringens* in the list of important organisms of the butyrate fermentation group in Table I was not done because this potential pathogen has any direct industrial value. Rather, we shall see how it plays an important role in the development of a genetic system in the clostridia. Also a new mesophilic cellulolytic organism, *C. cellulovorans*, has recently been added to this group of potentially usable bacteria, where cellulosic biomass is to be converted to organic acids (Sleat *et al.*, 1984). This strain together with *C. lochheadi*, isolated from rumen (Hungate, 1957), are the only cellulolytic clostridia known that produce butyrate as a major fermentation product.

C. kluyveri is the only example listed (Table I) among the clostridia that produce the higher chain length organic acids. During the ethanol–acetate fermentation, it was shown some time ago that, in addition to caproate and butyrate, a small amount of molecular hydrogen is also produced (Thauer *et al.*, 1968). This indicated that *C. kluyveri* obtains energy entirely from conversion of acetyl-CoA to acetate while the remainder of the fermentation producing butyrate and caproate balances the electrons arising from oxidation of ethanol to acetyl-CoA (Thauer *et al.*, 1977). Although not investigated, it well may be that a phosphotransbutyrylase and/or transcaproylase and a butyrate and/or caproate kinase play a role in additional ATP production in this fermentation. This organism has recently yielded a selenomethionine-containing thiolase (Sliwkowski and Stadtman, 1983). The global intermediary metabolism of *C. kluyveri* is now under study using nuclear magnetic resonance (NMR) by mixing cells with [1-^{13}C]ethanol and [2-^{13}C]acetate or with $^{31}PO_4$ (Smith and Roberts, 1984). This sort of approach may serve as a model for deriving information on regulation of fermentation pathways in other clostridia, particularly new strains that are to be genetically engineered for specific industrial processes.

D. The Ethanol Fermentations

There is an excellent recent review that deals exclusively with the thermophilic anaerobic cellulolytic bacteria, in which *C. thermocellum* and its close relatives are the main characters (Duong *et al.*, 1983). Two additional reviews of the thermophilic ethanol fermentations (Wiegel, 1980; Zeikus *et al.*, 1981) have appeared, in which work with *C. thermocellum* and *C. thermohydrosulfuricum* was emphasized. Finally, the third organism listed in Table I, *C. saccharolyticum*, was first described by Murray and Khan in 1982. It grows on 20 different sugars and sugar alcohols including cellobiose but does not utilize cellulose or starch and is mesophilic (37°C optimum growth) (Murray and Khan, 1983a). The ethanologenic clostridia convert sugars to pyruvate via the fructose-biphosphate pathway producing two ATPs and two NADHs per mole of hexose. Most of the pyruvate is converted to acetyl-CoA, reduced ferredoxin (Fd_{red}), and CO_2 with a minor amount reduced by lactate dehydrogenase to lactate (Fig. 6). The majority of acetyl-CoA is reduced to acetyldehyde and then to ethanol catalyzed by NAD-linked dehydrogenases. Acetate is formed via another branch from acetyl-CoA yielding a stoichiometric amount of ATP as with other clostridial fermentations. *C. thermocellum* normally produces a good deal of acetate in contrast to the other two ethanologens where acetate is a relatively minor component (Table I). Brener and Johnson (1984) studied growth and ethanol production by *C. thermocellum* at a series of cellobiose concentrations. Maximum ethanol was formed when 0.8% cellobiose was added. No mechanism for this effect has been proposed. In both the *C. thermohydrosulfuricum* and *C. saccharolyticum* fermentations, 1.8–1.9 mol of ethanol are produced per mole of hexose, classing them as excellent ethanol producers (see Table I).

The biochemical basis for the different reduced end product ratios in the three different strains that possess essentially the same glycolytic pathways and the same branched pathways has been investigated indirectly. Note that all of the ethanologens listed in Table I make excess ethanol and that the levels of molecular hydrogen produced are low. Referring to Fig. 6, these organisms must all transfer a significant quantity of electrons from reduced ferredoxin via ferredoxin:NAD oxidoreductase to NADH which then permits additional ethanol formation beyond that needed to reoxidize the two NADHs formed by oxidation of sugars to pyruvate. The importance of this

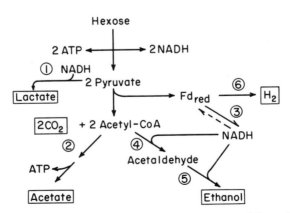

Fɪɢ. 6. Ethanol fermentations. The numbers indicate enzymes as follows: (1) Lactate dehydrogenase; (2) phosphotransacetylase + acetate kinase; (3) NADH:ferredoxin oxidoreductase; (4) acetaldehyde dehydrogenase; (5) ethanol dehydrogenase; (6) hydrogenase.

bypass for electrons during formation of extra ethanol was demonstrated by Lamed and Zeikus (1980). A strain (LQR1) of *C. thermocellum* was shown to conform approximately to Reaction (11) during fermentation of cellobiose:

$$1 \text{ Hexose} \rightarrow 1 \text{ acetate} + 1 \text{ ethanol} + 2H_2 + 2CO_2 + \text{ some lactate} \tag{11}$$

Thus, this strain of *C. thermocellum* produces neither excess ethanol nor low hydrogen as is typical of other strains such as AS39 (see Table I). In contrast to *C. thermocellum*, strain AS39, and *Thermoanaerobium brockii*, crude extracts of strain LQR1 did not have detectable ferredoxin:NAD oxidoreductase activity. Also, in contrast to *C. pasteurianum* and *C. butyricum* (Jungermann, 1973), no NADH-ferredoxin reductase activity was determined for either strain of *C. thermocellum* studied. Apparently, in the ethanologenic clostridia, transfer of electrons is restricted toward the direction $Fd_{red} \rightarrow$ NADH, which is in contrast to that found for the butyrate fermenters where the direction NADH $\rightarrow Fd_{red}$ is favored (Lamed and Zeikus, 1980). The possibility that high partial pressures of H_2 might affect the flow of electrons into ethanol has been tested. It is interesting that growth of *T. brockii* but not *C. thermocellum* is inhibited by molecular hydrogen (Zeikus *et al.*, 1981). There is also a minor effect of H_2 in increasing the yield of ethanol by strain AS39 of *C. thermocellum* but not by strain LQR1 (Lamed and Zeikus, 1980).

Although the addition of exogenous H_2 in the head space did not inhibit the growth of *C. saccharolyticum*, an increase in ethanol production was observed (Murray and Khan, 1983b). Stationary cultures also produced more ethanol and less hydrogen and acetate than shake cultures. In general, it appears that molecular hydrogen does not inhibit growth of the clostridia. The increase in ethanol observed in the ethanol fermentors suggests a partial inhibition of either hydrogenase or ferredoxin-NAD oxidoreductase activities by molecular hydrogen. However, this inhibition is not as extreme as that shown for the H_2 sensitive acidogenic anaerobes (Peck and Odom, 1981). In order to obtain strains of *C. saccharolyticum* with improved conversion of hexose to ethanol, pyruvate-negative mutants were sought following *N*-methyl-*N'*-nitro-*N*-nitrosoguanidine mutagenesis and penicillin enrichment (Murray *et al.*, 1983). Growth of clostridium on pyruvate requires that both phosphotransacetylase and acetate kinase be intact for synthesis of ATP. But pyruvate-negative cells not able to synthesize these enzymes (Fig. 6) should be able to grow on sugars, producing ethanol and perhaps lactate. Three mutants were found that produced 80 to 90% of theroretical yield of ethanol. One strain also showed a tolerance to 6.5% ethanol compared to 3.5% for the parent strain (Murray *et al.*, 1983).

The control of the lactate branch of the fermentation probably involves the intracellular concentration of fructose-biphosphate, since, in cell-free extracts of *C. thermocellum*, lactate dehydrogenase activity is activated by this intermediate (Lamed and Zeikus, 1980). A comparison of this lactate dehydrogenase with those of *C. thermohydrosulfuricum* or *C. saccharolyticum*, both of which produce much less lactate, has not been carried out. It may be that genetic programming of *C. thermocellum* in the future would involve substitution of or alteration of this enzyme. *C. thermohydrosulfuricum*, fermenting xylose in a nitrogen-limited chemostat, was shown to produce more lactate and less ethanol and acetate with increasing dilution rate (Ng and Zeikus, 1982); and *C. saccharolyticum*, fermenting sugars in limiting amounts of yeast extract in the medium, produced relatively more lactate (Murray and Khan, 1983b). The biochemical basis for these observations has not been investigated.

All three of these ethanol-producing clostridia are of great interest as candidates for use in future bioconversion processes. *C. thermocellum* and *C. thermohydrosulfuricum* can grow and ferment at 62–75°C making them attractive, since both whole cells and their enzymes are very stable, allow for increased production rates, and facilitate reactant activity and product recovery (Ljungdahl, 1979). *C. thermocellum* produces a complex of cellulose- and hemicellulose-degrading enzymes, discussed in the next section. Otherwise this organism is rather limited to glucose and cellobiose as substrates, producing a mixture of acetate and ethanol. In contrast, *C. thermohydrosulfuricum* and *C. saccharolyticum* are much more versatile, fermenting a number of hexoses, pentoses, and sugar alcohols to mostly ethanol and CO_2; but they do not degrade cellulose (Zeikus *et al.*, 1981; Murray and Khan, 1983b). A consequence of these properties has been the development of cocultures such as *C. thermocellum–C. thermohydrosulfuricum* (Zeikus *et al.*, 1981), *C. thermocellum–C. thermosaccharolyticum* (Avgerinos, 1982), *C. thermocellum–Thermoanaerobacter ethanoliticus* (Wiegel *et al.*, 1984), *C. strain H10–C. acetobutylicum* (Fond *et al.*, 1984; Petitdemange *et al.*, 1983), *C. saccharolyticum*–unknown cellulolytic anaerobe (Murray and Khan, 1982a). In the first two well-defined cases the coculture produces a substantially higher yield of ethanol from cellulose than the monoculture of *C. thermocellum*. A metabolic explanation for the phenomenon of a stable coculture between *C. thermocellum* and *C. thermohydrosulfuricum* fermenting cellulose has been proposed (Ng and Zeikus, 1982). Glucose and cellobiose are produced from degradation of cellulose by cellulases excreted by *C. thermocellum*. Glucose metabolism and uptake are constitutive in *C. thermohydrosulfuricum*, while both glucose uptake and hexokinase activity are repressed during growth with cellobiose in *C. thermocellum* (Ng and Zeikus, 1982). Cellobiose is consumed via cellobiose phosphorylase in *C. thermocellum* at a 30% slower rate than by *C. thermohydrosulfuricum*, which uses an intracellular cellobiase (Ng and Zeikus, 1982). So the two fermentation products produced by *C. thermocellum* are apparently more rapidly utilized by *C. thermohydrosulfuricum,* which then allows a balance to develop in a coculture of the two organisms. There is also evidence that cellulases of *C. thermocellum* are inhibited by cellobiose (Johnson *et al.,* 1982). Fermentation of cellulose in this balanced coculture results in a much higher yield of ethanol than when *C. thermocellum* ferments cellulose in monoculture. The coculture strategy is depicted in Fig. 7.

It is important to emphasize the significance of these experimental cocultures as precursor models for future industrial scale fermentations. This is best illustrated with reference to Table II where the primary bacterium is the cellulose fermentor (strain B) and the ancillary bacteria are the cellobiose fermentors (strains 3 and 4) (Peck and Odom, 1981). During the time of the experiment the cellulose fermentor did not grow significantly. These cellulolytic strains often grow very slowly and require 4–6 weeks to produce significant amounts of products from cellulose. However, this organism grew well on cellobiose and glucose, producing mostly acetate and H_2. It is only in association with an ancillary organism that growth on cellulose is enhanced. In fact, these associations are often isolated from nature as symbiotic cultures of two organisms (Khan and Murray, 1982b). The key to the association is that the ancillary organism must utilize cellobiose. Note that the cellobiose fermentor, strain 3, produces ethanol and acetate in a 2:1 ratio (Table II). This strain cannot grow on cellulose without strain B. The striking feature of the coculture fermentation is that the distribution of products formed tends to mimic the fermentation of the cellobiose fermentor alone. The same is

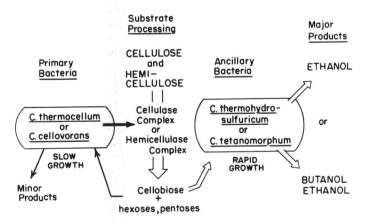

FIG. 7. Coculture strategy for solvent production from cellulose. Two alternative primary bacteria that produce the cellulase and hemicellulase enzyme complexes are shown. Two alternative ancillary bacteria are shown that utilize the major portion of the sugars to produce the major products in the fermentation.

TABLE II

The Effect of the Ancillary Bacteria on the Fermentation Products
of Successful Associations[a]

Bacterial types	Substrate	Products (μmol/ml)			
		H_2	Ethanol	Acetate	Formate
Cellulose fermentor (B)	Cellulose[b]	0	0	0	0
Cellulose fermentor (B)	Glucose	58.5	9.6	54.2	0
Cellobiose fermentor (3)	Cellobiose	12.0	41.0	21.4	10.0
Cellulose (B) plus cellobiose (3) fermentor	Cellulose	29.0	37.8	18.7	1.0
Cellobiose fermentor (4)	Cellobiose	1.0	1.1	68	0
Cellulose (B) plus cellobiose (4) fermentor	Cellulose	3.2	0.4	65.6	0

[a] From Peck and Odom (1981, p. 388).
[b] Significant growth of the cellulose fermentor was not observed during the time of the experiment.

true of a second cellobiose fermentor (strain 4) that produces primarily acetate in monoculture and in coculture with the cellulose fermentor B (Table II).

The reason for the increased rate of cellulose utilization by the cocultures versus the monoculture of the cellulose fermentor strains, appears to be the inhibition of the activity of cellulase complex of clostridium by cellobiose (Peck and Odom, 1981). This inhibition is released by the presence of an efficient cellobiose-utilizing ancillary bacterium, and then cellulose is more rapidly degraded. Clearly, it should be possible to design cellulose fermentations tailored to yield any desired single product under desired conditions as long as the chosen ancillary bacteria can efficiently use cellobiose. For example, in Fig. 7 a coculture with *C. cellovorans* as the primary bacterium and *C. tetanomorphum* as an ancillary bacterium is proposed for production of butanol and ethanol from cellulose.

E. The Butanol Fermentations

Following the renewed interest in developing microbial systems for bioconversion of biomass to commodity chemicals, there has been a flurry of new physiological and biochemical research activity centered upon the butanol-forming saccharolytic clostridia. This revival is largely due to two extrinsic factors. First, there was a successful use of *Clostridium acetobutylicum* strains in commercial production of acetone and butanol in the U.S.

and Canada early in this century. Second, the total production of butanol, acetone, and isopropanol taken together was about 4 billion pounds in the U.S. alone in 1982 (Office of Technology Assessment, 1984). Indeed, there is today a great deal of developmental research activity aimed at butanol, acetone, and isopropanol production, mostly employing the organisms *C. acetobutylicum* and *C. beijerinckii (butylicum)*. New reactor techniques, such as immobilization of cells (Häggström and Enfors, 1982; Häggström and Molin, 1980; Krouwell *et al.*, 1980; Langier *et al.*, 1985), continuous culture (Monot and Engasser, 1983; Monot *et al.*, 1984; Krouwel *et al.*, 1983; Jobses and Roels, 1983) and aqueous two-phase systems (Mattiasson *et al.*, 1981; Griffith *et al.*, 1983; Mattiasson, 1983) are being applied to these fermentations. A variety of natural carbon sources such as hemicellulose hydrolysates (Mes-Hartree and Saddler, 1982), whey filtrates (Maddox, 1980), and wood hydrolysates (Maddox and Murray, 1983) are under study for butanol production. In addition, there is an intrinsic value to the study of these clostridial butanol fermentations. From the standpoint of the research biochemist or microbiologist, a knowledge of the molecular basis of regulation of this relatively complex multibranched catabolic pathway offers a model system for understanding how pathway-switching mechanisms are integrated into the clostridial patterns for vegetative growth, survival under stress, and sporulation. Emerging from this work will be a clearer vision of workable approaches to genetic or physiological programming of these organisms to produce the desired solvents under specified conditions in commercial reactors. There have been a number of recent reviews that concentrate largely upon the production of solvents and the fermentation process with only limited attention to the biochemistry of the butanol fermentation (Gottschalk and Bahl, 1981; Linden *et al.*, 1986; Linden *et al.*, 1984; Moreira, 1983; Walton and Martin, 1979).

There are four species of *Clostridium* classed as butanol fermentors listed in Table I. The classical Weitzmann bacterium, *C. acetobutylicum*, produces a 6:3:1 ratio of the solvents butanol:acetone:ethanol (Prescott and Dunn, 1959). *C. beijerinckii (C. butylicum)* produces the same product ratios, but isopropanol is formed instead of acetone, and *C. aurantibutyricum* produces both acetone and isopropanol together with butanol (George *et al.*, 1983). Finally, a newly described bacterium, *C. tetanomorphum*, produces an almost equimolar amount of ethanol and butanol but no other solvents (Gottwald *et al.*, 1984). All of these strains ferment common hexoses, pentoses, and cellobiose, but *C. tetanomorphum* (MG1) cannot utilize starch (see Table I).

Figure 8 summarizes the proposed pathways leading to the major products of this fermentation. Clearly, there are three major branch points, at acetyl-CoA, acetoacetyl-CoA, and butyryl-CoA. Both *C. acetobutylicum* and *C.*

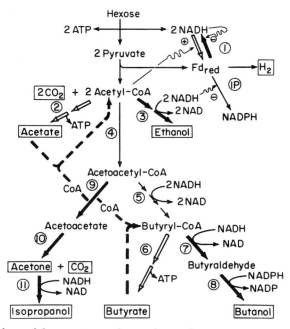

FIG. 8. The butanol fermentations. The numbers indicate enzymes as follows: (1 and 1P) NAD: and NADP:ferredoxin oxidoreductases; (2) phosphotransacetylase + acetate kinase; (3) acetaldehyde dehydrogenase + ethanol dehydrogenase; (4) acetyl-CoA acetyl transferase (or thiolase); (5) three enzymes producing butyryl-CoA (β-hydroxy butyryl-CoA dehydrogenase, + crotonase + butyryl-CoA dehydrogenase); (6) phosphotransbutyrylase + butyrate kinase; (7) butyraldehyde dehydrogenase; (8) butanol dehydrogenase; (9) CoA-transferase; (10) acetoace-tate decarboxylase; (11) isopropanol dehydrogenase. The wavy arrows indicate allosteric ef-fectors that activate (+) or inhibit (−) enzyme activity. The thick, heavy arrows emphasize reactions that predominate during the acidogenic phase of fermentation (⇒) or during the solventogenic phase of the fermentation (→). Proposed recycling of acetate and butyrate during solventogenesis is shown by broken arrows (··▸).

beijerinckii display an interesting shift in their fermentation which was known for some time by the "brewmasters" running the early fermentations (Beesch, 1953; Prescott and Dunn, 1959). In quantitative studies Davies and Stephen-son (1941) first showed that these clostridia produce acetate and butyrate during exponential growth and then switch to butanol, acetone, and ethanol during the deceleration phase of growth in a batch culture. One of our experiments is shown in Fig. 9, where the early accumulation of acids during the acidogenic phase is followed by the solventogenic phase, depicted here by butanol formation and a reutilization of some of the acetate and butyrate. The consequences of this dual fermentation strategy are best illustrated in Table III. During the acidogenic phase, essentially a butyric fermentation obtains.

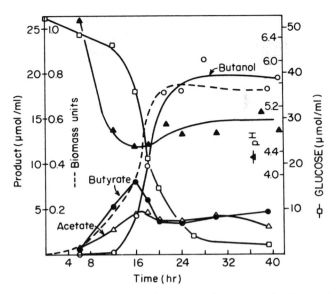

Fig. 9. The switch from acidogenic to solventogenic fermentation by *C. acetobutylicum* in batch culture. *C. acetobutylicum* was grown in a yeast-extract, amino acid medium (Davies and Stephenson, 1941) at 37°C with glucose as an energy source. Samples withdrawn at times shown were analyzed for biomass (- - -), glucose, (□), pH (▲), and products in the broth were determined using gas chromatography: acetate (△), butyrate (●), and butanol (○). Acetone and ethanol determinations were omitted.

Per mole of hexose, 2 mol of CO_2 and 2.5 mol of molecular hydrogen are produced; the excess molecular hydrogen is balanced by the amount of acetate made. This phase is estimated to yield about 3.25 mol of ATP per mole of glucose. In contrast, the solventogenic phase of the fermentation always yields less than 2 mol of molecular hydrogen and, in this case, 2.35 mol of CO_2; the excess CO_2 is balanced by the amount of acetone made. In Table III the imaginary balances are included so that the consequences of running each branch of the fermentation on electron flow and ATP production can be seen. The heavy open arrows and heavy black arrows in Fig. 8 depict the branches of this catabolic network that are believed to be open during the acidogenic and solventogenic phases, respectively. The pathway of butyrate production from butyryl-CoA has been known for some time to require two enzymes, phosphotransbutyrylase and butyrate kinase, identified in a number of butyrate-forming clostridia to be proteins separate from, but analogous to, the similar enzymes for acetate formation from acetyl-CoA in the same bacteria (Twarog and Wolfe, 1962; Valentine and Wolfe, 1960). Since both butyrate kinase and acetate kinase produce extra ATP during the acidogenic phase of the fermentation, about 3.25 mol ATP/mole glucose are predicted (Table III), while only

TABLE III

Clostridium acetobutylicum BALANCES[a]

Products	Fermentation phase[b]		Imaginary balances					
	Acidogenic	Solventogenic	Butyrate	Acetate	Butanol	Acetone	Ethanol	Isopropanol[c]
H_2	2.5	1.4	2	4	0	4	0	3
CO_2	2.0	2.3	2	2	2	3	2	3
Acetate	0.5			2				
Butyrate	0.75		1					
Butanol		0.65			1			
Acetone		0.3				1		
Ethanol		0.1					2	
(Isopropanol)[c]								1
ATP/glucose	3.25	2.0	3	4	2	2	2	2

[a] Products formed (mol/1 mol glucose).

[b] The products formed were calculated from yields determined during fermentation of a batch culture grown on a defined medium containing glucose, yeast extract, and amino acids.

[c] Isopropanol is formed by C. beijerinckii in place of acetone.

2 mol ATP/mol glucose are available in the solventogenic fermentation. Experiments with *C. acetobutylicum* growing in glucose-limited chemostats show surprisingly high biomass yields of 40–46 g dry wt/mol glucose suggesting that 3–4 ATPs/glucose were produced (Bahl *et al.*, 1982b). Cell extracts from acid-producing *C. acetobutylicum* contained two- to sixfold higher specific activities of all four terminal enzymes [Fig. 8, Reactions (2) and (6)] catalyzing butyrate and acetate formation compared to specific activities found in extracts of solvent-producing cells (Andersch *et al.*, 1983). In contrast, Hartmanis and Gatenback (1984) observed that butyrate kinase activity [Fig. 8, Reaction (6)] increased only slightly during solvent formation. The specific activities (units/mg protein) of the four enzymes that catalyze the conversion of 2 acetyl-CoA units to butyryl-CoA [Fig. 8, Reactions (4) and (5)] also increased two- to threefold during both phases of the batch fermentation and then dropped during the final 10 hours after growth has stopped (Hartmanis and Gatenbeck, 1984). These authors calculated the *in vivo* metabolic flux in the whole cells for all of the enzymes by measuring the rates of product formation. The conclusion is that *in vitro* enzyme activities are 10–1000 times higher than the rates of flux of product would demand. These preliminary data predict that allosteric modification of some or all of these enzymes by metabolic intermediates may well regulate the pathway to butyrate and acetate.

The terminal enzymes catalyzing solvent production are now under intensive study since their specific activities have recently been found to be from 10- to 70-fold higher in solvent-producing cells than in acid-producing cells. Acetoacetyl-CoA is most likely converted to acetoacetate in *C. acetobutylicum* by a acetoacetyl CoA:coenzyme A transferase that was found recently to utilize either acetate or butyrate as the CoA acceptor (Andersch *et al.*, 1983). The specific activity of this enzyme increases 10-fold in solvent-producing cells. A similar acetoacetyl-CoA:butyrate CoA-transferase was purified earlier from a lysine-fermenting *Clostridium SB4* (Barker *et al.*, 1978).The acetoacetate is decarboxylated by acetoacetate decarboxylase to form acetone. This enzyme was studied and purified earlier from *C. acetobutylicum* (Davies, 1943; Westheimer, 1969), and recently it was found that the specific activity rises about 40-fold in solvent-producing cells of *C. acetobutylicum* (Andersch *et al.*, 1983). As shown in Fig. 8, this process may allow conservation of the CoA thioester during acetone synthesis and account for the recycling of accumulated acetate and butyrate to form solvents in batch cultures during the solvent-producing phase. Recently, the pathway for uptake of acetate and butyrate by *C. acetobutylicum* has been shown by ^{13}C NMR investigations to proceed directly through an acyl-CoA intermediate to ethanol and butanol (Hartmanis and Gatenbeck, 1983). Apparently, the key enzyme in the uptake of acids is acetoacetate decarboxylase, which, after induction late in the fermentation, pulls the transferase reaction toward formation of acetoacetate. Based on an analysis of the data from 13 acetone-butanol fermentations in which the uptake of acetate and butyrate and production of acetone were measured (Hartmanis *et al.*, 1984) it was found that

$$[\text{acetate}] + [\text{butyrate}] \text{ utilized} = [\text{acetone}] \text{ produced}$$

These results explain the many observations that added acetate and buty-rate enhance both butanol and acetone formation (Bahl *et al.*, 1982; Gott-schall and Morris, 1981; J. R. Martin *et al.*, 1983; Nakmanovich and Shcheblykina, 1960; Yu and Saddler, 1983). The stoichiometry also permits the hypothesis that the two enzymes of the acetone pathway are produced as a metabolic device of the organism designed to couple removal of potentially toxic acids with less toxic solvent production. In any event, it is clear that in batch fermentation with *C. acetobutylicum*, a high yield of butanol is impos-sible unless coupled to acetone production.

The new pathway from butyryl-CoA to butanol requires two dehydrogen-ases both of which have been recently demonstrated in *C. acetobutylicum* extracts in our laboratory (Rogers and Hansen, 1983). The butyraldehyde dehydrogenase is assayed in the reverse direction at pH 8.5 and is CoA and NADH dependent (legend, Fig. 10). In contrast, the butanol dehydrogenase is an NADPH-dependent enzyme in both *C. acetobutylicum* and *C. bei-jerinckii* (Rogers and Hansen, 1983; George and Chen, 1983). Both enzymes require added sulfhydryl reagents to demonstrate activity. Extracts from solvent-producing cells show a 70- to 90-fold higher specific activity of both of these enzymes than acid-producing cells (Rogers and Hansen, 1983). Figure 10 shows that the increase in butyraldehyde dehydrogenase activity occurs in batch culture just prior to butanol production and then decays rapidly when cells cease biomass increase. Butanol dehydrogenase also shows this behavior. Whether the two dehydrogenase activities producing isopropanol or ethanol [Fig. 8, Reactions (11) and (3)] are the same or differ-ent proteins than those producing butanol has not been determined. The coordinated increase in activity of the sets of enzymes leading to acetone, butanol, and probably ethanol formation suggests a common regulatory sig-nal system. Recently we found that the increase in butyraldehyde dehydro-genase activity during the fermentation shift in batch culture (Fig. 10) re-quires new protein synthesis, since added rifampicin or chloramphenicol immediately halts the increase in enzyme activity (Palosaari and Rogers, unpublished observations). The nature of the rapid deactivation of this en-zyme at the end of the fermentation has not been determined. The regula-

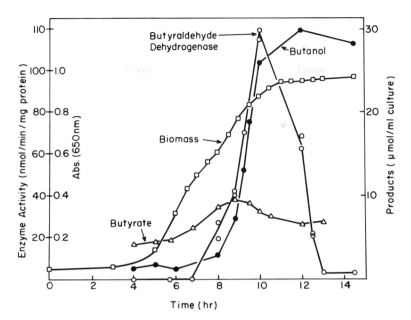

FIG. 10. Induction and decay of CoA-dependent butyraldehyde dehydrogenase of *C. acetobutylicum* during batch fermentation. *C. acetobutylicum* was grown as described for Fig. 9. Samples were analyzed for biomass (□), butyrate (△), and butanol (●). Butyraldehyde dehydrogenase (○) was assayed as described recently (Rogers and Hansen, 1983). In 1.0 ml: glycylglycine, pH 8.5, 50 μmol; NAD, 0.5 μmol; CoA, 0.1 μmol; dithiothreitol, 80 μmol; extract *C. acetobutylicum*, 50–100 μg protein. Preincubate 15 minutes, add butyraldehyde, 45 μmol. Incubate at 25°C and measure NADH at 340 nm.

tion of some of these enzymes, then, must involve induction or derepression of enzyme biosynthesis and the molecularity of this process is unknown.

As pointed out recently by Kim and Zeikus (1985), the electron flow is different in the acidogenic and solventogenic phases of growth of *C. acetobutylicum*. As in other clostridia, the electron distribution system shown in Fig. 8 consists of hydrogenase and the NADH: and NADPH:ferredoxin oxidoreductases. It was shown by Petitdemange *et al.* (1976) that extracts of *C. acetobutylicum* contain a NADH:ferredoxin oxidoreductase that both oxidizes NADH and reduces NAD, the former reaction being strongly activated by acetyl-CoA and the latter reaction being inhibited by NADH very much like the situation in *C. butyricum* (Jungermann *et al.*, 1973). The NADPH: ferredoxin oxidoreductase apparently works only to produce NADPH for biosynthesis. As shown in Fig. 8, there is a high net flow of electrons from both NADH and pyruvate through ferredoxin to molecular hydrogen during acid-

ogenesis, whereas during solvent formation extra electrons flow into the
NADH pool to the more reduced end products such as butanol and ethanol.
This results in a reduction in molecular hydrogen produced during solvent
formation below 2 mol of H_2/mol glucose as is shown in Table III. Hydrogenase activity measured recently in whole cells from acidogenic cultures was
about 2.2 times higher than that measured in solventogenic cells (Kim and
Zeikus, 1985). The assay result was entirely independent of assay pH,
whether pH 5.8 or 4.5, and added acetate or butyrate was without effect (Kim
et al., 1984). It appears, then, that a small regulated reduction in hydrogenase
occurs during growth of cells in a batch culture. Other workers measuring
hydrogenase at pH 8.5 showed no difference in cells from the two phases of
fermentation (Andersch et al., 1983).

How are the carbon flow and electron flow changes regulated and coordinated in these organisms? Davies and Stephenson (1941) as well as other
early workers (Spivey, 1978) pointed out that solvents are formed by C.
acetobutylicum when the pH drops to about pH 4.5 and that low phosphate
promotes solvent production. Using continuous culture, Bahl et al. (1982a)
confirmed that maximum butanol and acetone yields were obtained in two-
stage limiting phosphate chemostats at pH 4.3, operating at low dilution
rates (0.03–0.04 hour^{-1} at 33°C. Continuous culture with sulfate limitation
(Bahl and Gottschal, 1984) or nitrogen source limitation (Monot and Engasser, 1983) were also shown to increase solvent production at low dilution
rates as long as a high $[H^+]$ of pH 5.0 or less was maintained. Glucose
limitation gave poor solvent yields (Bahl et al., 1982). George and Chen
(1983) reported that batch cultures of C. beijerinckii maintained automatical-
ly at either pH 6.8 or 5.0 passed normally from the acidogenic to solventogenic phase producing the same yields of solvents. In contrast, batch
cultures of C. acetobutylicum controlled to pH 5.8 or above do not make
solvents, but cells poised at pH 4.5 appear to pass through an acidogenic
phase prior to producing solvents (Kim and Zeikus, 1985). Thus, acid pH
may be one of the environmental signals permitting the switch to a solvent
fermentation at least in C. acetobutylicum but not in C. beijerinckii. In
addition, a limitation for a nutrient, such a phosphate, provides a strong
signal for solventogenesis. How these signals are translated into molecular
events which adjust specific enzyme activities is still unknown. A model for
control of the entire system has been proposed as emanating from the hydrogenase step (Kim and Zeikus, 1985; Kim et al., 1984). The observation is that
adding 15% carbon monoxide to the headspace reduced whole cell hydrogenase activity while there was a significant increase in butyrate consumption and butanol and acetone production (Datta and Zeikus, 1985). Thus,
since solventogenesis is associated with a decrease in hydrogen production
and hydrogenase activity, it is proposed that a search for the control mecha-

nism at this point in the electron flow system will lead to an understanding of the initiating factors controlling the carbon flow system as well (Kim and Zeikus, 1985).

III. Clostridial Degradation of Polymers

Cellulose degradation using low-temperature biocatalysis is of interest because of the potentially low-cost conversion of major sources of renewable biomass reserves to a high yield of sugars. These sugars can then be further converted into commodity chemicals by subsequent fermentation. Cellulases from a number of bacterial and fungal sources have been purified and partially characterized (Duong *et al.*, 1983). The cellulase system consists of a combination of three general classes of enzymes: 1,4-β-D-glucan-cellobiohydrolases or exoglucanases (EC 3.2.1.91), which specifically cleave cellobiose units from the nonreducing end of cellulose chains; endo-1,4-β-D-glucanases (EC 3.2.1.4), which cleave internal cellulosic bonds; and 1,4-β-D-glucosidases, or cellobiases (EC 3.2.1.21) which specifically cleave glucosyl units from the nonreducing ends of cellooligosaccharides. Often these activities are produced in multiple forms by the same organism; they seem to act synergistically; and at the present time, the mechanism of action of these multienzyme cooperative systems remains unknown. There are a limited number of the clostridia that are able to utilize cellulose and thus produce their own cellulase complex. These include five mesophilic cellulose degraders: *C. cellobioparum* (Chung, 1976), *C. lochheadii* (Hungate, 1957), *C. papyrosolvens* (Madden *et al.*, 1982), *C. cellovorans* (Sleat *et al.*, 1984), and *Clostridium* sp. strain H10 (Giallo *et al.*, 1983). Perhaps more interesting are the two thermophilic cellulolytic organisms, *C. thermocellum* (Duong *et al.*, 1983), and *C. stercorarium* (Madden, 1983). Most of these species either have been isolated from nature only recently or have not been carefully studied, so that the properties of their cellulase complexes are virtually unknown. For example, two strains of *C. acetobutylicum* were shown recently to produce cell-bound and extracellular endogluconase and cellobiase activities during growth on cellobiose (Lee *et al.*, 1985). Curiously, however, these bacteria do not grow on cellulose as a sole source of carbohydrate, although the enzymes from these strains degrade acid-swollen cellulose slowly to glucose. In contrast, the properties of the thermophilic *C. thermocellum* cellulase system have been under intense study over the past 5 years. Much of this work was reviewed recently by Duong *et al.*, (1983). The other heavily studied cellulase complex is secreted by the aerobic fungus, *Trichoderma reesei* (Ladish *et al.*, 1981; Montenecourt *et al.*, 1981). Although cultures of *C. thermocellum* grow faster on native cellulose (cotton, Avicel) than *T. reesei*, standard assays show 100 times less cellulase activity in the bacterial broth than in the fungal growth medium (Zeikus *et al.*, 1981). However, after adjusting the assay buffer, adding Ca^{2+} and dithiothreitol (DTT), the *C. thermocellum* broth, which contains about 0.2 mg/ml protein, had equal activity to *T. reesei* culture filtrates (9 mg/ml protein) (Johnson *et al.*, 1982). The *C. thermocellum* complex works better than the *T. reesei* cellulase complex on cotton and other crystalline cellulose sources (Ng and Zeikus, 1981a). The overall *C. thermocellum* cellulase activity is stable in the absence of substrate at 70 but not at 80°C. However, using carboxymethyl cellulose (CMC) to measure endogluconase activity, Johnson *et al.* (1982) showed this component of cellulase to be much more sensitive to heat, losing 87% of activity at 70°C for 5 hours. This and other reports indicate that *C. thermocellum* makes several endo-1,4-β-gluconases with different properties. Purification of components of the cellulase complex by two different laboratories yielded two different endogluconases. One enzyme, with a M_r of 56,000 obtained after a 100-fold purification, was stable to O_2 and unaffected by sulfhydryl reagents such as dithiothreitol or mercaptoethanol or inhibitors such as iodoacetate or *N*-ethylmaleimide (Petre *et al.*, 1981). The second endogluconase was isolated from broth of *C. thermocellum* grown on cellobiose. It has a M_r of 88,000, has different pI and pH characteristics than the 56,000-M_r enzyme, and also contains no cysteine and is insensitive to O_2 (Ng and Zeikus, 1981b). As with other cellulolytic organisms, there are most likely other endogluconases in the cellulase complex of *C. thermocellum* that have not yet been characterized.

Also involved with the metabolism of cellulose by *C. thermocellum* are two cell-bound or intracellular enzymes, β-glucosidase (cellobiase) and cellobiose-phosphorylase, that metabolize cellobiose, the main product of cellulose digestion. Cellobiose phosphorylase was isolated and purified from *C. thermocellum* some time ago (Duong *et al.*, 1983). It carries out a phosphorolysis of cellobiose by Reaction (12).

$$\text{Cellobiose} + \text{HOPO}_3^{2-} \xrightarrow{\substack{\text{cellobiose} \\ \text{phosphorylase}}} \alpha\text{-D-glucose-1-PO}_4 + \text{glucose} \tag{12}$$

A cellodextrin phosphorylase is also present in this organism which phosphorolyzes β-1-4 oligoglycans to form glucose 1-phosphate similar to Reaction (12). These phosphorolytic cleavages of cellobiose and oligoglycans which yield 1 mol of glucose 1-phosphate per mole substrate utilized are peculiar to the cellulolytic bacteria. *C. thermocellum*, most cellulolytic fungi, such as *T. reesei* (Montenecourt *et al.*, 1981) and other clostridia, such as *C. thermohydrosulfuricum*, that can grow on cellobiose (Ng and Zeikus, 1982), produce a β-glucosidase (cellobiase) which carries out Reaction (13).

$$\text{Cellobiose} + \text{H}_2\text{O} \xrightarrow{\text{β-glucosidase}} 2 \text{ glucose} \tag{13}$$

Here there is no conservation of the bond energy, so that both free glucose molecules must be phosphorylated during transport in order to enter the metabolic pathways. However, following phosphorolysis of 1 mol of cellobiose by *C. thermocellum,* glucose 1-phosphate can enter the cell metabolism via phosphoglucomutase as glucose 6-phosphate (Ng and Zeikus, 1982). Apparently, the second glucose is converted to glucose-6-phosphate at the membrane by a membrane-bound hexokinase. It is interesting that *C. thermocellum* also contains an intracellular β-glucosidase activity, although at a 10-fold lower specific activity than that found intracellularly in cells of *C. thermohydrosulfuricum.* The cell associated β-glucosidase probably plays a minor role *in vivo* in *C. thermocellum* relative to cellobiose phosphorylase because its apparent K_m for cellobiose is 10-fold higher (Ng and Zeikus, 1982). This enzyme has been purified 940-fold and the isolated enzyme is stable at 60°C for 7 hours (Ait *et al.,* 1982).

The third component of the cellulase complex of *C. thermocellum,* the exo-β-gluconase, has not been successfully purified or characterized (Duong *et al.,* 1983). The unusual ability of *C. thermocellum* cellulase complex to successfully cleave highly crystalline cellulose may well involve this component. *C. thermocellum* also degrades xylan (Zeikus *et al.,* 1981) which makes up the backbone of the polymer, hemicellulose. Thus, this bacterium must make a hemicellulase complex of enzymes, which have not been studied to date. In contrast to glucose and cellobiose, many strains of *C. thermocellum* do not readily use xylose or xylobiose, the products of xylan degradation. *C. thermocellum* strains selected or engineered to rapidly utilize these products will be of value for future mixed-polymer fermentations.

The frustration with trying to isolate these various cellulase-complex components from broths containing cellulose, along with the observation that *C. thermocellum* cells as well as other cellulolytic bacteria bind tightly to cellulose substrates early in fermentation, have led recently to a characterization of the nature of cell-binding and cellulase complex-binding to the cellulose substrate. Adherence of *C. thermocellum* to insoluble cellulose substrate is due to an unusual cell-associated cellulose-binding factor (CBF) or factors (Bayer *et al.,* 1983; Lamed *et al.,* 1983). Adherence defective mutant strains have been isolated that lack the cell-associated CBF, but these cells form a modified CBF and excrete it into the broth. After complete hydrolysis of cellulose by a culture of *C. thermocellum* and removal of cells, the CBF could be isolated from the broth by attachment and elution from cellulose. Biochemical characterization of this CBF revealed a particulate multiprotein complex with a M_r of about 2.1×10^6. This fascinating complex appears in electron microscopy to be 18 nm in size and is not broken apart by urea treatment. However, SDS gel electrophoresis yields 14 distinct polypeptides ranging in M_r from 48,000 to 210,000 (Lamed *et al.,* 1984). Using a gel overlay assay with CMC, eight of these polypeptides were found to have cellulolytic activity. Interestingly, only the M_r 210,000 subunit was found antigenically active to immune serum prepared from whole cells of C. thermocellum (Bayeret al., 1983). It appears that the CBF is not only responsible for cell adherence to cellulose but also contains a major portion of this cellulolytic enzyme consortium.

Ljungdahl et al. (1983), using another approach, showed that there is a yellow substance (YS) made by C. thermocellum that attaches to cellulose early in the fermentation. This YS was shown to bind C. thermocellum endogluconase to the cellulose fibers and perhaps it plays a role in the cellulolytic process. The YS is a water-insoluble, acetone-soluble, M_r 1500 material that has a main absorption peak at 460 nm but is bleached to a 375 nm peak in O_2. Wiegel and Dykstra (1984) recently published electron micrographs of C. thermocellum showing adhesion to cellulose and hemicellulose fibers of wood stained with ruthenium red. The pictures and staining suggest multiple thin fibrils containing mucopolysaccharides which connect the C. thermocellum cells to the fibrous substrate. This tight attachment allowed sporulation to occur on the fiber surfaces and permits isolation and purification of these and many other cellulolytic organisms from artificial mixed cultures or from natural sources. The relationships between the CBF, the YS, and the red-staining fibrils in attachment of cells and cellulolytic activity are still unknown. These observations all serve to point out that these clostridial cellulose-degrading complexes differ markedly from the well-studied processes in fungi involving synergistic action between exo- and endogluconases of Trichoderma reesei (Ladish et al., 1981). If naturally organized multienzyme particles such as CBF can be produced and collected it may be that these can serve as enzyme cassettes for a first stage in bioprocessing of crystalline cellulose biomass in the future.

Pectinolytic activity is also important in biodegradation of plant material, since pectins are the glues that hold plant cells together. A number of fungi and bacteria including the clostridia produce the pectinolytic enzymes, polygalacturonate hydrolase (EC 3.2.1.15) and pectin methylesterase (EC 3.1.1.11). Studies of the extracellular pectic enzymes of C. multifermentans show that both polygalacturonate lyase and methyl esterase activities are organized into one functional complex (Lee et al., 1970; MacMillen et al., 1970; Miller et al., 1970). These enzymes have also been reported in broths from C. felsineum and C. roseum cultures (Lund and Brocklehurst, 1978). Recently, both pectinolytic enzymes have been isolated from cultures of a recently described thermophic strain, C. thermosulfurogenes (Schink and Zeikus, 1983a). Both of the pectin-degrading enzyme activities are distributed evenly between the cells and the culture fluid and both are induced by growth on polygalacturonate but not on glucose (Schink and Zeikus, 1983b). These enzymes from C. thermosulfurogenes function at 60°C and are stable at that temperature. This thermostable pectinolytic activity may have application in the food processing industry or in processing other agricultural and forest products, where pectin gums are a problem.

IV. Genetics of *Clostridium*

A. MUTANTS AND MUTAGENESIS

Until very recently, there were virtually no genetic studies with the clostridia. However, Walker (1983) in his review on the genetics of bacterial fermentations, and Snedecor and Gomez (1983) in their review on the genetics of strict anaerobes, predicted that *Clostridium* genetics would be on the move in the 1980s; they were correct.

Table IV lists some of the few mutants of various clostridia reported recently. It appears that *C. thermocellum* is resistant to mutagenesis by N-methyl-N'-nitro-N-nitrosoguanidine (NNG); mutagenesis by UV is only

TABLE IV

MUTANTS OF CLOSTRIDIA

Clostridium species	Mutagenesis method[a]	Mutant characteristics	Reference
C. thermocellum	UV	5-Fluorouracil resistant, rifampicin resistant	Gomez *et al.* (1980)
C. thermocellum	UV, Spon.	Leu⁻, Ade⁻	Mendez and Gomez (1982)
C. thermocellum	Spon.	Ethanol tolerant	Herrero and Gomez (1980)
C. thermocellum	Gamma rays	Low acid producer	Duong *et al.* (1983)
C. thermoaceticum	EMS	Acid tolerant	Schwartz and Keller (1982)
C. acetobutylicum	EMS	Autolysis deficient	Allcock *et al.* (1981)
C. acetobutylicum	Spon.	Butanol tolerant	Lin and Blaschek (1983b)
C. acetobutylicum	EMS	Arg⁻, Tyr⁻, His⁻, Met⁻	Jones *et al.* (1985)
C. acetobutylicum	EMS, Spon.	Streptomycin resistant, rifampicin resistant, allyl-alcohol resistant	Rogers (unpublished)
C. acetobutylicum	EMS	Rifampicin resistant and sporulation negative	Jones *et al.* (1982)
C. saccharolyticum	NNG	Pyruvate negative	Murray *et al.* (1983)
C. perfringens	NNG	λ-Toxin⁻, K-toxin⁻, hemagglutinin⁻	Tatsuki *et al.* (1981)
C. perfringens	NNG	DNase negative	Blaschek and Klacik (1984)

[a] Abbreviations: UV, ultraviolet light; Spon., spontaneous mutation; EMS, ethylmethanesulfonate; NNG, N-methyl-N'-nitro-N-nitrosoguanidine.

modestly effective. Perhaps *C. thermocellum* contains very weak "error-prone-repair" enzymes (Walker, 1983). It is interesting that all reports show that mutagenesis in *C. acetobutylicum* was obtained with ethyl-methanesulfonate (EMS), which does not require error-prone repair processing. In our laboratory we were unable to induce either streptomycin-resistant mutants or rifampicin-resistant mutants with UV, although EMS was an effective mutagen (Table IV). A few mutants with defective fermentation pathways have been obtained. One strategy is to select pyruvate-negative mutants of clostridia that normally can grow on pyruvate as indicated in Section II,D above. Using dye selection (brom-cresol purple) low acid-producing strains of *C. thermocellum* were isolated and found to accumulate an 8:1 ratio of ethanol:acetate during fermentation of cellobiose, instead of the 1:1 ratio found in parental strains (Duong *et al.*, 1983). We have isolated allyl-alcohol-resistant mutants of *C. acetobutylicum*, some of which make very little butanol but produce excess butyrate (Table IV). Allyl alcohol is oxidized by alcohol dehydrogenases to a toxic aldehyde, acrolein. Thus, mutant organisms that synthesize altered forms of butanol and ethanol dehydrogenases will survive selection in the presence of allyl alcohol (Lutsdorf and Megnet, 1968). These sorts of mutants affecting the fermentation pathways of the clostridia will be useful for detailed genetic studies on the mechanisms of regulation of specific branches of a fermentation pathway.

Solvent tolerance may very well be developed and studied by collecting mutants such as those studied in *C. thermocellum* and *C. acetobutylicum* (Table IV), especially when genetic transfer systems for these bacteria are further developed. Similar to the results in *Bacillus subtilis*, selection for rifampicin-resistant mutants included some mutant strains of *C. acetobutylicum* that were deficient in a stage of sporulation (Jones *et al.*, 1982). As is discussed in the next section, some of these strains were also deficient in solvent production.

Simple minimal defined media have now been developed for *C. acetobutylicum* (Long *et al.*, 1983; Monot *et al.*, 1982; O'Brien and Morris, 1971), *C. thermocellum* (Johnson *et al.*, 1981), and *C. thermoaceticum* (Lundie and Drake, 1984). This has allowed isolation of a few auxotrophic mutants (Table IV) which already have proved important in development and study of a protoplast fusion system in *C. acetobutylicum* (Jones *et al.*, 1985).

B. Plasmids

Plasmids are playing an important role in the development of the emerging genetics of the clostridia. Traditional gene transfer systems such as transformation, or uptake of extracellular DNA by cells, and conjugation, requiring DNA exchange during cell–cell contact, are both under study utilizing clostridial plasmids. Also, the development of modern genetic tools such as cloning vectors, expression vectors, and shuttle vectors is dependent upon clostridial plasmids or parts of them. Table V lists most of the plasmids of clostridia described recently. The plasmids of *C. acetobutylicum, C. beijerinckii,* and *C. butyricum* are all as yet cryptic so that they cannot be identified easily by a loss or gain of function such as resistance to an antibiotic. However, the 9.4-MDa plasmid of *C. butylicum* may serve as a vector for gene cloning in clostridia sometime in the future, since it has a single site for digestion by each of the restriction endonucleases, *Eco*RI, *Pst*1, and *Sal*1 (Urano *et al.,* 1983). *C. perfringens* is an important pathogen which causes clostridial myonecrosis (gas gangrene), uterine infections, bacteremia, and food poisoning. Because of the discovery of clinical isolates with multiple antibiotic resistances, attention has been directed to the characterization of the plasmids of *C. perfringens* that have been found to carry and perhaps spread the capability to mount resistance to antibiotics from strain to strain (Brefort *et al.,* 1977). A number of these plasmids carrying antibiotic resistance markers, as well as other characters, such as caseinase activity and bacteriocin production, have been identified, isolated, and sized (Table V). *C. difficile,* another human pathogen, causes a severe illness, antibiotic-associated pseudomembraneous enterocolitis. These organisms contain a number of plasmids of various sizes that have been identified and probably carry antibiotic-resistance markers (Table V). Recently, strains of *C. tetani* that produce the potent neurotoxin, tetanoplasmin, have been shown to carry a large plasmid (49.5 MDa) that has been extensively studied, mapped by restriction endonucleases, and the position of the neurotoxin gene identified (Finn *et al.,* 1984). Isolation of large plasmids as covalently closed circular DNA from some strains such as *C. perfringens* has been variable and irreproducible because of both extracellular and intracellular DNases. Blaschek and Klacik (1984) report that a significant improvement in isolation of intact large plasmids is obtained by adding 0.2% diethyl pyrocarbonate before protoplast disruption.

C. GENETIC TRANSFER IN THE CLOSTRIDIA

1. Conjugation

It appears that natural conjugal transfer of genetic material does occur in the clostridia. However, the reports are very limited; they involve only specific strains of *C. perfringens* and *C. difficile,* and the molecular and physiologic details of these transfers remain to be clarified. Strains of *C.*

TABLE V

Plasmids of the Clostridia

Species	Plasmid characteristics	Reference
C. acetobutylicum	One or two plasmids per strain, 5.2, 6.7, 11, 50 MDa, all cryptic (in 5 out of 17 strains examined)	Truffant and Sebald (1983)
C. beijerinckii (*C. butylicum*)	One plasmid per strain, 2.6 MDa, cryptic	Truffant and Sebald (1983)
C. butyricum	One or two plasmids per strain, 3.9, 4.3, 5.2 MDa, all cryptic (in 3 out of 7 strains examined)	Minton and Morris (1981)
C. butyricum	Strain IF03847, three plasmids, 9.4, 32, 51 MDa, all unknown functions	Urano *et al.* (1983)
C. cochlearium	One plasmid, decomposition of methyl mercury	Pan *et al.* (1980)
C. perfringens		
Strain CNP50	pIP404, 5.5 MDa, bacteriocin	Brefort *et al.* (1977)
Strain CP590	pIP401, 37 MDa, chloramphenicol and tetracycline resistance	
	pIP402, 41 MDa, erythromycin and clindomycin resistance	
	pI403, 6.5 MDa, bacteriocin	
Various strains	One to three plasmids/strain, 1.9 to 74.9 KDa resistance to one or more antibiotics, bacteriocin or cryptic	Li *et al.* (1980); Mihelc *et al.* (1978); Rokos *et al.* (1978); Rood *et al.* (1978a,b)
ATCC 3626B	Three plasmids, 45, 52, and 68 MDa, functions unknown	Blaschek and Klacik (1984)
ATCC 10543	Two plasmids, 9.4 and 30 MDa, functions unknown	
ATCC 3626	pHB101, 2.1 MDa, casinase	Blaschek and Solberg (1981)
	pHB102, 9.4 MDa, cryptic	
Strain 12502	pJU121-pJU123, 2.1, 2.6, 11.2 MDa, cryptic	Squires *et al.* (1984)
	pJU124, 25.6 MDa, tetracycline resistant	
Strain 11268	pCW3, 28.2 MDa, tetracycline resistant	
Cl perfringens–E. coli shuttle vectors	pJU12 and four others, 6.6 to 8 MDa, tetracycline resistance, ampicillin resistance	Squires *et al.* (1984)
C. difficile	One to four plasmids per strain, 2.7 to 60 MDa, some carry resistance to one or more antibiotics	Ionesco (1980); Muldrow *et al.* (1982)
C. tetani (toxogenic strains)	pCL1, 3, 4, and 5, 49.5 MDa, production of tetanus toxin	Finn *et al.* (1984); Laird *et al.* (1980)
C. botulinum	Cryptic plasmids, one or more per strain, 2.1 to 81 MDa (in 38 of 68 strains)	Scott and Duncan (1978); Strom *et al.* (1984)

perfringens have been isolated that are multiply resistant to clindamycin (Cl), chloramphenicol (CM), erythromycin (Em), and tetracycline (Tc). These strains were found to carry two different plasmids, pIP401 and pIP402, which, from curing experiments, were shown to carry markers for Tc,Cm-resistance and Em,Cl-resistance, respectively (Table V). *In vitro* mating on plates demonstrated transfer of Tc and Cm plasmid markers together to sensitive strains of *C. perfringens*. In contrast, transfer of Em, Cl plasmid genes did not occur (Brefort *et al.*, 1977). In addition, transfer of a bacteriocinogenic gene carried by plasmid pIP404, harbored by another *C. perfringens* strain, was demonstrated. Brefort *et al.* (1977) reported kinetic data showing transfer of the Tc and Cm markers at a frequency of about 0.2–1.2×10^{-4}/donor/hour using a plating technique for mating, while in liquid mating an insignificant level of 10^{-8}/donor/hour was obtained. Millipore filter matings and experiments designed to demonstrate a role for DNA or bacteriophages appear to rule against transduction or transformation mechanisms and favor cell–cell contact for this presumed plasmid exchange system. In a second report, a conjugation-like mechanism was described which was resistant to DNase, was not mediated by donor filtrates, and required cell–cell contact (Smith *et al.*, 1981). The Tc-resistant transconjugants contained no plasmid DNA although the donor of Tc resistance contained both 5.1- and 22-MDa plasmids. Similarly, another recent report of mixed culture matings of *C. difficile* on filters showed transfer of Tc, Cl, Em, and streptogramin resistance to sensitive strains at low frequencies (0.1 to 5×10^{-7} per donor cell) (Wust and Hardegger, 1983). Again the mechanism of transfer seems to be a conjugation-like phenomenon requiring close cell–cell contact but not any detectable plasmid transfer. In summary, the low level of these transfer events, together with the lack of good evidence involving specific plasmids or other mechanisms in the conjugation process, make this system difficult to use at the moment. Further work on natural conjugation mechanisms in the clostridia should be pursued.

2. Transformation

Development of a DNA transformation technique for *Clostridium* species is apparently a more promising approach to a "user-friendly" genetic transfer system. This technique is an artificial one and is based upon the ability to prepare protoplasts and then permit them to regenerate into vegetative cells. Interposed between preparation and regeneration is a process of protoplast fusion between protoplasts of two strains where the genomes of both cells end up in the same hybrid cell where they are free to recombine. Alternatively, the protoplast may be a recipient of exogenous plasmid or bacteriophage DNA, thus forming an efficient transformation or transfection system. These techniques of transformation and cell fusion requiring protoplast intermediates have been studied extensively for gram-positive organisms such as *Bacillus* and *Streptomyces* and are reviewed by Hopwood (1981).

Three types of wall-less clostridia have been found useful in genetic studies. L-Phase variants of *C. perfringens* can be generated in the presence of penicillin, grown in broths and they form colonies on plates (Kawatomari, 1958). However, these L-forms could not be regenerated into walled rods following transformation (Heffner *et al.*, 1984). Autoplasts, which are protoplasts that form as a result of autolytic activity, can be formed in *C. perfringens* (Heffner *et al.*, 1984), *C. botulinum* (Kawata *et al.*, 1968), *C. acetobutylicum* (Allcock *et al.*, 1981), and "*C. saccharoperbutylacetonium*" (Ogata *et al.*, 1981). Cells of these bacteria undergo autolysis in media or buffer, but form autoplasts rather than lyse when an osmotic stabilizer such as sucrose (0.4 *M*) is added. These autoplasts have been successfully regenerated into walled cells (Heffner *et al.*, 1984; Kawata *et al.*, 1968; Ogata *et al.*, 1981).

Lysozyme treatment of early exponential phase cells yields almost complete protoplasting in *C. acetobutylicum* and *C. pasteurianum*. Regeneration to walled cells following the published procedure of Allcock *et al.* (1982) was about 1.0 to 5% of the original cells used to produce lysozyme-protoplasts for *C. acetobutylicum* in our laboratory. A reversion frequency of 5.8 to 10% for *C. pasteuranium* was obtained after optimizing the growth medium and the regeneration medium (Minton and Morris, 1983).

The present meager reports of genetic transformation and cell fusion are limited to *C. acetobutylicum* and *C. perfringens* "wall-less cells." *C. acetobutylicum* lysozyme-protoplasts were transfected with DNA from a clostridium bacteriophage, CA1, and in the absence of polyethylene glycol (PEG) (Reid *et al.*, 1983). Transfection was sensitive to DNase but no careful study of conditions, and no quantitative data were given. Lin and Blaschek (1983a, 1984) reported transformation of *C. acetobutylicum* protoplasts by DNA from a kanamycin-resistant plasmid, pUB110, from *Staphylococcus*. Transformation to kanamycin resistance (Km^R) required PEG 6000 and a preheating of the protoplasts at 55°C for 15 minutes to inactivate DNase. The plasmid, pUB110, was isolated from the Km^R transformants.

Mixtures of protoplasts from two double auxotroph strains of *C. acetobutylicum* have been induced to fuse using PEG for 3 to 15 minutes. Regenerated colonies gave recombinants and biparental diploid cells at a frequency of 0.3 to 2.6% and 1.4 to 8.5%, respectively (Jones *et al.*, 1985). Electric field-mediated fusion has not been reported for clostridia (Zimmerman, 1983). However, our preliminary experiments with electric field fusion of *C. acetobutylicum* protoplasts show excellent chain formation, and many pairs that remain together following a direct current pulse. This method has the advantage of avoiding the toxic effects of PEG treatment of protoplasts.

Both autoplasts and L-phase variants of *C. perfringens* have been transformed to Tc resistance with the plasmids pCW3 and pUV124 (see Table V), and this transformation event requires PEG (Heffner *et al.*, 1984). Restriction analysis revealed that the transformants contained the same plasmids as those in the transforming DNA. The transforming frequency was low, about 1 to 4 transformants per 10^6 viable cells. Using this transforming system it is possible to test shuttle vectors that replicate both in *E. coli* and in *C. perfringens*. Squires *et al.* (1984) have constructed small shuttle plasmids by recombining *E. coli* plasmid pBR322 with three different small plasmids, pJU121, pJU122, and pJU281, of *C. perfringens*. Then two tetracycline-resistant genes (*tet*) were cloned into the recombinant plasmids from two other *C. perfringens* plasmids. Both *tet* genes made *E. coli* resistant to tetracycline. These shuttle plasmids (see Table V) can be transformed into both bacterial species and are maintained in both by virtue of the antibiotic resistance markers. These plasmids have restriction sites that will make them very useful as cloning vehicles in the future.

D. CLONING AND EXPRESSION OF *Clostridium* GENES IN *E. coli*

The DNA of *C. thermocellum* has been cleaved with restriction endonucleases, ligated into cosmid vectors, and cloned into *E. coli* (Cornet *et al.*, 1983a,b). It is apparent that the *C. thermocellum* genes coding for amino acid biosynthesis and cellulose hydrolysis are expressed in the *E. coli* host (Snedecor and Gomez, 1983). Also, cloned genes for β-isopropyl-malate dehydrogenase and hydrogenase from *C. butyricum* are expressed in *E. coli* (Ishii *et al.*, 1983; Karube *et al.*, 1983).

Because of the importance of cellulases in biomass conversion projects, there has been a great deal of recent research on molecular cloning, expression, and sequencing of the genes from cellulase-producing organisms. Fragments of DNA carrying the endogluconase gene from the thermophilic filamentous bacterium, *Thermomonospora* XY, have been cloned into plasmid pBR322. Restriction maps have been prepared and expression of the *Thermomonospora* cellulase is under study (Collmer and Wilson, 1983). Likewise, the enzymes and major genes of the cellulase complex of the fungus *Trichoderma reesei* are under intensive study. Shoemaker *et al.* (1983a,b) have purified and characterized most of the major enzymes in the cellulase complex of this organism. They have cloned, completely sequenced, and expressed in *E. coli* the exo-cellobiohydrolase I of *Trichoder-*

ma ressei. Terri *et al.* (1983) reported cloning and partial sequencing of the same gene. Cornet *et al.* (1983a) subcloned two structural genes, *celA* and *celB*, coding for endoglucanase A and B of *C. thermocellum* from two cosmids into the smaller pBR322 plasmid of *E. coli.* Both genes were carried on 1.8-MDa DNA segments, but[32]P-labeled probes showed no homology between them. The *celA* gene product made in *E. coli* cross-reacts with antiserum raised against the M_r 56,000 endogluconase from *C. thermocellum.* The *celB* gene product was purified from *E. coli* and antisera prepared allowed identification of new endoglucanase B in *C. thermocellum* broth of M_r 66,000 not formerly discovered (Beguin *et al.*, 1983). It is interesting that the *celB* and *celA* genes are expressed in *E. coli* when incorporated into pBR322 in either direction, and expression did not inhibit growth of *E. coli.* In contrast to *C. thermocellum,* where the endogluconases are exported into the growth medium, these enzymes collect in the cytoplasm and periplasmic space of *E. coli.* It is not clear why this is, but it may be that the outer membrane of the latter organism provides a barrier to export.

A recombinant plasmid pTC1 with a 1.5-MDa sequence of *C. thermocellum* DNA has been reported to generate genetic variants in *E. coli* rather than complement for defects (Snedecor and Gomez, 1983). Analysis of the cloned *C. thermocellum* DNA piece suggests that it has the properties of an insertion sequence. Such elements may be engineered into a utilizable transposon which would have importance in genetic analysis of the clostridia. Similar well-characterized enteric bacterial elements and the *Streptococcus faecalis* transposon, Tn917 (Youngman *et al.*, 1983) have been successfully used for insertional mutagenesis in gram-negative and gram-positive bacteria, respectively.

These initial easy successes in cloning and expression in *E. coli* of enzymes important to the fermentation process, such as cellulases and hydrogenase, open the opportunity to apply all of the flexibility and power of the *E. coli* molecular cloning systems to analyze metabolic pathways and other aspects of *Clostridium* biology, such as solvent tolerance, involved in fermentation processes.

V. The Coupling of Nutritional Stress, Sporulation, and Solventogenesis

It has been known for some time that a conventional batch fermentation of *Clostridium acetobutylicum* begins with a phase of rapid vegetative growth and an acidogenic fermentation and then following the "pH breakpoint" switches into a slow-growing phase or "ripe culture" (Davies and Stephenson, 1941; Spivey, 1978) in which solvents are produced. The relationship between solvent formation and sporulation was recognized and utilized by early workers who used cycles of sporulation, heating, and outgrowth to ensure the selection of a high solvent-producing strain (Beesch, 1953).

Low solvent-yielding strains usually sporulated poorly. Only recently has the connection between the loss of spore-forming potential and the loss of solvent formation been investigated in both batch cultures (Jones *et al.*, 1982; Long *et al.*, 1983a,b) and chemostat cultures (Gottschal and Morris, 1981a,b). In 1970 it was shown that *C. thermosaccharolyticum*, cultured under conditions of restricted growth, by slow feeding of glucose or using starch as a carbon source, shifted into a sporulation phase (Hsu and Ordal, 1970). This sporulating *C. thermosaccharolyticum* goes through an intermediate "enlongated cell stage" that produces mostly ethanol, instead of acetate, butyrate, and lactate typical of the vegetative cells (Hoffmann *et al.*, 1978; Hsu and Ordal, 1970). The normal strains of *C. thermosaccharrolyticum* initially form elongated cells when grown continuously on xylan and form mostly ethanol (Landuyt *et al.*, 1983). However, they either revert back to vegetative cells again becoming acidogenic organisms, or go on and sporulate. A mutant strain, SD105, grown similarly, continued to produce ethanol for 26 hours and remained as elongate cells (Landuyt and Hsu, 1985). Similarly, during batch fermentation, *C. acetobutylicum* changes morphologically to a cigar-shaped, granulose-containing "clostridial form" which has been correlated with the shift from an acid fermentation to a solvent fermentation (Jones *et al.*, 1982; Long *et al.*, 1983). This correlation is strengthened by selection of two classes of mutants. The *cls* mutants are unable to form a clostridial stage, make granulose, form capsules, or produce any endospore stage; *spo* mutants are blocked at a later stage in sporogenesis (Jones *et al.*, 1982). The *cls* mutants are unable to switch to solventogenesis whereas the *spo* mutants or mutants blocked in granulose formation or capsule formation all can produce solvents (Long *et al.*, 1984b). These results indicate that these two events share some regulatory features in common. From the study of the above-mentioned species of clostridia it appears that, in contrast to the "signal" of nutrient deprivation that starts the sporulation program running in the aerobic *Bacillus*, in the clostridia there are a different set of signals. In the case of *C. thermosaccharolyticum*, significant ethanol production and elongate cell formation both require a combined signal of specific carbon source (L-arabinose or L-xylose), lower pH, and a restricted rate of supply of energy source (Landuyt *et al.*, 1983). In the case of *C. acetobutylicum*, switching to the solventogenic phase and differentiation to the clostridial stage requires a threshold concentration of acetate and butyrate, the correct pH (4.3 to 5.0), as well as enough glucose and nitrogen to induce the new program and end the old one (Long *et al.*, 1984a).

Careful studies employing chemostats have revealed that restriction of growth by phosphate or sulfate limitation at pH 4.3 but not by nitrogen or carbohydrate limitation are conditions for continuous solvent formation (Bahl et al., 1982a,b; Bahl and Gottschalk, 1984; Gottschal and Morris, 1981b). High solvent yield depends upon adjusting the growth-limiting factor in a range which allows some growth but still high substrate consumption. Solventogenesis, then, is apparently a metabolic response to a condition of unbalanced growth where the utilizable energy source remains in excess but growth is restricted by other limiting factors or growth inhibitors. Suprisingly, sulfate limitation of C. acetobutylicum in chemostats above pH 5.0 caused production of L-lactate as a major fermentation product (Bahl and Gottschalk, 1984). This is probably not due to a significant lowering of intracellular concentrations of compunds like coenzyme-A since it participates in both solvent and acid production from acetyl-CoA at low pH as well. Bahl and Gottschalk (1984) suggested that sulfate limitation may influence hydrogen evolution and that at high pH the reduction of pyruvate to lactate is an outlet for excess reducing equivalents. As suggested in a previous section, the production of the solvents butanol, acetone, and ethanol (or lactate) may be correlated with or a response to decreased activity of hydrogenase, thus providing a relief valve for the need to reduce excess pools of NADH and NADPH (Datta and Zeikus, 1985; Kim et al., 1984). The molecular interlocks and internal signals turned on by growth restriction that result in these changes in solvent production remain unknown.

The shift into a presporulation phase by C. acetobutylicum requires glucose and ammonia in contrast to the aerobic bacillus, and the presence of acid endproducts butyrate and acetate, whose function remains unknown. It appears that the low pH, solvent production, and a clostridial stage are all intimately connected to endospore formation (Long et al., 1984a).

But if this is true, how can phosphate-limited chemostat cultures of C. acetobutylicum be maintained that produce a high yield of solvents? Recently, as mentioned above, Bahl and Gottschalk (1984) ran chemostats with low dilution rates ($D = 0.1$ hour^{-1}) low pH (4.3), excess substrate, threshold level of butyrate and/or acetate and a suitable growth-limiting factor (e.g., phosphate) with high yields of acetone and butanol from glucose. These steady state chemostats have been maintained for 1 year (Bahl et al., 1982a). However, a steady state condition is theoretically excluded if a shift to a sporulation program is connected with the onset of solvent formation. The difficulty has now been solved by the recent finding that asporogenous mutant strains of C. acetobutylicum are selected during the chemostat operation (Meinecke et al., 1984). After the first 12 days of operation, stable asporogenous strains can be isolated showing lack of granulose formation in colonies; after 35 days the asporogenous strain predominates in the continuous culture and sporulating strains cannot be found (Meinecke et al., 1984). Following this finding, Largier et al. (1985) investigated solvent production

by asporogenous mutants of *C. acetobutylicum* immobilized in calcium alginate beads. An early-sporulation mutant, *spoA2*, which forms a clostridial
stage but does not produce granulose, capsule, or forespore septum was a
markedly better solvent producer than either a later sporulation mutant,
spoB, or the wild type. Using a continuous fluidized bed reactor, these
workers demonstrated the superiority of the *spoA2* mutant for solvent production as evaluated by a number of parameters, such as productivity, yield
coefficient, and solvent concentration.

Although the isolation of these mutants demonstrate that sporulation is
not a *prerequisite* for solvent production, the view remains that the signals
for initiation of sporulation and of solvent production are tightly connected.
But are these two events coupled or do they just happen at the same time?
In *Bacillus subtilis*, initiation of sporulation is dependent upon chromosomal
replication (Young and Mandelstam, 1980). Since inhibition of DNA synthesis blocks sporulation this has been used to differentiate between sporula-
tion-specific events and events which may be essential to the sporulation
process but are not sporulation specific. Using this strategy with *C. acetobutylicum*, low concentrations of the DNA synthesis inhibitors, ethidium bromide, novobiocin, and 6-(*p*-hydroxyphenylazo)-uracil inhibited septum and
spore formation but not the clostridial stage, granulose and capsule formation. Ethidium bromide (1 μg/ml), which blocked sporulation, actually allowed a higher than normal level of solvent production (Long *et al.*, 1984b).
These results appear to mimic the observations with the sporulation (*spo*)
and clostridial stage (*cls*) mutants mentioned above. We can conclude that
inhibition of spore formation does not affect solvent production; however,
since *cls* mutants produce neither endospores nor solvents, these events
share common regulatory features. The molecular nature of the common and
separate control devices for these two processes is a central facinating subject for future investigation.

VI. Product Tolerance

For development of any commercial solventogenic or acidogenic microbial
fermentation, the challenge is to utilize the conditions and microorganisms
that produce the desired product at the lowest possible cost. Among other
parameters such as the rate and efficiency of conversion of substrate to
product, is the difficult problem of product recovery from water. Typically,
fermentations for chemicals produce low concentrations of the desired acids
and solvents dissolved in the broth. It has been calculated that the energy
requirements for recovery of ethanol, at less than 6–8%, or *n*-butanol, at less
than 2–3% (w/v) from fermentation beers using standard distillation methods exceeds the recovery of net energy necessary for a reasonable commer-

cial process (Phillips and Humphrey, 1983). Calculations show that to increase the fermentor concentration of butanol from 1.2 to 2% w/v would
halve the energy consumption for distillation (Linden *et al.*, 1984). The
limiting factor in most fermentations is the fact that the microorganism
carrying out the bioconversion is intolerant to low concentrations of its own
solvent or acid products so that toxic and inhibitory effects cause growth and
the fermentation to stop. There is both an engineering approach and a
biological approach to minimizing this problem. In the former, the toxic
products and their inhibitory effects could be eliminated by continuous
extraction of the product from the fermentation broth. These technologies
are being developed and include vacuum fermentation and extractive fermentation and some are claimed to reduce the overall energy cost for extraction by 50% (Phillips and Humphrey, 1983; Griffith *et al.*, 1983; Mattiasson, 1983). The second approach, the biological one, has been essentially
neglected until recently, as pointed out elsewhere (Linden *et al.*, 1984;
Rabson and Rogers, 1981). This approach first entails research leading to an
understanding of the physiologic and molecular nature of solvent or acid
toxicity and tolerance. As we gain knowledge of how tolerance works for an
organism, developmental researchers will be in a better position to produce
solvent- or acid-tolerant strains by genetic manipulation or by optimizing
tolerance by changes in media or fermentation conditions. The seriousness
of the problem of product toxicity and tolerance in the case of the clostridia
becomes even more apparent when one compares them to yield capabilities
in fermentations by other microorganisms. Yeasts produce a single product,
ethanol, often at relatively high yields; for example, *Saccharomyces cerevisiae* and *S. carlsbergensis* produce 12% ethanol while the slower fermen-
tating *S. sake* yields 20% (Rose and Beavan, 1981). The bacterium *Thermoanaerobacter ethanolicus*, a thermophile, is tolerant to 6–10% ethanol
and yields 4% ethanol in a mixed fermentation (Wiegel and Ljungdahl, 1981)
while strains of *Zymomonas mobilis* can yield 12% alcohol as a single product
(Eveleigh *et al.*, 1983). The yeasts and zymomonads convert sugars to an
ethanol concentration high enough (10–20%) so that the energy required for
solvent separation from the fermentation beer is low enough to permit industrial production of alcohol. In contrast, cocultures of *C. thermocellum*
and *C. thermohydrosulfuricum* that ferment delignified wood to ethanol
produce a low final solvent yield of less than 2%, a concentration above
which growth of both of these bacteria is inhibited (Zeikus *et al.*, 1981).
Likewise, the butanol–acetone fermentation by *C. acetobutylicum* stops
when only 2% w/v of total solvents is reached, primarily due to the toxicity of
butanol which is the main product (Linden and Moreira, 1983). Similarly,
Acetobacter suboxydans oxidizes ethanol to acetic acid in a process yielding
food-grade vinegar at a concentration of about 120 g/liter (Sinsky, 1983)

while a fermentation using *C. thermoaceticum*, which theoretically yields 100% acetic acid from hexose, is inhibited by acetic acid at only about 15 g/liter (Ljungdahl, 1983). With a mutant strain selected to grow in continuous culture at pH 4.5, a yield of 4.5 g/liter of acetic acid was obtained (Schwartz and Keller, 1982). In a careful study of growth inhibition and acetic acid production by *C. thermoaceticum* in batch cultures with pH control, Wang and Wang (1984) showed that at pH 6.9 a yield of 56 g/liter acetic acid is produced instead of 15 g/liter in uncontrolled conditions. The combined results show that the undissociated acid completely inhibits growth at 2–3 g/liters while the acetate ion inhibits at 48 g/liter. Thus at low pH the undissociated acid inhibits growth, while at pH 6.0 or over it is the ionized acetate that is responsible. It is important to consider these findings when selecting high-growth rate, acetic acid-tolerant mutants of *C. thermoaceticum*. Thus, unless product tolerance can be built into these *Clostridium* species, the potentially valuable microbial process will remain in limited use for commodity chemical production.

In summarizing the work on ethanol tolerance in yeast, Rose and Beavan (1981) proposed the view that the most probable molecular basis for this property involves the well-known denaturing effect of ethanol on proteins. Thus, strains with proteins that resist the denaturing effects of ethanol can survive, metabolize, and grow in higher concentrations of ethanol and are thus more tolerant. The outcome of this view would be that the basis of tolerance to acids and solvents would resemble the basis of thermophily and support the contention that solvent tolerance is under complex genetic control. Furthermore, like with thermophily, the isolation of tolerant mutants would be unlikely. As an aside, it is noteworthy that the thermophilic clostridia are not obviously more or less resistant to solvents and acids than their mesophilic cousins. Thus, the protein structures leading to thermal tolerance do not translate into the structural needs for solvent tolerance. If tolerance level depends upon structure of a number of proteins, the question is, are there certain critical proteins whose function is more affected by specific solvents than others? Are there sensitive targets that might be changed or replaced by genetic or physiologic manipulation to cause increased tolerance? The two groups of proteins or processes studied most heavily in yeast are the fermentation pathway and the various protein-connected functions involved with cell membranes (Rose and Beavan, 1981). For the most part the small amount of research reported on the basis of tolerance in the clostridia focuses on cell membranes and membrane functions and has involved the three organisms: *C. thermocellum*, *C. thermohydrosulfuricans*, and *C. acetobutylicum*. The effect of growth of *C. thermocellum* at low concentrations of ethanol was to increase the ratio of unsaturated/saturated fatty acids consistent with work in *E. coli* (Herrero *et al.*, 1982). Using a mutant strain C9 adapted to growth in higher levels of ethanol (20 g/liter) (Herrero and Gomez, 1980), even more extreme changes were seen. The view is that ethanol enters the membrane causing a decrease in fluidity, and microorganisms attempt to adapt by increasing the membrane fluidity. In *S. cerevisiae*, enrichment of plasma membranes with sterols having unsaturated side chains and with unsaturated fatty acids increased survival and resistance of transport function inhibited by ethanol (Rose and Beavan, 1981). The low ethanol tolerance of *C. thermocellum* (5 g/liter) was also ascribed to specific inhibition of some glycolytic enzymes for conversion of glucose to glyceraldehyde-3-phosphate (Herrero and Gomez, 1981). In contrast, *C. thermohydrosulfuricum* was found to be tolerant to methanol or acetone at 5% (w/v), although glucose fermentation was inhibited by 0.5 to 2.0% ethanol (Lovitt *et al.*, 1984). A mutant strain (39EA) grew in 8.0% ethanol at 45°C and up to 3.3% ethanol at 68°C. In contrast to the parent strain (39E) that produces an ethanol:lactate ratio of 8:1, the mutant strain produced equal amounts of lactate and ethanol (Lovitt *et al.*, 1984). The ethanol-tolerant mutant also lost the ability to utilize starch or pyruvate for growth which is not understood. Apparently, the mechanism of solvent tolerance in *C. thermohydrosulfuricum* does not involve either disruption of membrane fluidity or glycolytic enzyme activity, although a change in the fermentation pattern was found.

Most of the research on product tolerance in *C. acetobutylicum* has been reviewed recently (Linden and Moreira, 1983; Linden *et al.*, 1986). The effects of butanol appear to be opposite to that observed with ethanol in that butanol increases membrane fluidity. The expected response of cells adapting to butanol would be to mimic adaptation to high growth temperatures, that is, an increased ratio of saturated to unsaturated fatty acids in the membrane should be observed. Thus, one might expect to find that a thermophilic butanol-producing strain would be more butanol tolerant. Recently, Vollherbst-Schneck *et al.* (1984) found that growth of *C. acetobutylicum* in 1% butanol increased levels of saturated acyl chains at the expense of unsaturated acyl chains. Also, butanol caused a 20–30% increase in fluidity of lipid dispersions from *C. acetobutylicum.* It is interesting that growth of *C. acetobutylicum* is inhibited 50% by 6.0–8.0 g/liter of butyrate and acetate while 11, 51, and 43 g/liter of butanol, ethanol, and acetone, respectively, are required for the same toxic effect (Linden and Moreira, 1983). Thus, the observed shift in this fermentation to solvents discussed above may be viewed as a metabolic detoxification. However, the fermentation is limited by the major solvent, butanol. The effects of alcohols on membrane-associated functions in *C. acetobutylicum* have been studied. Membrane-bound ATPase activity, alanine and 3-*O*-methyl glucose uptake were all found to show similar greater sensitivity (8-fold) to butanol compared to ethanol than was found for growth (Linden and Moreira, 1983). Goldfine and Johnston (1980) showed that in biotin-free media, fatty acid biosynthesis by *C. butyricum* is blocked. Using this medium incorporation of added oleate and elaidate into both glycerol-ester linkages and glycerol-ether linkages of phospholipids of *C. butyricum* was enriched more than 90%. In this organism, the fluidity of the cell membrane was increased by oleic acid (*cis* 18:1) and decreased by elaidic acid (*trans* 18:1). Linden and Moreira (1983) showed that growth of *C. acetobutylicum* to produce enriched oleic- and elaidic-cell membranes yielded bacteria about twice as tolerant to butanol inhibition of both growth and membrane-ATPase. Since *trans*- and *cis*-fatty acid-enriched cell membranes both yielded an increase in butanol tolerance the mechanism of butanol action remains unclear. Another approach to understanding the mechanism of butanol tolerance is to produce mutants and study their properties. From a high butanol-producing strain ATCC 824 (7.9 g/liter on extruded corn) Lin and Blaschek (1983) selected a butanol-tolerant mutant that grew well at 15 g/liter butanol and produced 14 g/liter. Allcock *et al.* (1981) isolated an autolysis resistant, autolysin-negative mutant of *C. acetobutylicum* (Table IV). This mutant was more butanol tolerant than the parent strain and grew with 16 g/liter butanol but during fermentation it produced only 20 g/liter total solvent. The protein or membrane alternations in these mutants have not been investigated. However, these results encourage the view that membrane proteins are the proper targets for further research on the tolerance mechanism and for development of butanol-tolerant bacteria.

VII. Summary and Conclusions

This review presents five major areas of investigation that are now yielding new knowledge upon which future applications of the clostridia to the production of commodity chemicals can be based. It should also be apparent that in none of these areas is our understanding complete. Indeed, in each section some of the obvious gaps in our knowledge were presented. Again, in this summary, the focus is on those areas of research or directions in the study of the biology of the clostridia that seem to be of the highest priority for the immediate future.

First, with respect to the biochemistry of clostridial fermentations, three rather urgent problems emerge:

1. With the possible exception of the homoacetogenic fermentation, our knowledge of the basic enzymology of most of these pathways is rather primitive. Molecular structure, stability characteristics, substrate specificity, and kinetic constants for most of these biocatalysts are hardly known and

work in this area will be vital for reprogramming these pathways for human uses.

2. Even if we know the general enzymic steps in these pathways of the clostridia, we are almost entirely ignorant of even the general mechanisms of their regulation, to say nothing of the sophisticated molecular mechanisms involved. Application of the emerging genetic methods for the clostridia to defining the regulatory networks for these fermentations will allow a clearer picture of how these metabolic pathways operate.

3. The biology of defined cocultures involving two or more organisms for specific fermentations is just emerging. Already it is clear from the few studies presented that this approach when combined with a growing knowledge of the species interaction can yield new efficient fermentations now impossible for one organism. The strategy of physiologic combination of two or more natural germplasms as an alternative to the recombinant DNA method of putting it all in one package should be carefully investigated.

Second, in the realm of infant genetics of the clostridia it is not hard to find areas of research that are essential:

1. Both natural genetic transfer mechanisms such as conjugation or transduction and artificial genetic transfer methods must be further investigated and improved to make them usable for genetic study of at least two or three key species of *Clostridium*.

2. Basic mutagenesis methods such as transposon mutagenesis should be intensively studied and when available, they will allow production of mutants for study of many processes. Careful study of mutation mechanism, recombination, DNA repair, and restriction enzymes in the clostridia are essential for effective use of these organisms.

3. Research to further our understanding of clostridial plasmids as natural replicative units should be encouraged. Development of expression vectors and cloning vehicles designed for clostridia should be continued. Future programming of the production of individual enzymes or the regulation of entire pathways will depend upon these genetic tools.

Third, the area of clostridial enzyme complexes that degrade biopolymers should be of great future interest:

1. The multi enzyme complex of *C. thermocellum* should be studied intensively in order to understand its many functions in binding cells and substrates, catalytic functions, and stability.

2. Biochemical studies on secretion, activation, and inhibition of extracellular enzymes should be carried out along with genetic methods to aid in establishing the mechanisms of all of these processes.

3. Detailed studies on the hemicellulase complex and pectinase complex

of the clostridia should be considered, since at the present time very little is known concerning these potentially important polymer-degrading systems.

Fourth, research on morphogenesis and endospore formation in the clostridia should reveal similar mechanisms of global regulatory changes as shown in the aerobic bacilli. This knowledge already seems important to an understanding of the major internal mechanisms and environmental signals controlling solvent production in *C. acetobutylicum*, as reviewed above.

Finally, the prospect of utilizing the clostridia in any cost-effective process in the future will depend heavily upon improving their tolerance to toxic solvent and acid products. Research in this area is still very rudimentary. With the new tools of genetics applied to this problem in the clostridia, perhaps some progress in our understanding of tolerance will be made.

Surely the work reviewed here will act as a platform from which the above suggested work will be launched. Indeed, the recent findings presented in this article develop a structure of understanding that already provides a glimpse into the molecular nature of clostridial fermentation programs. This knowledge also provides a rational approach to the conversion and domestication of these organisms for technical uses as biocatalysts in chemical production in the future.

On the other hand, the microbiologist delights in simply understanding the molecular biology of this curious group of bacteria.

ACKNOWLEDGMENTS

I am most grateful to all those generous colleagues who shared their ideas and their most recent unpublished results for inclusion in this review. Also special thanks to Jan Smith for help in editing and preparing the manuscript.

Some of the work reported from our laboratory was supported by Grant No. DOE/DE-AC02-83ER13068 from the U.S. Department of Energy.

REFERENCES

Ait, N., Creuzet, N., and Cattaneo, J. (1982). *J. Gen. Microbiol.* **128**, 569–577.
Akedo, M., Cooney, C. L., and Sinskey, A. J. (1983). *Abstr. Annu. Meet ASM* p. 240.
Allocck, E. R., Reid, S. J., Jones, D. T., and Woods, D. R. (1981). *Appl. Environ. Microbiol.* **42**, 929–935.
Allcock, E. R., Reid, S. J., Jones, D. T., and Woods, D. R. (1982). *Appl. Environ. Microbiol.* **43**, 719–721.
Andersch, W., Hubert, B., and Gottschalk, G. (1983). *Eur. J. Appl. Microbiol. Biotechtol.* **18**, 327–332.
Andreesen, J. R., and Ljungdahl, L. G. (1973). *J. Bacteriol.* **114**, 743–751.
Andreesen, J. R., Gottschalk, G., and Schlegel, H. G. (1970). *Arch. Mikrobiol.* **72**, 154–174.

Avgerinos, G. C. (1982). Ph.D. thesis, MIT, Cambridge, Massachusetts.

Avgerinos, G. C., Faug, H. Y., Biocio, I., and Wang, D. I. C. (1981). *In* "Advances in Biotechnology" (M. Moo Young and C. W. Robinson, eds.), Vol. II, pp. 119–124. Pergamon, New York.

Bahl, H., and Gottschalk, G. (1984). *In* "Sixth Symposium on Biotechnology for Fuels and Chemicals" (D. I. C. Wang and C. D. Scott, eds.), Vol. 14, pp. 215–223. Wiley, New York.

Bahl, H., Andersch, W., and Gottschalk, G. (1982a). *Eur. J. Appl. Microbiol. Biotechnol.* **15,** 201–205.

Bahl, H., Andersch, W., Braun, K., and Gottschalk, G. (1982b). *Eur. J. Appl. Microbiol. Biotechnol.* **14,** 17–20.

Barker, H. A., Jeng, I. M., Neff, N., Robertson, J. M., Tam, F. K., and Hosaka, S. (1978). *J. Biol. Chem.* **253,** 1219–1225.

Bayer, E. A., Kenig, R., and Lamed, R. (1983). *J. Bacteriol.* **156,** 818–827.

Beesch, S. C. (1953). *Appl. Microbiol.* **1,** 85–95.

Beguin, P., Cornet, P., and Millet, J. (1983). *Biochimie (Paris)* **65,** 495–500.

Blaschek, H. P., and Klacik, M. A. (1984). *Appl. Environ. Microbiol.* **48,** 178–181.

Blaschek, H. P., and Solberg, M. (1981). *J. Bacteriol.* **147,** 262–266.

Braun, K., and Gottschalk, G. (1981). *Arch. Microbiol.* **128,** 294–298.

Braun, K., Mayer, F., and Gottschalk, G. (1981). *Arch. Microbiol.* **128,** 288–293.

Brefort, G., Magot, M., Ionesco, H., and Sebald, M. (1977). *Plasmid* **1,** 52–66.

Brener, D., and Johnson, B. F. (1984). *Appl. Environ. Microbiol.* **47,** 1126–1129.

Chen, J. S. (1978). *In* "Hydrogenases—their Catalytic Activity, Structure and Function" (H. G. Schlegel and K. Schneider, eds.), pp. 57–81. Goltz, Göttingen, West Germany.

Chen, J. S., and Blanchard, D. K. (1984). *Biochem. Biophys. Res. Commun.* **122,** 9–16.

Chung, K. T. (1976). *Appl. Environ. Microbiol.* **31,** 342–348.

Clark, J. E., Ragsdale, S. W., Ljungdahl, L. G., and Wiegel, J. (1982). *J. Bacteriol.* **151,** 507–509.

Collmer, A., and Wilson, D. B. (1983). *Bio/Technology,* **1,** 594–601.

Cornet, P., Millet, J., Beguin, P., and Aubert, J. P. (1983a). *Bio/Technology* **1,** 589–594.

Cornet, P., Tronik, D., Millet, J., and J. P. Aubert. (1983b). *FEMS Microbiol. Lett.* **16,** 137–141.

Datta, R., and Zeikus, J. G. (1985). *Appl. Environ. Microbiol.* **49,** 522–529.

Davies, R. (1943). *Biochem. J.* **37,** 230–238.

Davies, R., and Stephenson, M. (1941). *Biochem. J.* **35,** 1320–1331.

Dorn, M., Andreesen, J. R., and Gottschalk, G. (1978). *J. Bacteriol.* **133,** 26–32.

Drake, H. L. (1982a). *J. Bacteriol.* **150,** 702–709.

Drake, H. L. (1982b). *J. Bacteriol.* **149,** 561–566.

Drake, H. L., Hu, S. I., and Wood, H. G. (1980). *J. Biol. Chem.* **255,** 7174–7180.

Drake, H. L., Hu, S. I., and Wood, H. G. (1981). *J. Biol. Chem.* **256,** 11137–11144.

Duong, T. V. C., Johnson, E. A., and Demain, A. L. (1983). *In* "Topics in Enzyme and Fermentation Biotechnology 7" (A. Weisman, ed.), pp. 156–195. Wiley, New York.

Eveleigh, D. E., Stokes, H. W., and Dally, E. L. (1983). *In* "Organic Chemicals from Biomass" (D. L. Wise, ed.), pp. 69–91. Cummings, Menlo Park, California.

Finn, C. W., Silver, R. P., Habig, W. H., Hardegree, M. C., Zon, G., and Garon, C. F. (1984). *Science* **224,** 881–884.

Fond, O., Petitdemange, E., Petitdemange, H., and Engasser, J. M. (1984). *Biotechnol. Bioeng. Symp.* (13), 217–224.

Fontaine, F. E., Peterson, W. H., McCoy, E., Johnson, M. J., and Ritter, G. J. (1942). *J. Bacteriol.* **43,** 701–715.

George, H. A., and Chen, J. S. (1983). *Appl. Environ. Microbiol.* **46**, 321–327.

George, H. A., Johnson, J. L., Moore, W. E. C., Holdeman, L. V., and Chen, J. S. (1983). *Appl. Environ. Microbiol.* **45**, 1160–1163.

Giallo, J. C., Gaudin, J., Belaich, P., Petitdemange, E., and Caillet-Mangin, F. (1983). *Appl. Environ. Microbiol.* **45**, 843–849.

Goldfine, H., and Johnston, N. C. (1980). *In* "Membrane Fluidity: Biophysical Techniques and Cellular Recognition" (M. Kates and M. Kuksi, eds.), pp. 365–380. Human Press, Clifton, New Jersey.

Gomez, R. F., Snedecor, B., and Mendez, B. (1980). *Dev. Ind. Microbiol.* **22**, 87–96.

Gottschal, J. C., and Morris, J. G. (1981a). *FEMS Microbiol. Lett.* **12**, 385–389.

Gottschal, J. C., and Morris, J. G. (1981b). *Biotechnol. Lett.* **3**, 525–530.

Gottschalk, G. (1979). "Bacterial Metabolism." Springer-Verlag, New York.

Gottschalk, G., and Bahl, H. (1981). *In* "Trends in the Biology of Fermentations for Fuels and Chemicals" (A. Hollaender, ed.), pp. 463–471. Plenum, New York.

Gottwald, M., Hippe, H., and Gottschalk, G. (1984). *Appl. Environ. Microbiol.* **48**, 573–576.

Griffith, W. L., Compere, A. L., and Googin, J. M. (1983). *Dev. Ind. Microbiol.* **24**, 347–352.

Häggström, L., and Enfors, S. (1982). *Appl. Biochem. Biotechnol.* **7**, 35–37.

Häggström, L., and Molin, N. (1980). *Biotech. Lett.* **3**, 241–246.

Hartmanis, M. G. N., and Gatenbeck, S. (1983). *Fed. Proc.* **42**, 1981.

Hartmanis, M. G. N., and Gatenbeck, S. (1984). *Appl. Environ. Microbiol.* **47**, 1277–1283.

Hartmanis, M. G. N., Klason, T., and Gatenbeck, S. (1984). *Appl. Microbiol. Biotechnol.* **20**, 66–71.

Heffner, D. L., Squires, C. H., Evans, R. J., Kopp, B. J., and Yarus, M. J. (1984). *J. Bacteriol.* **159**, 460–464.

Herrero, A. A., and Gomez, R. F. (1980). *Appl. Environ. Microbiol.* **40**, 571–577.

Herrero, A. A., and Gomez, R. F. (1981). *In* "Advances in Biotechnology" (M. Moo-Young and C. W. Robinson, eds.), Vol. II, pp. 213–218. Pergamon, New York.

Herrero, A. A., Gomez, R. F., and Roberts, M. F. (1982). *Biochim. Biophys. Acta* **693**, 195–204.

Hoffman, J. W., Chang, E. K., and Hsu, E. J. (1978). *In* "Spores" (G. Chambliss and J. C. Vary, eds.), Vol. VII, pp. 312–318. Am. Soc. Microbiol., Washington, D.C.

Hollaender, A. (1982). "Genetic Engineering of Microorganisms for Chemicals." Plenum, New York.

Hopwood, D. A. (1981). *Annu. Rev. Microbiol.* **35**, 237–272.

Hsu, E. J., and Ordal, Z. J. (1970). *J. Bacteriol.* **102**, 369–376.

Hu, S. I., Drake, H. L., and Wood, H. G. (1982). *J. Bacteriol.* **149**, 440–448.

Hungate, R. E. (1957). *Can. J. Microbiol.* **3**, 289–311.

Ionesco, H. (1980). *Ann. Microbiol.* **131**, 171–179.

Ishii, K., Kudo, T., Honda, H., and Horikoshi, K. (1983). *Agric. Biol. Chem.* **47**, 2313–2318.

Jobses, I. M. L., and Roels, J. A. (1983). *Biotechnol. Bioeng.* **25**, 1187–1194.

Johnson, E. A., Madia, A., and Demain, A. L. (1981). *Appl. Environ. Microbiol.* **41**, 1060–1062.

Johnson, E. A., Sakajoh, M., Halliewell, G., Madia, A., and Demain, A. L. (1982a). *Appl. Environ. Microbiol.* **43**, 1125–1132.

Johnson, E. A., Reese, E. T., and Demain, A. S. (1982b). *J. Appl. Biochem.* **4**, 64–71.

Jones, D. T., van der Westhuizen, A., Long, S., Allcock, E. R., Reid, S. R., and Woods, D. R. (1982). *Appl. Environ. Microbiol.* **43**, 1434–1439.

Jones, D. T., Jones, W. A., and Woods, D. R. (1985). *J. Gen. Microbiol.* **131**, 1213–1216.

Jungermann, K., Thauer, R. K., Leimenstoll, G., and Decker, K. (1973). *Biochim. Biophys. Acta* **305**, 268–280.

Karube, I., Urano, N., Yamada, T., Hirochika, H., and Sakaguchi, K. (1983). *FEBS Lett.* **158,** 119–122.

Kawata, T., Takumi, K., Sato, S., and Yamashita, H. (1968). *Jpn. J. Microbiol.* **12,** 445–455.

Kawatomari, T. (1958). *J. Bacteriol.* **76,** 227–232.

Kellum, R., and Drake, H. L. (1984). *J. Bacteriol.* **160,** 466–469.

Kerby, R., and Zeikus, J. G. (1983). *Curr. Microb.* **8,** 27–30.

Khan, A. W., and Murray, W. D. (1982a). *Biotechnol. Lett.* **4,** 177–180.

Khan, A. W., and Murray, W. D. (1982b). *FEMS Microbiol. Lett.* **13,** 377–381.

Kim, B. H., and Zeikus, J. G. (1985). *Dev. Ind. Microbiol.* **26,** 549–556.

Kim, B. H., Bellows, P., Datta, R., and Zeikus, J. G. (1984). *Appl. Environ. Microbiol.* **48,** 764–770.

Krouwel, P. G., van der Laan, W. F. M., and Kossen, N. W. F. (1980). *Biotech. Lett.* **2,** 253–258.

Krouwel, P. G., Groot, W. R., and Kossen, N. W. F. (1983). *Biotechnol. Bioeng.* **25,** 281–299.

Ladisch, M. R., Hong, J., Voloch, M., and Tsao, G. T. (1981). *In* "Trends in the Biology of Fermentations for Fuels and Chemicals" (A. Hollaender, ed.), pp. 55–83. Plenum, New York.

Laird, W. J., Asronson, W., Silver, R. P., Habig, W. H., and Hardegree, M. C. (1980). *J. Infect. Dis.* **142,** 623.

Lamed, R., and Zeikus, J. G. (1980). *J. Bacteriol.* **144,** 569–578.

Lamed, R., Setter, E., and Bayer, E. A. (1983). *J. Bacteriol.* **156,** 828–836.

Lamed, R., Setter, E., Kenig, R., and Bayer, E. A. (1984). *Biotechnol. Bioeng. Symp.* (13), 163–182.

Landuyt, S. L., and Hsu, E. J. (1985). *In* "Fundamental and Applied Aspects of Bacterial Spores" (G. J. Ding, G. W. Gould, and D. J. Ellar, eds.), pp. 477–493. Academic Press, New York.

Landuyt, S. L., Hsu, E. J., and Lu, M. (1983). *Ann. N.Y. Acad. Sci.* **413,** 474–478.

Largier, S. T., Long, S., Santangelo, J. D., Jones, D. T., and Wood, D. R. (1985). *Appl. Environ. Microbiol.* **50,** 477–481.

Leaver, F. W., Wood, H. G., and Stjerholm, R. (1955). *J. Bacteriol.* **70,** 521–530.

Lee, M., Miller, L., and MacMillan, J. D. (1970). *J. Bacteriol.* **103,** 595–600.

Lee, S. F., Forsberg, C. W., and Gibbins, L. N. (1985). *Appl. Environ. Microbiol.* **50,** 220–228.

LeGall, J., Moura, J. J. G., Peck, H. D., and Xavier, A. V. (1982). *In* "Iron Sulfur Proteins" (T. Spiro, ed.), Vol. IV, pp. 177–248. Wiley, New York.

Li, A. W., Krell, P. J., and Mahony, D. E. (1980). *Can. J. Microbiol.* **26,** 1018–1022.

Lin, Y. L., and Blaschek, H. P. (1983a). *Abstr. Annu. Meet. Am. Soc. Microbiol.* **H79,** p. 119.

Lin, Y. L., and Blaschek, H. P. (1983b). *Appl. Environ. Microbiol.* **45,** 966–973.

Lin, Y. L., and Blaschek, H. P. (1984). *Appl. Environ. Microbiol.* **48,** 737–742.

Linden, J. C., and Moreira, A. R. (1983). *In* "Biological Basis of New Developments in Biotechnology" (A. Hollaender, A. I. Laskin, and P. Rogers, eds.), pp. 377–404. Plenum, New York.

Linden, J. C., Moreira, A. R., and Lenz, T. G. (1986). *In* "Comprehensive Biotechnology" (M. Moo Young, ed.), Vol. 3. Pergamon, Oxford, England (In press).

Lipinsky, E. S. (1981). *Science* **212,** 1465–1471.

Liu, C. L., and Mortenson, L. E. (1983). *Fed. Proc.* **42,** 2266.

Ljungdahl, L. G. (1979). *Adv. Microbial. Physiol.* **19,** 150–243.

Ljungdahl, L. G. (1983). *In* "Organic Chemicals from Biomass" (D. L. Wise, ed.), pp. 219–248. Benjamin/Cummings, Menlo Park, California.

Ljungdahl, L. G., and Wood, H. G. (1982). In "B_{12}" (D. Dolphin, ed.), Vol. 2, pp. 166–202. Wiley, New York.

Ljungdahl, L. G., Pettersson, B., Eriksson, K. E., and Wiegel, J. (1983). *Curr. Microbiol.* **9**, 195–200.

Long, S., Jones, D. T., and Woods, D. R. (1983). *Appl. Environ. Microbiol.* **45**, 1389–1393.

Long, S., Jones, D. T., and Woods, D. R. (1984a). *Appl. Microbiol. Biotechnol.* **20**, 256–261.

Long, S., Jones, D. T., and Woods, D. R. (1984b). *Biotechnol. Lett.* **6**, 529–535.

Lovitt, R. W., Longin, R., and Zeikus, J. G. (1984). *Appl. Environ. Microbiol.* **48**, 171–177.

Lund, B. M., and Brocklehurst, T. F. (1978). *J. Gen. Microbiol.* **104**, 59–66.

Lundie, L. L., and Drake, H. L. (1984). *J. Bacteriol.* **159**, 700–703.

Lutsdorf, U., and Megnet, R. (1968). *Arch. Biochem. Biophys.* **126**, 933–944.

MacMillen, J. D., Phaff, H. T., and Vaughn, R. H. (1964). *Biochemistry* **3**, 572–578.

Madden, R. H. (1983). *Int. J. Syst. Bacteriol.* **33**, 837–840.

Madden, R. H., Bryder, M. J., and Poole, N. J. (1982). *Int. J. Syst. Bacteriol.* **32**, 87–91.

Maddox, I. S. (1980). *Biotechnol. Lett.* **2**, 493–498.

Maddox, I. S., and Murray, A. E. (1983). *Biotechnol. Lett.* **5**, 175–178.

Martin, D. R., Lundie, L. L., Kellum, R., and Drake, H. L. (1983). *Curr. Microbiol.* **8**, 337–340.

Martin, J. R., Petitdemange, H., Ballogue, J., and Gay, R. (1983). *Biotechnol. Lett.* **5**, 89–94.

Mattiasson, B. (1983). *Trends Biotechnol.* **1**, 16–20.

Mattiasson, B., Suominen, M., Andersson, E., Haggstrom, L., and Albertsson, P. A. (1981). In "Enzyme Engineering" (I. Chibata, S. Fukui, and L. B. Wingard, eds.), Vol. 6, pp. 153–155. Plenum, New York.

Meinecke, B., Bahl, H., and Gottschalk, G. (1984). *Appl. Environ. Microbiol.* **48**, 1064–1065.

Mendez, B. S., and Gomez, R. F. (1982). *Appl. Environ. Microbiol.* **43**, 496–498.

Mes-Hartyee, and Saddler, J. N. (1982). *Biotechnol. Lett.* **4**, 247–252.

Mihelc, V. A., Duncan, C. L., and Chambliss, G. H. (1978). *Antimicrob. Agents Chemother.* **14**, 771–779.

Miller, L., and MacMillan, J. D. (1970). *J. Bacteriol.* **102**, 72–78.

Minton, N., and Morris, J. G. (1981). *J. Gen. Microbiol.* **127**, 325–331.

Minton, N. P., and Morris, J. G. (1983). *J. Bacteriol.* **155**, 432–434.

Monot, F., and Engasser, J. M. (1983). *Biotechnol. Lett.* **5**, 213–218.

Monot, F., Martin, J., Petitdemange, H., and Gay, R. (1982). *Appl. Environ. Microbiol.* **44**, 1318–1324.

Monot, F., Engasser, J. M., and Petitdemange, H. (1984). *Biotechnol. Bioeng. Symp.* (13), 207–216.

Montenecourt, B. S., Nhlapo, S. D., Trimino-Vazquez, H., Cuskey, S., Schambart, D. H. J., and Eveleigh, D. E. (1981). In "Trends in the Biology of Fermentations for Fuels and Chemicals" (A. Hollaender, ed.), pp. 33–53. Plenum, New York.

Moreira, A. R. (1983). In "Organic Chemicals from Biomass" (D. L. Wise, ed.), pp. 385–406. Benjamin/Cummings, Menlo Park, California.

Muldrow, L. L., Archibald, E. R., Nunez-Montiel, O. L., and Scheehy, R. J. (1982). *J. Clin. Microbiol.* **16**, 637–640.

Murray, W. D., and Khan, A. W. (1983a). *Can. J. Microbiol.* **29**, 348–353.

Murray, W. D., and Khan, A. W. (1983b). *Can. J. Microbiol.* **29**, 342–347.

Murray, W. D., Khan, A. W., and VandenBerg, L. (1982). *Int. J. Syst. Bacteriol.* **32**, 132–135.

Murray, W. D., Wemyss, K. B., and Khan, A. W. (1983). *Eur. J. Appl. Microbiol. Biotechnol.* **18**, 71–74.

Nakmanovich, B. M., and Shcheblykina, N. A. (1960). *Mikrobiologiya* **29,** 67–72.

Ng, T. K., and Zeikus, J. G. (1981a). *Appl. Environ. Microbiol.* **42,** 231–240.

Ng, T. K., and Zeikus, J. G. (1981b). *Biochem. J.* **199,** 341–350.

Ng, T. K., and Zeikus, J. G. (1982). *J. Bacteriol.* **150,** 1391–1399.

Ng, T. K., Busche, R. M., McDonald, C. C., and Hardy, R. W. F. (1983). *Science* **219,** 733–740.

O'Brien, R. W., and Morris, J. G. (1971). *J. Gen. Microbiol.* **68,** 307–318.

Odom, J. M., and Peck, H. D. (1984). *Annu. Rev. Microbiol.* **38,** 551–592.

Office of Technology Assessment (1984). *In* "Commercial Biotechnology: An International Analysis" pp. 237–250. (U.S. Congress, OTA-BA-218), U.S. Gov. Printing Office, Washington, D.C.

Ogata, S. K., Choi, H., Yoshino, S., and Hayashida, S. (1981). *J. Fac. Agric. Kyushu Univ.* **25,** 201–222.

Pan-Hou, H. S. K., Hosono, M., and Imura, N. (1980). *Appl. Environ. Microbiol.* **40,** 1007–1011.

Papoutsakis, E. T. (1984). *Biotechnol. Bioeng.* **26,** 174–187.

Peck, H. D., and Odom, M. (1981). *In* "Trends in Biology of Fermentations for Fuels and Chemicals" (A. Hollaender, ed.), pp. 375–395. Plenum, New York.

Peck, H. D., Liu, C. L., Varma, A. K., Ljungdahl, L. G., Szulezynski, M., Bryant, F., and Carreira, L. (1983). *In* "Basic Biology of New Developments in Biotechnology" (A. Hollaender, A. I. Laskin, and P. Rogers, eds.), pp. 317–348. Plenum, New York.

Petitdemange, H., Cherrier, C., Raval, G., and Gay, R. (1976). *Biochim. Biophys. Acta* **421,** 334–347.

Petitdemange, E., Fond, O., Caillet, F., Petitdemange, H., and Gay, R. (1983). *Biotechnol. Lett.* **5,** 119–124.

Petre, J., Longin, R., and Millet, J. (1981). *Biochemie* **7,** 629–639.

Pezacka, E., and Wood, H. G. (1983). *Fed. Proc.* **42,** 1981.

Pezacka, E., and Wood, H. G. (1984). *Arch. Microbiol.* **137,** 63–69.

Phillips, J. A., and Humphrey, A. E. (1983). *In* "Organic Chemicals from Biomass" (D. L. Wise, ed.), pp. 249–304. Benjamin/Cummings, Menlo Park, California.

Prescott, S. C., and Dunn, C. G. (1959). "Industrial Microbiology" 3rd Ed., pp. 240–284. McGraw-Hill, New York.

Prevot, A. R., Turpin, A., and Kaiser, P. (1967). "Les Bacteries Anaerobies," pp. 928–931. Dunod, Paris.

Rabson, R., and Rogers, P. (1981). *Biomass* **1,** 17–37.

Ragsdale, S. W., and Ljungdahl, L. G. (1984). *J. Bacteriol.* **157,** 1–6.

Ragsdale, S. W., Clark, J. E., Ljungdahl, L. G., Lundie, L. L., and Drake, H. L. (1983a). *J. Biol. Chem.* **258,** 2364–2369.

Ragsdale, S. W., Ljungdahl, L. G., and DerVantanian, D. V. (1983b). *J. Bacteriol.* **155,** 1224–1237.

Reid, S. J., Allcock, E. R., Jones, D. T., and Woods, D. R. (1983). *Appl. Environ. Microbiol.* **45,** 305–307.

Rogers, P., and Hansen, W. R. (1983). *Abstr. Annu. Meet. Am. Soc. Microbiol.* p. 239.

Rokos, E. A., Rood, J. I., and Duncan, C. L. (1978). *FEMS Microbiol. Lett.* **4,** 323–326.

Rood, J. I., Mahler, E. A., Somers, E. B., Campos, E., and Duncan, C. L. (1978a). *Antimicrob. Agents Chemother.* **13,** 871–880.

Rood, J. I., Scott, V. N., and Duncan, C. L. (1978b). *Plasmid* **1,** 563–570.

Rose, A. H., and Beavan, M. J. (1981). *In* "Trends in the Biology of Fermentation for Fuels and Chemicals" (A. Hollaender, ed.), pp. 513–532. Plenum, New York.

Schink, B., and Zeikus, J. G. (1983a). *J. Gen. Microbiol.* **129,** 1149–1158.

Schink, B., and Zeikus, J. G. (1983b). *FEMS Microbiol. Lett.* **17**, 295–298.

Schwartz, R. D., and Keller, F. A. (1982). *Appl. Environ. Microbiol.* **43**, 117–123.

Schweiger, G., and Buckel, W. (1984). *FEBS Biochem. Lett.* **171**, 79–84.

Scott, V. N., and Duncan, C. L. (1978). *FEMS Microbiol. Lett.* **4**, 55–58.

Shoemaker, S., Schweickart, V., Laduer, M., Gelfand, D., Kwok, S., Myambo, K., and Innis, M. (1983a). *Bio/Technology* **1**, 691–696.

Shoemaker, S., Watt, K., Tsitovsky, G., and Cox, R. (1983b). *Bio/Technology* **1**, 687–690.

Sinskey, A. J. (1983). *In* "Organic Chemicals from Biomass" (D. L. Wise, ed.), pp. 1–68. Benjamin/Cummings, Menlo Park, California.

Sinskey, A. J., Akedo, M., and Cooney, C. L. (1981). *In* "Trends in the Biology of Fermentations for Fuels and Chemicals" (A. Hollaender, ed.), pp. 473–492. Plenum, New York.

Sjolander, N. O. (1937). *J. Bacteriol.* **34**, 419–428.

Sleat, B., Mah, R. A., and Robinson, R. (1984). *Appl. Environ. Microbiol.* **48**, 88–93.

Sliwkowski, M. X., and Stadtman, T. C. (1983). *Fed. Proc.* **42**, 1974.

Smith, C. J., Markowitz, S. M., and Macrina, F. L. (1981). *Antimicrob. Agents Chemother.* **19**, 997–1003.

Smith, G. M., and Roberts, J. D. (1984). *Fed. Proc.* **43**, 1865.

Snedecor, B. R., and Gomez, R. F. (1983). *In* "Organic Chemicals from Biomass" (D. L. Wise, ed.), pp. 93–108. Benjamin/Cummings, Menlo Park, California.

Spivey, M. J. (1978). *Process Biochem.* **13**, 2–5, and 25.

Squires, C. H., Heffner, D. L., Evans, R. J., Kopp, B. J., and Yarus, M. J. (1984). *J. Bacteriol.* **159**, 465–471.

Strom, M. S., Eklund, M. W., and Poysky, F. T. (1984). *Appl. Environ. Microbiol.* **48**, 956–963.

Tatsuki, T., Imagawa, T., Kitajma, H., Higashi, Y., Chazono, M., Yanagase, Y., and Amano, T. (1981). *Bikens J.* **24**, 1–11.

Teeri, T., Salovuori, I., and Knowles, J. (1983). *Bio/Technology* **1**, 696–699.

Thauer, R. K., Jungermann, K., Henninger, H., Wenning, H., and Decker, K. (1968). *Eur. J. Biochem.* **4**, 173–180.

Thauer, R. K., Jungermann, K., and Decker, K. (1977). *Bacteriol. Rev.* **41**, 100–180.

Truffaut, N., and Sebald, M. (1983). *Mol. Gen. Genet.* **189**, 178–180.

Twarog, R., and Wolfe, R. S. (1962). *J. Biol. Chem.* **237**, 2474–2477.

Urano, N., Karube, I., Suzuki, S., Yamada, T., Hirochika, H., and Sakaguchi, K. (1983). *Eur. J. Appl. Microbiol. Biotechnol.* **17**, 349–354.

Valentine, R. C., and Wolfe, R. S. (1960). *J. Biol. Chem.* **235**, 1948–1952.

Vollherbst-Schneck, K., Sands, J. A., and Montenecourt, B. S. (1984). *Appl. Environ. Microbiol.* **47**, 193–194.

Walker, G. C. (1983). *In* "Basic Biology of New Developments in Biotechnology" (A. Hollaender, A. I. Laskin, and P. Rogers, eds.), pp. 349–376. Plenum, New York.

Walton, M. T., and Martin, J. L. (1979). *In* "Microbial Technology" (H. J. Peppler and D. Perlman, eds.), 2nd Ed., Vol. I, pp. 187–209. Academic Press, New York.

Wang, G., and Wang, D. I. C. (1984). *Appl. Environ. Microbiol.* **47**, 294–298.

Westheimer, F. H. (1969). *In* "Methods in Enzymology" (J. M. Lowenstein, ed.), Vol. 14, pp. 231–241. Academic Press, New York.

Westhuizen, A., van der, Jones, D. T., and Woods, D. R. (1982). *Appl. Environ. Microbiol.* **44**, 1277–1281.

Wiegel, J. (1980). *Experientia* **36**, 1434–1446.

Wiegel, J. (1982). *Abstr. Annu. Meet. Am. Soc. Microbiol.* p. 112.

Wiegel, J., and Dykstra, M. (1984). *Appl. Microbiol. Biotechnol.* **20**, 59–65.

Wiegel, J., and Ljungdahl, L. G. (1981). *Arch. Microbiol.* **128**, 343–348.

Wiegel, J., Ljungdahl, L. G., and Rawson, J. R. (1979). *J. Bacteriol.* **139**, 800–810.

Wiegel, J., Braun, M., and Gottschalk, G. (1981). *Curr. Microbiol.* **5**, 255–260.

Wiegel, J., Carreira, L. H., Mothershed, C. P., and Puls, J. (1984). *Biotechnol. Bioeng. Symp.* (13), 193–205.

Wood, W. A. (1961). *In* "The Bacteria: A Treatise on Structure and Function" (I. C. Gunsalus and R. Y. Stanier, eds.), Vol. 2, pp. 59–150. Academic Press, New York.

Wood, W. A. (1981). *In* "Trends in the Biology of Fermentations for Fuels and Chemicals" (A. Hollaender, ed.), pp. 3–17. Plenum, New York.

Wust, J., and Hardegger, V. (1983). *Antimicrob. Agents Chemother.* **23**, 784–786.

Yamamoto, I., Saiki, T., Liu, S. M., and Ljungdahl, L. G. (1983). *J. Biol. Chem.* **258**, 1826–1832.

Young, M., and Mandelstam, J. (1980). *Adv. Microb. Physiol.* **20**, 103–162.

Youngman, P. J., Perkins, J. B., and Losick, R. (1983). *Proc. Natl. Acad. Sci. U.S.A.* **80**, 2305–2309.

Yu, E. K. C., and Saddler, J. N. (1983). *FEMS Microbiol. Lett.* **18**, 103–109.

Zeikus, J. G. (1980). *Annu. Rev. Microbiol.* **34**, 423–464.

Zeikus, J. G. (1983). *Adv. Microb. Physiol.* **24**, 215–299.

Zeikus, J. G., Ben-Bassat, A., Ng, T. K., and Lamed, R. J. (1981). *In* "Trends in the Biology of Fermentation for Fuels and Chemicals" (A. Hollaender ed.), pp. 441–462. Plenum, New York.

Zimmermann, V. (1982). *Biochim. Biophys. Acta* **694**, 227–277.

The Acetone Butanol Fermentation

B. McNeil and B. Kristiansen

Department of Bioscience and Biotechnology
Applied Microbiology Division
University of Strathclyde
Glasgow, Scotland

I. Introduction

The production of chemical feedstocks from carbohydrates by the action of microorganisms encompasses a wide range of microbes and processes, both aerobic and anaerobic. Essentially, it involves the use of pure cultures of microorganisms to produce basic feedstock chemicals of a relatively simple structure. By virtue of their prospective role such chemicals must be produced inexpensively. This basic fact, coupled with the difficulty of recovering an often small amount of product from a large volume of process water,

ADVANCES IN APPLIED MICROBIOLOGY, VOLUME 31

has been the major factor in determining industrial progress in these solvent fermentations throughout this century.

The list of products produced by the fermentation route is wide, and includes, or has included, glycerol, ethanol, 2,3-butanediol, isopropanol, lactic acid, acetone, and butanol.

With the exception of the ethanol fermentation, the most important of these fermentations has been the acetone/butanol fermentation carried out by microbes of the genus *Clostridium*. This process, once of great industrial significance, is now carried out on an industrial scale only where special conditions permit it to rival feedstock produced from a petrochemical source.

Pasteur was probably the first person to recognize butanol as a product of microbial action (1861). Some of the most eminent names in early microbiology worked on the process, including Beijerinck, Duclaux, Schardinger, and Prazmowski, who proposed the name *Clostridium* for this genus.

Much of the early interest focused on the possible use of butanol for conversion to butadiene, which would then be polymerized to give synthetic rubber. However, with the onset of the first World War, the demand for acetone, for use in cordite manufacture and as a general solvent, led to the setting up of a number of large plants in the United Kingdom, then Canada and the United States, producing acetone and butanol by the fermentation of starchy materials, usually grains. The process actually used was that pioneered by Weizmann (Weizmann, 1915, 1919). The ratio of butanol to acetone was approximately two to one, and a major difficulty arose with regard to the storage and disposal of the butanol which was largely regarded as a waste product.

As the war ended, these plants were gradually run down as demand for acetone fell sharply. Many plants were reopened, however, when a use was found for butanol as a solvent in the manufacture of nitrocellulose lacquers, which were important to the growing automobile industry. Subsequently, the acetone/butanol fermentation expanded rapidly, again using the classical Weizmann process, with grain as the raw material. Due to the difficult nature of grain as a fermentation substrate, investigations into the use of sugars were carried out. These culminated in the early 1930s in the development of a process for the fermentation of molasses which rapidly superseded the previous grain process. In 1945, the process was still flourishing, although competition from synthetic butanol (from acetaldehyde) was increasing steadily as the petrochemical industry expanded. In that year, 66% of United States butanol and 10% of the United States requirement for acetone was being met by fermentation.

Following this period, the fermentation route went into sharp decline for several reasons, including steep rises in the price of all types of molasses and grain, severe competition from the petrochemical industry, which at this period had an abundance of a very cheap raw material, and perhaps most relevant, the still relatively low yield of solvents produced by the microbiological process. Despite many attempts to improve the yield of the fermentation, it never rose above 33% on sugar fermented.

Economic pressure on the fermentation led to the necessity to recover all possible products of the fermentation; thus, the dried biomass (being high in riboflavin), was used as a feed additive for ruminants, and the fermentation gases hydrogen and carbon dioxide were recovered. Such integrated plants achieved a high degree of efficiency in operation. Similarly, rising molasses prices led to investigations into alternative, cheaper sources of fermentable carbohydrates, all of which proved successful at least to the pilot plant stage.

Following the quadrupling of the price of crude oil in 1973, the synthetic route to butanol became much more expensive, and this triggered a greatly revived interest in the potential of the fermentation. This interest has, so far, been sustained only at the research scale. The earlier empirical work had left many questions unanswered regarding the physiology and biochemistry of *Clostridium acetobutylicum*. For this reason, much of the current interest centers round these vital areas, seeking a more reasoned approach to the control of the fermentation.

II. The Organisms

Although considerable research had been carried out on the production of acetone and butanol from starchy materials by others previously, it is usual to describe the process of Weizmann (Weizmann, 1915, 1919) as being the first successful method of producing acetone and butanol by fermentation. He described the isolation, from natural sources, soil, or cereals, of a heat-resisting bacterium capable of fermenting starch or other carbohydrates largely to acetone and butanol. The fermentation could be carried out either aerobically, which he preferred, or anaerobically.

Weizmann's method of isolation involving successive heat shocking is still the method in current use (Calam, 1980; Spivey, 1978). Normally a potato-based medium is used for the serial subculturing of isolates obtained from soils or muds, cereals, potatoes, or the roots of leguminous plants, where the organism is in loose association with nitrogen fixers.

A systematic study of the "butyl alcohol organisms" was carried out (McCoy *et al.*, 1926) studying 11 different industrial producer strains from separate sources. The life cycle of the organisms was described in detail, placing them as members of the low acid/high alcohol-producing butyric acid bacteria. Morphologically, they are motile (by peritrichous flagella), gram-positive bacilli, bearing granulose (a storage carbohydrate) and swelling to clostridial form at sporulation. A single oval spore is located subterminally. Dimensions of the organism were stated to vary considerably with strain, cultural conditions and age, but typical dimensions are given in Table I.

It was concluded that since no organism previously described possessed the primary characteristics of this organism, none of the names of the earlier butanol producers was applicable. They therefore proposed the name *Clostridium acetobutylicum* (Weizmann).

When attempts were made to utilize molasses as a carbohydrate source, this original organism was found to be capable of fermenting a mixture of 50% starch and up to 50% molasses, though with greatly lengthened fermentation time (Beesch, 1953). Methods of isolating clostridia capable of using saccharine materials were developed, essentially analogous to those of Weizmann (Prescott and Dunn, 1959) with the substitution of invert molasses for starch, plus 4% ammonium sulfate, 5% calcium carbonate, and 0.3% phosphorous pentoxide (based on weight of the sugar). In 1938, a patent was taken out on a process involving fermentation of a 4 to 6% monosaccharide solution to give a 25 to 30% yield of solvents, mostly butanol, based on original sugar concentration (Muller, 1938). This organism was designated *Clostridium propylbutyricum* alpha.

There followed a period in which many saccharolytic solventogenic clostridia were isolated. These produced a varied ratio of neutral products, and had varied abilities to utilize a range of carbohydrate substances. Many patents were issued in respect to these naturally isolated strains. It has been suggested, however, that these organisms should be considered merely to be strains of *Clostridium acetobutylicum* (Beesch, 1953). Most such organisms, or strains, possess many of the characteristics of *C. acetobutylicum*, as listed in Bergey's eighth edition (Buchanan and Gibbons, eds., 1974).

Recent work suggests that taxonomy based on end product analysis is an

TABLE I

CELL TYPE AND DIMENSIONS OF *Clostridium acetobutylicum*
AT VARIOUS TIMES IN THE ACETONE–BUTANOL FERMENTATION[a]

Time after inoculation	Cell type	Dimensions (μm)
3 hours	Young vegetative rods, granulose negative	4.7 × 0.72
27 hours	Vegetative rods	3.4 × 0.66
	Clostridia, granulose positive	4.7 × 1.6
75 hours	Vegetative rods	2.6 × 0.6
	Clostridia	5.5 × 1.3
	Free spores	2.4 × 1.2

[a] From McCoy *et al.* (1926).

unreliable means of speciation in the solventogenic clostridia, since solvent production can be a very variable trait (George *et al.*, 1983). However, an organism isolated from soil producing relatively more butanol and less acetone than comparable cultures of *C. acetobutylicum*, has been named *Clostridium saccharoperbutylacetonicum* (Hongo and Murata, 1965). This organism is claimed to be a distinct species mainly on the basis of differences in types of infecting bacteriophages.

Some workers in the field distinguish between two fermentation types, viz. the acetone/butanol fermentation, carried out by *C. acetobutylicum* and closely related saccharolytic strains, and the butanol/isopropanol fermentation, of much less importance industrially, carried out by *Clostridium butylicum* and related organisms (Prescott and Dunn, 1959). The exact taxonomic status of many butanol-producing bacteria still awaits clarification.

III. The Fermentation Process

A. Inoculum Preparation

The inoculum procedure in current industrial use builds up in stages of progressively larger volume (Spivey, 1978). A heat-shocked spore suspension is added to 150 ml of potato glucose medium and grown for 12 hours. This is then used to inoculate 500 ml of molasses medium, grown for 6 hours, and added to 3.5 liters of the same medium for a further 6 hours incubation before being added to 9 liters of molasses medium and incubated for 9 hours. Finally, three of these 9-liter cultures are used as inoculum for 90,000 liters of medium. Such progressive increase in inoculum volume is routine in the acetone/butanol process though final inoculum size varies. Mass inoculation techniques allowing fermentation of up to 10% sugar solutions have also been developed (Weizmann, 1945).

The process of inoculum preparation in traditional grain and molasses fermentations was essentially the same (Beesch, 1952, 1953). In each case three stages, of increasing volume, led to a 4000-liter prefermenter stage. The final inoculum level was between 0.5 and 3%.

B. Production of Solvents from Starches

Starch-bearing grains and potatoes were the first sources of carbohydrates used in the acetone/butanol process. In industrial practice the grain of choice was maize. A flow diagram indicating what a typical fermentation plant using starch products may incorporate is shown in Fig. 1 (Beesch, 1952). The culture is usually added to the fermentor at the start, but in "deferred filling" processes, the fermentor is partially filled with corn mash, inoculated and the rest of the mash added later. Advantages claimed with

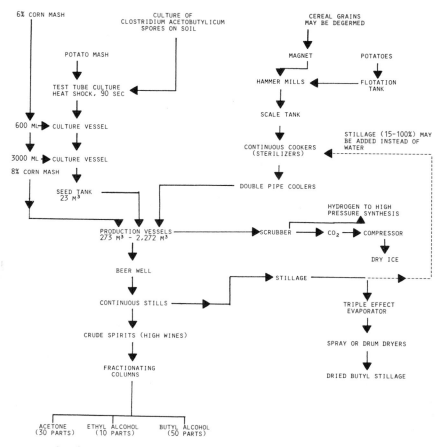

Fɪɢ. 1. Flow diagram of acetone–butanol fermentation using starch products (Beesch, 1952).

this method include the speeding up of the final fermentation, and the suppression of contamination.

Optimum temperature of the fermentation is 36 to 37°C though the fermentation is normal in the range 34 to 41°C. There is usually a loss of acetone at the higher temperatures, and a change in the solvent ratio.

The pH at the start of fermentation is usually 6 to 6.5 and at the end 4.2 to 4.4. The addition of calcium carbonate, to act as a buffer, causes a decrease in the production of acetone and butanol in direct proportion to the amount of calcium carbonate added, and a corresponding increase in the amount of butyric and acetic acids found.

The starch process does not require the use of medium supplements as the cooked grain supplies all necessary nutrients for growth of the organism. The use of butyl stillage in the fermentation as a replacement for part or all of the water used in diluting the mash has been proposed (Legg and Stiles, 1938). Advantages cited include a better than normal fermentation, a saving in water and steam, and less foaming in the fermentors. Stillage or "slop" may comprise up to 40% of the total medium volume. Others, however, considered continued use of stillage undesirable due to build-up of solid residues in the plant (Beesch, 1953).

The grain fermentation is complete in 50 to 60 hours with an overall solvent yield of 38%, based on starch. A typical product balance per 100 kg of starch used is butanol, 22 kg; acetone, 10.5 kg; ethanol, 5.3 kg; hydrogen, 1.7 kg; carbon dioxide, 62.4 kg (Beesch, 1953). Peterson and Fred (1932) obtained a similar ratio of butanol/acetone/ethanol of 6:3:1 in their starch-based fermentation.

The gases, hydrogen and CO_2, weigh almost one and a half times as much as the solvents. Total gas evolved contains, by volume, 60% CO_2 and 40% hydrogen. Once the hydrogen is separated from the CO_2, by passage through a scrubber, it may be used to synthesize ammonia. Alternatively, by passing the mixed exhaust gases over heated carbon, a mixture of carbon monoxide and hydrogen is produced which could be used in methanol synthesis. Occasionally the exhaust gases were burnt, after scrubbing to remove all solvents (approximately 1% of total solvent content was recovered, in the ratio 65% acetone, 30% butanol, 5% ethanol). The fermentation gas was sometimes used to maintain an anaerobic blanket over the culture in the early stage, until the culture began producing sufficient volume of gas to ensure anaerobiosis (Prescott and Dunn, 1959; Beesch, 1953).

The dried stillage, containing between 33 and 40% protein, and up to 40 to 70 μg of riboflavin is used extensively as a valuable feed additive for livestock (Beesch, 1953; Spivey, 1978).

Other Products

Other than residual acids (acetic and butyric acids), the products of the industrial fermentation include acetyl methylcarbinol, formic acid, and yellow oil (Wilson *et al.*, 1927). In a normal fermentation from 295 kg of starch, 5 kg of residual acids remain, and acetyl methylcarbinol was produced up to concentration of 0.3 to 0.4 g/liter (Reilly *et al.*, 1920). Yellow oil, composing 0.5 to 1% of total solvents, was shown to be a mixture of *n*-butanol, amyl alcohol, isoamyl alcohol, *n*-hexyl alcohol and the butyric, caprylic, and capric esters of these alcohols (Marvel and Broderick, 1925).

C. PRODUCTION OF SOLVENTS FROM MOLASSES

When the acetone/butanol fermentation first achieved industrial importance attempts were made to use commercial blackstrap molasses as a carbohydrate source using the Weizmann process. This, however, proved unsuccessful, though it was possible to substitute 50% of the starch with molasses. This procedure, using the original Weizmann-type cultures, produced a much slower fermentation (Beesch, 1952). It was not until saccharolytic strains of the organism were isolated in the late 1930s that the molasses fermentation took off (Arroyo, 1938; Woodruff et al., 1937). At first, it was found essential to invert the molasses, but sucrose-fermenting strains were rapidly isolated.

Despite the wide range of sugars which could be utilized by the new saccharolytic strains, in practice, the industrial processes developed using either invert or "high test" molasses, or blackstrap molasses. Typical composition of each type is shown below in Table II (Meade, 1963).

Molasses media were usually made up to a concentration of 4 to 6% sugar, but, unlike grain mash which constituted a complete medium, it was found necessary to supplement the molasses medium. Figure 2 shows a flow diagram of a typical acetone/butanol fermentation plant using sugar products as a raw material (Beesch, 1952). Additional nitrogen was supplied either by the addition of corn steep liquor, yeast water, or, by adding NH_3 or its salts. If ammonium sulfate was used, normally 4 to 6% of the weight of sugar was added (Rose, 1961). Ammonia was sometimes added to raise initial pH to 5.7 to 6.7 and it was also used as a titrant to keep the pH above 5.3 to 5.6. Most of the ammonia was added before the "break point" when the culture changes from being acidogenic to solventogenic. About 1 to 1.4% NH_3 was required, though this depended on the weight of sugar in the medium. If ammonium salts were used, calcium carbonate was used as a buffer, though, as in the grain process, this increases the proportion of acetone (Prescott and Dunn, 1959).

In general, nitrogen requirements vary with strain of organism and fermentation conditions (Rose, 1961). It was noted that NH_3 or its salts as sole nitrogen supplements did not give best yields. Thus, generally, a mixture of complex (e.g., corn steep liquor) and inorganic nitrogen sources were used, e.g., $(NH_4)_2SO_4$ (Beesch, 1952; Spivey, 1978).

TABLE II

COMPOSITION OF INVERT
AND BLACKSTRAP MOLASSES

	Invert molasses	Blackstrap molasses
Solids	84.5%	77–84%
Sucrose	27%	25–40%
Reducing sugars	50%	12–35%
Total sugars	77%	>50%

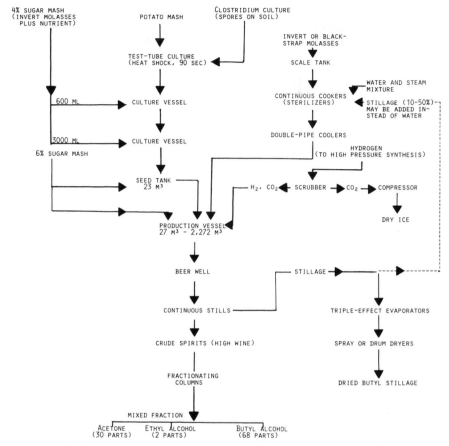

FIG. 2. Flow diagram of typical acetone–butanol fermentation using sugar products (Beesch, 1953).

Phosphate requirements also vary with type of raw material. Blackstrap molasses requires less additional phosphate than other types (Beesch, 1952). Phosphate was added in the form of monoammonium phosphate, calcium acid phosphate, or other soluble phosphates to a concentration of 0.2 to 0.4% (based on sugar).

The fermentation proceeded at a pH between 5 and 7, but best results were achieved with an initial pH in the region 5.5 to 6.5. After 16 to 18 hours pH fell to 5.3 to 5.5, with the final pH in the range 5.2 to 6.2. A pH lower than this indicated the likelihood of a *Lactobacillus* contaminant (Spivey, 1978).

The optimum temperature for the fermentation was around 30°C, lower than the starch process optimum. The demands for accurate temperature control were not stringent, as the fermentations were normally carried out in the range 29 to 35°C, though, as with the grain process, some acetone was lost at higher temperatures due to increased evaporation. Despite the loss of acetone, raising the temperature increased the proportion of acetone in the final product. It was noted that cooling a molasses fermentation to about 24–25°C about 16 hours after inoculation resulted in a significant increase in the amount of butanol at the expense of the other solvents (Carnarius, 1940).

Molasses fermentations were not normally agitated by mechanical means. Any agitation in the fermentors was achieved by sparging with CO_2 or by production of gas by the culture itself (Beesch, 1952; Spivey, 1978).

To maximize fermentor usage, the employment of finishing tanks has been advocated whereby the partially complete fermentation could be pumped from the pressure vessel to a nonsterile finishing tank, thus freeing the fermentor at an earlier stage. This could be done at any point after the "break point," the drawbacks being a slight reduction in yield, and a more acidic mash (Beech, 1953). The use of nonsterile mashes in the fermentation has been patented (Gavronsky, 1945). This involved development of a massive inoculum such that the clostridial culture outgrew all contaminants.

Duration of the molasses fermentation was normally 40 to 50 hours; however, in one modern process, the fermentation was said to be complete by 34 hours (Spivey, 1978). Solvent yields on sugar utilized ranged from 29 to 33% (Beesch, 1952). The ratio of solvents was different from that of the grain process, the ratio normally being of the order of 75% butanol:20% acetone:5% ethanol, but great variation in the ratio exists (Beesch, 1952; Prescott and Dunn, 1959; Rose, 1961). The ratio of solvents was very dependent upon strain of organism, but also depended on the origin, type, and nature of molasses. Under the same conditions, molasses of the same type (invert) from three different locations gave distinctly different solvent ratios. A typical material balance for the acetone/butanol fermentation using blackstrap molasses was given as follows (Beesch, 1952): Starting material: blackstrap molasses, 45.5 kg; total solids, 37 kg; sucrose and invert sugar, 26 kg; and protein, 1.4 kg. Yields were: butanol, 5.2 kg; acetone, 2.2 kg; ethanol, 0.23 kg; carbon dioxide, 14.6 kg; hydrogen, 0.36 kg; and dry feed, 13 kg.

Normally the limiting factor in the fermentation has been the final butanol concentration in the "beer." It was found that a concentration of butanol above 13.5 g liter^{-1} completely stopped fermentation, and with a 30% conversion to solvents, the sugar concentration in the medium was limited to around 6%. Higher concentrations of sugar were fermented less efficiently (i.e., resulting in higher residual sugar concentrations), though molasses media containing sugars at up to 10% have been fermented by the method of Weizmann (Weizmann, 1945).

The CO_2 produced was normally converted to dry ice, while the H_2 was either burnt (Spivey, 1978) or utilized in synthesis (Beesch, 1952, 1953). Normal dried stillage from the molasses fermentation contains various B group vitamins, including B_2 in amounts of 40 to 80 $\mu g/g$. Despite many attempts it proved impossible to raise these levels to those of the grain fermentation. The stillage was used as an animal feed additive, or recycled into the fermentation.

The advantages of using a molasses-based medium were said to include (Beesch, 1952) the following: (1) The mash is sterilized at lower temperature; 107°C for molasses as opposed to 126°C for starch. (2) A higher proportion of butanol is produced. (3) The fermentation is run around 33°C as opposed to 37°C for starch. (4) Equipment is more readily cleaned, stills less frequently blocked. (5) Molasses is normally cheaper than starches. (6) The fermentation is of shorter duration.

D. Production of Solvents from Other Raw Materials

Due to economic pressures, investigations into the utilization of a wide range of carbohydrate-containing wastes for solvent production have been carried out. *C. acetobutylicum* is versatile in its ability to utilize carbohydrates, being capable of using sugars such as xylose and arabinose, as well as hexoses.

Among the wastes proposed for use in acetone/butanol production are whey (essentially 4 to 5% lactose), which is readily available and easily transportable (Frey *et al.*, 1939; Meade *et al.*, 1945; Solomons, 1976), waste sulfite liquor (I. G. Farbenindustrie, 1938; Vierling, 1938; Wiley *et al.*, 1941; McCarthy, 1954), containing up to 2% monosaccharides, agricultural wastes such as wood shavings, bagasse, and rice straw (Leonard and Peterson, 1947; Langlykke *et al.*, 1948; Soni *et al.*, 1982). Both the agricultural and domestic wastes normally require pretreatment, notably acid or enzymatic hydrolysis, increasing their cost as fermentation raw material (Mes-Hartree and Saddler, 1982; Saddler *et al.*, 1983). However, in one study using *C. felsinae* the necessity for hydrolysis was avoided. It was possible to ferment garbage to give 45 liters of mixed solvents per ton of raw material (Jean, 1939).

E. Growth Rate in Batch Culture

The rate of cell growth of solventogenic species varies considerably and is influenced by medium composition, temperature, strain of organism, etc. After a short lag phase (2–3 hours at most) the organism quickly achieves its fastest vegetative growth rate. The maximum specific growth rate of *C.*

acetobutylicum in a synthetic medium has been estimated to be around 0.20 hour^{-1}, at 6–8 hours, thereafter declining slowly to around 0.15 hour^{-1} at about 11.5 hours. After this phase the proportion of cells showing signs of sporulation increases, while specific growth rate declines (Gottschal and Morris, 1981a).

IV. Factors Affecting the Fermentation

A. VITAMIN REQUIREMENTS

In general, the vitamin requirements of the solvent-producing clostridia were for many years obscured by the use of complex media containing materials such as yeast extract or corn steep liquor as sources of nitrogen and vitamins. In 1940, however, a synthetic medium was used to study the requirements of *C. acetobutylicum* (Oxford *et al.*, 1940). These studies showed that biotin and para-aminobenzoic acid were essential growth factors for the organism. It was further noted that, although all strains of *C. acetobutylicum* showed an absolute requirement for biotin, not all required p-aminobenzoic acid for growth (Lampen and Peterson, 1943). Other workers confirmed the vitamin requirements, usually using glucose as the carbon and energy source (Davies, 1942).

The knowledge of the organism's vitamin requirements allowed formulation of the synthetic media in common use nowadays, as shown in Table III (Andersch *et al.*, 1982).

TABLE III

GROWTH MEDIUM FOR *C. acetobutylicum*

Glucose	54 g liter^{-1}
$(NH_4)_2SO_4$	2 g liter^{-1}
KH_2PO_4	1 g liter^{-1}
K_2HPO_4	1 g liter^{-1}
$MgSO_4 \cdot 7H_2O$	0.1 g liter^{-1}
NaCl	0.01 g liter^{-1}
$Na_2MoO_4 \cdot 2H_2O$	0.01 g liter^{-1}
$CaCl_2 \cdot 6H_2O$	0.01 g liter^{-1}
$MnSo_4 \cdot 4H_2O$	0.015 g liter^{-1}
$FeSO_4$	0.015 g liter^{-1}
Sodium dithionite	0.035 g liter^{-1}
p-Aminobenzoic acid	2 mg liter^{-1}
Thiamine HCl	2 mg liter^{-1}
Resazurin	1 mg liter^{-1}
Biotin	0.1 μg liter^{-1}

B. Sugar Concentration

Using glucose or sucrose, cultures of solvent-producing clostridia were able to ferment media with concentrations of sugar ranging from 3 to 10% (Spivey, 1978). The ability of each strain to ferment various levels of sugar has to be determined experimentally. Similarly, organisms vary in their ability to ferment various sugars, the fermentability of sugars such as xylose and arabinose being much lower than glucose (Leonard and Petersen, 1947).

The initial sugar concentration used is normally 60 g liter^{-1} (as sucrose/glucose Eqs.) which yields a 20 g liter^{-1} solution of mixed solvents; the toxicity of butanol to the producing organism determines the substrate concentration that can be used economically, as shown in the next section. Using a synthetic medium Monot *et al.* (1982) found that at levels of glucose of 5 to 10 g liter^{-1} growth was poor, fermentation incomplete, and only 8 and 14%, respectively, of sugar consumed was converted to solvents. At a concentration of 20 g liter^{-1}, 27% of the glucose was converted to solvents with 13% being converted to acids (acetic and butyric), while at a glucose concentration of 80 g liter^{-1}, 31% of glucose converted went to solvent production, but 13 g liter^{-1} glucose remained unfermented. Best fermentation and yield of solvents was at a glucose concentration of 40 g liter^{-1}.

C. Butanol Toxicity

Strains vary in their resistance to butanol quite considerably. The toxicity of butanol is believed to be exerted on the lipoprotein envelope of the cell. Many attempts have been made to develop strains of *C. acetobutylicum* capable of resisting a butanol concentration of greater than 13 g liter^{-1}. These early attempts were almost totally unsuccessful. The experiments usually involved ultraviolet irradiation of cultures, then selection of mutants. Slightly increased tolerance was found when inocula containing large numbers of active bacteria were used (Ryden, 1958). It was demonstrated that butanol concentrations between 7 and 16 g liter^{-1}, which are within the range obtained in industrial fermentations, increase the rate of autolysis of clostridia, but had no effect on an autolysis-deficient mutant which could grow in higher concentrations of butanol. Acetone at concentrations up to 20 g liter^{-1} had no effect on the rate of autolysis (van der Westhuizen *et al.*, 1982).

There have been several recent attempts to produce butanol-tolerant strains. One method used was that of serial enrichment whereby growing cultures of *C. acetobutylicum* were challenged by butanol (5 g liter^{-1}) and the fastest growing culture was then subcultured into fresh media containing increasing butanol concentrations. After 12 such transfers a strain capable of growing in the presence of high butanol concentrations was produced. The parent strain showed no growth at 15 g liter^{-1} butanol while the mutant grew at 66% of the uninhibited control. The mutant also produced 5 to 14% more butanol than its parent (Lin and Blaschek, 1983).

Another method involved plate cultivation. A culture of *C. acetobutylicum* was plated onto media containing 5, 7, 10, and 12 g liter^{-1} butanol. The cells were mutagenized by addition of a small crystal of nitrosoguanidine (NG) to the center of a plate. After 3 days growth, no inhibition was seen on the plates with 5 and 7 g liter^{-1} butanol whereas the 12 g liter^{-1} butanol plates showed a clear background of no growth. Butanol-resistant mutants were picked from around the zone of inhibition on the 10 g liter^{-1} butanol-containing plates, with one mutant producing 24 g liter^{-1} solvents, as opposed to 17 g liter^{-1} for the wild type. The mutant was also more resistant to butanol, ethanol, and methanol.

The butanol-resistant mutant also showed anomalous sporulation. It was proposed that mutation affected the membrane and the fact that the average chain length of fatty acids in the membrane was higher in the mutant, as was the percentage of plasmalogens in the phospholipids confirmed this (Hermann *et al.*, 1982).

Different methods to increase the butanol produced by the organism have been used. The addition of 6% activated carbon to a 12% sugar solution gave rise to a final butanol concentration of around 28.5 g liter^{-1} (Hongo and Nagata, 1961).

D. Oxygen

C. acetobutylicum is an obligate anaerobe; thus, many steps have been taken to ensure anaerobic conditions, ranging from use of blankets of fermentation gas on the industrial scale, to use of reducing agents, blankets of oxygen-free nitrogen, and gas-proof tubing on the experimental scale (van der Westhuizen *et al.*, 1982).

The mechanism of oxygen toxicity in *C. acetobutylicum* has been investigated in detail (O'Brien and Morris, 1971). Four main hypotheses for this toxicity were proposed: (1) Oxygen is itself a toxic agent. (2) Anaerobes require low redox potentials (E_h), e.g., around −200 mV or below, to grow well, but in the presence of O_2 these could not be achieved. (3) Organisms lacking catalase, e.g., the clostridia, are killed by the H_2O_2 formed by reducing some of the oxygen. (4) Oxygen is a more avid electron acceptor than the normal terminal oxidants so that anaerobes cannot maintain intracellular concentrations of electron donors such as NAD(P)H.

Growth of *C. acetobutylicum* was studied under three sets of conditions: (1) anaerobic (E_h, −400 to −370 mV), (2) aerated (E_h, −50 to 0 mV: dissolved O_2 < 1 μM), and (3) aerobic (E_h, +100 to +150 mV: dissolved O_2 40 to 50 μM). From Fig. 3 it can be seen that the specific growth rate of the organism under aerated and anaerobic conditions was very similar, at around 0.6 hour^{-1}, a value which differs significantly from that reported for a synthetic medium (Gottschal and Morris, 1981a). This difference may be accounted for by differences in pH (in the above experiments pH was main- tained at 7, hence no solvents were formed), the presence of 4 g liter^{-1} casein hydrolysate in the medium or differences in the regimes of culture maintenance and heat shocking.

Exposure of an anerobic culture (4 to 6 hours old) to oxygen (40 to 60 μM) for periods of up to 6 hours was not lethal, i.e., the culture could reestablish anaerobic conditions when oxygenation ceased. At high dissolved O_2 con- centrations (40 to 60 μM) (i.e., under aerobic conditions), the rate of glucose consumption fell, and growth was halted, as was net synthesis of DNA,

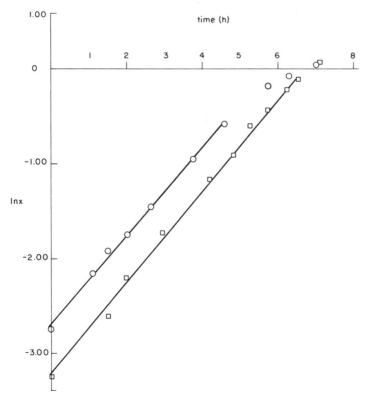

FIG. 3. lnx vs time under anaerobic and aerated conditions (adapted from O'Brien and Morris, 1971). ○, Anaerobic; □, aerated; x, biomass in g liter^{-1}.

RNA, and protein. The organism, under such aerobic conditions, was drained of reducing power and starved of energy, as evidenced by a cessation of butyrate (but not of acetate) formation and by a marked fall in intracellular ATP. These consequences of oxygenation were all reversible. There was no evidence to suggest formation of H_2O_2 nor could toxic effects be attributed merely to oxygen raising the culture E_h. Oxygen (40 μM) inhibited growth in a medium poised at -50 mV, whereas growth was normal in an anaerobic culture poised at $+370$ mV.

Thus, this work suggests oxygen itself to be directly toxic to the organism. On gradual oxygenation of a culture, the activity of NADH oxidase slowly increased, until it was five to six times that of an anaerobic culture. The enzyme was involved in reductive detoxification of the exogenous oxygen, and in doing so draining the cell of reducing power. It was noted that by short hourly periods of aeration in a fermenting rye flour mash the amount of butanol could be increased by 3.4 to 9.1% with concomitant increase in redox potential (Nakhmanovich and Kochkina, 1960).

E. Nitrogen

Molasses media were usually supplemented with inorganic sources of nitrogen. Tables IV and V show the effect of various inorganic nitrogen sources (Beesch, 1952).

Most synthetic media use 2 to 3 g liter^{-1} of $(NH_4)_2SO_4$ as the nitrogen source (Petitdemange et al., 1976; Gottschal and Morris, 1981a).

F. Butyric and Acetic Acids

In a typical sugar fermentation using C. acetobutylicum two distinct phases are seen. The first is a period of rapid growth associated with buty-

TABLE IV

EFFECT OF $(NH_4)_2SO_4$ CONCENTRATIONS
ON SOLVENTS[a]

$(NH_4)_2SO_4$ (as percent of added sucrose)	Solvent yield (%)
3.9	28.46
4.3	28.07
4.7	30.22
5.0	30.52
5.4	30.95
5.8	30.85

[a] From Beesch (1952).

TABLE V

Effect of Nitrogen Source on Solvent Yield

Nitrogen source	Solvent yield (%)	Acetone (percent of solvents)
$(NH_2)_2SO_4$	29.82	22.2
NH_4Cl	29.31	22.5
NH_4NO_3	23.68	40.9
$CH_3COO\ NH_4$	30.42	27.5
NH_4OH	31.58	22.4

rate, acetate, CO_2, and H_2 production. In the second phase, specific growth rate falls, acids are converted to solvents, and sporulation commences. What the underlying cause of this switch is has been under discussion for a long time.

Early work suggested the switch from acidogenesis to solventogenesis was merely due to the drop in pH caused by acid production (Ross, 1961) but when solvent production under controlled pH conditions (culture pH was held steady at 5.0) and the effect of acetate and butyrate (10 mM each) was examined, it was found that, instead of solvent production starting between 7 and 18 hours, it now commenced in between 1 to 3 hours and was associated with reduced specific growth rate and reduced H_2 evolution, while CO_2 evolution rate remained steady. All these features are characteristic of the solventogenic phase of culture. The culture was producing solvents at an earlier stage, and thus was essentially a vegetative culture; only 1 to 2% of cells showed signs of sporulation. This proved that reduced pH alone did not trigger the metabolic switch and that solventogenesis and sporogenesis were not causally linked (Gottschal and Morris, 1981b).

Addition of acetic acid at a concentration of 2 g liter^{-1} to a culture of *C. acetobutylicum* with a solvents yield on glucose of 32% and a butanol/acetone/ethanol ratio of 6:1.9:0.6 resulted in a yield of 34% and a butanol/acetone/ethanol ratio of 6:3:0.5. Addition of butyric acid (2 g liter^{-1}) increased production of all solvents to give an overall yield on glucose of 35% and a solvent ratio (butanol/acetone/ethanol) of 6:2.4:0.8. Addition of 4 g liter^{-1} butyric acid inhibited the culture. The addition of both butyric and acetic acids (2 g liter^{-1} each) gave a solvent yield of 34.7% with a ratio similar to that obtained by addition of acetic acid alone. This suggests that the initial concentration of the two acids strongly influences final solvent ratios (Martin *et al.*, 1983).

However, Yu and Saddler (1983) did not observe any beneficial effect on solvent production by addition of butyric and acetic acids (0.5 and 1.0 g liter^{-1}, respectively) to a culture grown on a glucose-based medium. Addition of 0.5 to 1.0 g liter^{-1} of acetic acid to xylose-based medium increased solvent levels to three to four times normal, while addition of butyric acid to a similar medium stimulated solvent production only when added prior to inoculation.

G. METALS

One of the most closely examined metals, in relation to its effect on the fermentation, is copper (Beesch, 1952, 1953; McCutchan and Hickey, 1954). The starch fermentation is inhibited by 40 ppm of copper and almost completely inhibited by 50 ppm. The maximum limit tolerated lies between 30 and 40 ppm of copper, though this depends on the strain. The molasses fermentation is more vulnerable, being inhibited by 2 ppm of the copper ion and completely inhibited by 5.0 ppm. The maximum tolerated level for strains of *C. saccharoaceto perbutylicum* was 1.0 to 2.0 ppm (Beesch, 1952).

The addition of aluminium, tin, iron, nickel, zinc, manganese, lead, cobalt, cadmium, chromium, thorium, thallium, and uranium ions at 50 ppm had no effect on the molasses fermentation (Beesch, 1952). Addition of mercury, from 7 to 50 ppm, delayed fermentation 24 hours while addition of antimony at 50 ppm seriously reduced yield.

The effect of adding various levels of metals was studied using a glucose-based synthetic medium. The following results were noted (Monot *et al.*, 1982):

1. MgSO₄

When no MgSO$_4$ was added, growth was very poor with poor utilization of sugar and low yield of solvents. Excess of MgSO$_4$ (0.35 g liter^{-1} and over) enhanced growth, but solvent yield decreased. At levels of 0.05 to 0.20 g liter^{-1} growth, solvent yield, and glucose utilization were optimal.

2. MnSO₄

The presence of MnSO$_4$ up to a concentration of 20 mg liter^{-1} had little effect on the fermentation. At 50 mg liter^{-1} the solvent yield decreased.

3. FeSO₄

Levels between 1 and 50 μg liter^{-1} of FeSO$_4$ gave similar results and optimal fermentation. In the absence of FeSO$_4$ growth was poor, and only 40% of sugar was utilized with a conversion to solvents of 25%.

4. KCl

Solvent levels increased with increases in potassium concentration from 0 to 60 mg liter^{-1}, then levelled off. The effect of very high levels of KCl (0.6

to 8 g liter^{-1}) was also studied. It was found that within this range substrate utilization was reduced but conversion of substrate to solvents was unaffected.

Other workers have confirmed that Mg and K play an important part in acetone formation by bacterial enzymes (Rosenfield and Simon, 1950). It was observed in one study that growth of *C. acetobutylicum* was K dependent (Davies, 1942).

H. Temperature

The temperature of the fermentation can affect overall yield, solvent ratios, and rate of solvent production.

In the molasses fermentation at 30°C yield of solvents was 31% with 23% of this acetone, at 33°C yield was 30% with acetone 26%, and at 37°C yield was around 24% with acetone being 38% of total solvents (McCutchan and Hickey, 1954). Similar results were obtained in a synthetic medium (see Table VI). The effect of temperature on the volumetric and specific rates of production are shown in Fig. 4.

I. Bacteriocin Production and Autolysis

C. acetobutylicum cells in an industrial fermentation were found to undergo autolysis in the stationary phase of the fermentation. A bacteriocin was isolated from these fermentations. This bacteriocin production had important implications in relation to solvent production in that it caused widespread lysis of the culture involving the spindle-shaped cells associated with solvent production; thus, its production decreased solvent yields (Barber *et al.*, 1979).

Release of the noninducible bacteriocin began late in the exponential

TABLE VI

Comparison of Temperature and Solvent Yield^a

Temperature of fermentation (°C)	Yield (%) $\left(\dfrac{\text{solvents (g liter}^{-1})}{\text{glucose (g liter}^{-1})} \times 100\%\right)$	Ratio (butanol/acetone)
25	29.1	3.48
30	28.4	3.70
37	25.5	4.73
40	24.9	5.67

^{a} From McNeil (1984).

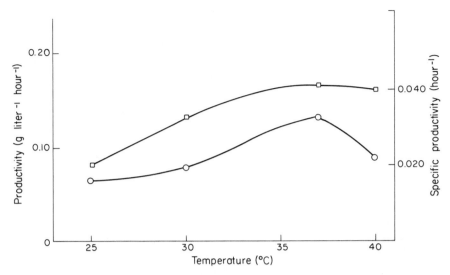

FIG. 4. Productivity, specific productivity of total solvents vs temperature in batch fermentations. □, Specific productivity; ○, productivity (McNeil, 1984).

phase (24 hours) and caused culture lysis and diminished solvent yield. The substance was characterized as a glycoprotein of M_r 28,000. The glycoprotein lysed cell wall preparations, and thus met some of the criteria of an autolysin. The autolysin gene appeared to be chromosomal since no plasmid DNA was detected in this *C. acetobutylicum* strain (Webster *et al.*, 1981). The relationship between high butanol levels and increased autolytic activity has been confirmed by van der Westhuizen *et al.* (1982).

The strain used in the above work, P262J, was later used in genetic transformation experiments. Attempts to transfer *Bacillus subtilis*, *Staphylococcus aureus*, and *Clostridum perfringens* plasmids to *C. acetobutylicum* had proved unsuccessful, but it was possible to transform protoplasts of strain P262J using phage CA I, which had been isolated from a phage infection of an acetone/butanol fermentation (Reid *et al.*, 1983). Such work, involving genetic manipulation, holds great promise for such a potentially important organism as *C. acetobutylicum*.

J. EFFECT OF REPEATED SUBCULTURING

As early as 1902, Winogradsky reported that continuous subculture of *Clostridium pasteurianum* brought about degenerative changes in the culture leading to decreased fermentation rate and absence of spores or clostridial forms in the culture.

TABLE VII

Effect of Subculture with Heat Shocking on Solvent Yield

Transfer	Total solvents (g liter^{-1})	Acetone (percent of total solvents)
2	15.08	35.45
4	16.03	32.40
6	18.40	28.00
8	19.30	30.50
10	19.98	28.50
12	19.85	25.80

Kutzenok and Aschner (1952), using a strain of *Clostridium butylicum*, found that direct subculturing could be carried out 6 to 10 times without any clear physiological change; after this, fermentation rate slowed down and the fermentation was incomplete. No slime was formed, and clostridial forms and spores were absent from the culture. Culture viability was low, and the product spectrum shifted from butanol with a little acetone to predominantly butyric acid with little solvents formed. Actively growing cultures and normal fermentations could be maintained only by heat shocking (immersion of a test tube containing the culture in boiling water for a period up to 90 seconds). It was stated that low pH (4.3) encouraged rapid degeneration (after three subcultures).

Maintaining a culture of *C. saccharoacetobutylicum* in the exponential phase of growth resulted in much less tendency to degenerate than when the culture was subjected to serial 24-hour transfers in the absence of heat shocking (Finn and Nowrey, 1959).

The effect of serial subculture with heat shocking at each stage is shown in Table VII (Beesch, 1953).

V. Biochemistry of the Fermentation

All accounts of the standard batch culture describe a series of metabolic stages in the acetone/butanol fermentation. The results of Peterson and Fred (1932) are typical. In this study a starch-based medium with an initial pH of 6.0 was used. The bacteria multiplied exponentially for up to 15 hours converting ca. 8 g liter^{-1} starch to give 1.3 g liter^{-1} acetic acid, 2.7 g liter^{-1} butyric acid, and very little solvents. During this time, culture pH fell. Total bacterial numbers remained steady for the next 15 hours or so and about 60% of accumulated acids were metabolized to form acetone and butanol. The proportion of spore-bearing cells increased. Starch utilization and solvent production continued with most solvents being formed in this period. Subsequently, bacterial numbers began to fall, solvent formation proceeded at a much slower rate, and some more acid accumulated as the remaining substrate was utilized. Evolution of H_2 and CO_2 occurred in a molar ratio of 2 to 1 in the first phase but fell to 1.5 to 1.

The organism breaks down glucose via the Embden–Meyerhof–Parnas pathway to give 2 mol of pyruvate, 2 mol of NADH + H^+, and 2 mol of ATP per mole of glucose fermented, followed by production of acetate from pyruvate via the "phosphoroclastic reaction" and the action of phosphotransacetylase and acetate kinase. Butyrate is also produced from pyruvate via the reoxidation of the NADH + H^+ produced by glycolysis, and the production of ATP during terminal conversion of butyryl-CoA to butyrate via phosphotransbutyrylase and butyrate kinase. Figure 5 shows the overall path-

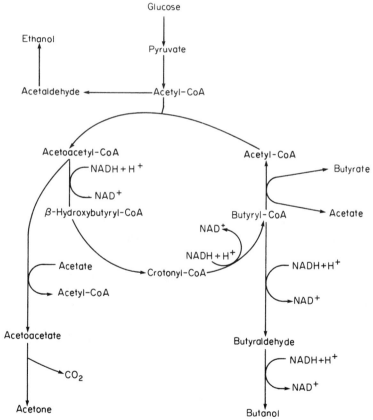

Fig. 5. Metabolic pathway leading to acetone–butanol formation (adapted from Doelle, 1975).

ways involved (Doelle, 1975). As can be seen, energy production is via substrate level phosphorylation. Substrate breakdown is associated with redox reactions. Electron transfers are effected by carriers such as NAD^+ and the reoxidation of such carriers is by means of ferredoxin. In the butyric clostridia production of acetyl-CoA and CO_2 from pyruvate is linked to reduction of ferredoxin. The redox potential of this coenzyme is -400 mV, i.e., low enough to allow reduction of protons to hydrogen via the enzyme hydrogenase (Dellweg, 1977). The overall effect of the pathways after the acetyl-CoA stage is to produce one extra molecule of ATP.

The reason for butyrate production via the cyclic system is that production of acetate as sole end product is unsatisfactory since it becomes more difficult to reoxidize $NADH + H^+$ as pH falls. Thus, the cyclic system producing the much less acidic end product butyrate comes into effect. An extra molecule of ATP can also be produced as follows (Valentine and Wolfe, 1960):

$$\text{Butyryl-CoA} + P \longrightarrow \text{butyryl}-P + \text{CoASH}$$
$$\text{Butyryl}-P + ADP \longrightarrow \text{butyrate} + ATP$$

The enzyme involved is phosphotransbutyrylase.

C. acetobutylicum differs from other butyric clostridia in that under the right conditions it can convert acids to neutral solvents. A transferase enzyme diverts acetoactyl-CoA from the normal cycling mechanism to give acetoacetate, which is then converted to acetone and CO_2 via the enzyme acetoacetate decarboxylase. This latter step is irreversible.

Diversion of the cyclic system to produce acetone stops further butyric acid formation. In addition to interruption of the cycle, two steps generating NAD^+ are eliminated; thus some other reduction process must be found. The reduction of butyrate to butanol does so in three steps (Doelle, 1975):

1. Acetyl CoA + butyrate $\longrightarrow$ acetate + butyryl-CoA
2. Butyryl CoA + NADH + H^+ $\longrightarrow$ butyraldehyde + NAD^+ + HS-CoA
3. Butyraldehyde + NADH + H^+ $\longrightarrow$ n-butanol + NAD^+

where (2) is catalyzed by aldehyde dehydrogenase, and (3) is carried out by NAD^+-linked alcohol dehydrogenase (the same enzyme used for ethanol formation).

The role of $NAD^+/NADH + H^+$ is vital throughout the series of reactions; thus, any mechanism regulating NAD^+ levels will affect solvent production. Regulation of NADH-ferredoxin oxidoreductase by acetyl-CoA (activator of NADH ferredoxin reductase activity) and by NADH (competitive inhibitor of ferredoxin:NAD^+ reductase activity) allows the enzymes to function correlatively with glycolytic enzymes to control levels of NAD^+ and $NADH + H^+$ in the cell. Since many reactions in acidogenesis/solventogenesis are NAD^+ linked they, too, are strongly influenced by this regulatory mechanism (Petitdemange et al., 1976).

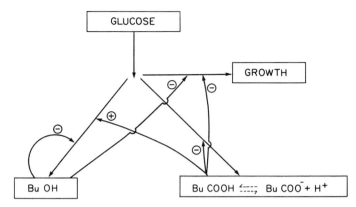

FIG. 6. Proposed regulation of growth and metabolism of *C. acetobutylicum* by its metabolites (Monot and Engasser, 1983a).

Monot and Engasser (1983a) proposed the scheme shown in Fig. 6 for the regulation of growth and metabolism in *C. acetobutylicum*.

VI. Economics

A. RAW MATERIALS

Essentially, acetone/butanol production was priced out of the market by rapidly increasing substrate prices and harsh competition based on very cheap acetone/butanol from petrochemical sources. Gibbs (1983) noted the inherent flexibility of a large petrochemical plant, which, unlike most fermentation plants, can vary the composition of product range to suit more closely market demand, making the synthetic route even more competitive.

It is relevant to consider the future of the process in the light of three areas (Solomons, 1976): (1) raw materials and transport costs, (2) processing costs, and (3) capital investments.

As regards raw materials, grain and other substrates, e.g., molasses, capable of food or feed use, will be so used and thus largely excluded from fermentation use. This leaves domestic, municipal, and agricultural wastes as the likely future feedstock. These have the disadvantages of being bulky, having low carbohydrate values and being, in many cases, seasonal. At present, with yields around 30%, Solomons (1976) estimated for plant outputs ranging from 50,000 to 500,000 tons per annum, that the use of more tractable wastes such as whey, straw, or wood hydrolysate would be required to show a profit.

The economics of an acetone/butanol plant utilizing wood chips as its raw

TABLE VIII

Comparison of Chemical vs Fermentation
Ethanol Production Cost Factors[a]

Cost factors	Ethylene chemical route	Anaerobic fermentation route
Raw materials	62	50
Utilities	18	15
Capital depreciation and overhead	13	25
Labor and maintenance	7	10

[a] From Tong (1978).

material has been illustrated by Gibbs (1983). The chips were derived from local wood-processing plants. Total value of products for 1 week's operation, producing 2 tonnes of acetone, 4 tonnes of ethanol, 24 tonnes of butanol, was £15,660, whereas total cost was estimated to be £10,683 per week, utilizing 200 tons of sawdust or wood chips. No allowance was made for H_2, CO_2, production and disposal or the use of the dried biomass and no clear mention was made of the cost of treating the massive volume of effluent.

It seems likely that any revival of the fermentation will depend on the availability of wastes and their efficient transport and utilization. In this context, however, it is worth noting that Tong (1978) estimated unit costs for feedstocks derived from n-paraffins at 38 (cents/kg of carbon), 28 from ethylene, and 27.5 from sugar in molasses (1977 values). Thus, some traditional feedstocks of the fermentation route were cheaper in 1977 than petroleum-derived feedstocks. This was illustrated by a consideration of the cost factors in chemical and fermentation ethanol production (Table VIII). The need to consider waste utilization in the fermentation route to acetone/butanol production was strongly emphasized (Gibbs, 1983; Zeikus, 1980).

Total crop utilization has been emphasized for solvent production; bagasse produced from maize or sugar cane cropping is used to generate steam (Tong, 1978) while the plant utilizes the sugar/starch produced. Improved strains of the organism capable of fermenting more sugar to butanol would be of double benefit in that productivity would increase while recovery costs would decrease (Hermann et al., 1982).

B. Product Recovery

One area in which improvement could lead to great benefits is that of product recovery. At present, as in the past, this is carried out by a distillation procedure, which when dealing with a maximum of 2% solvents has three major effects: (1) plants must be very large to be economical, (2) recovery is very expensive, and (3) a very large volume of process effluent has to be dealt with. Weizmann (1948) made many unsuccessful attempts to remove solvents by selective adsorption. Recent studies using silicalite, a zeolite analog capable of selectively adsorbing small organic molecules, show that this method of concentrating butanol may hold considerable promise for the future (Maddox, 1983). Continuous adsorption of butanol during the fermentation may also bring about, as an added advantage, reduced product inhibition. Alternatively, improvements in the areas of dialysis fermentation or ultrafiltration may allow solvent separation without distillation.

One of the major aims of future acetone/butanol plants must be to minimize capital installation costs. A major boost could be the use of much cheaper materials such as plastics, combined with nonsterile fermentation procedures. It has been estimated that almost two-thirds of direct energy costs in an acetone/butanol plant go into distillation, almost all the remainder being involved in sterilization (Walton and Martin, 1979). If a means of nonsterile fermentation could be effected it might be economically very worthwhile despite any possible reduced yield. The problems of phage contamination could be overcome by the use of immune strains.

The acetone/butanol fermentation has been singled out by many as having immediate potential (Solomons, 1976; Zeikus, 1980). Certainly, after many years of neglect there is much renewed interest in the fermentation. With careful planning the fermentation could become viable again, although initially, it seems to offer greatest promise for the third world or major agricultural countries.

VII. Continuous Culture

Industrially, perhaps the most promising aspect of continuous culture is the increased productivity often associated with this technology when compared to conventional batch fermentations. Despite the advantages offered, continuous culture has largely remained a research tool, possibly due to the very real difficulties inherent in such a technique, including the difficulty in preventing contamination, the possibility of mutation in the organism, the possibility of selecting unwanted strains of the organism by the imposed conditions and the capital cost of conversion from batch to continuous. Another major drawback is the product recovery stage, often the rate-limiting step in any fermentation. Despite these difficulties, however, the development of a system of continuous solvent production using *C. acetobutylicum* has been suggested (Bu'Lock, 1975). A continuous acetone/butanol fermentation was patented as long ago as 1932 but the process was never adopted on a commercial scale (Wheeler and Goodale, 1932).

The first clear report of continuous cultivation and solvent production is that of Dyr *et al.* (1958). The plant is described as having one seed tank and seven fermentors of 220–270 m³, with separately adjusted feeds of flour/molasses/wood hydrolysate mixtures to a concentration of 4 to 6% carbohydrate. The net flow rate through each was 20–28 m³/hour, and the system was said to give a 20% increase in productivity over the batch process, as well as substantial improvement in substrate economy.

Normal duration of continuous running was less than 90 hours because of infection problems. It is not clear whether productivity figures include any allowance for shut-down and start-up times.

Recent reports of glucose-limited cultures of *C. acetobutylicum* (Gottschal and Morris, 1981a; Bahl *et al.*, 1982a) have indicated that, under such conditions, the organism produces exclusively acetic and butyric acids as its end products, with production of solvents being almost totally absent in rigorously limited cultures. Gottschal and Morris (1981a) cultivated *C. acetobutylicum* NC1B8052 in a synthetic medium containing glucose, 2 g liter^{-1}; KH_2PO_4, 1 g liter^{-1}; K_2HPO, 1 g liter^{-1}; NH_4Cl, 0.4 g liter^{-1}; $MgSO_4 \cdot 7H_2O$, 0.4 g liter^{-1}; biotin, 10 μg liter^{-1}; *p*-aminobenzoic acid, 100 μg liter^{-1}; and trace elements.

Culture temperature was 35°C and strict anaerobic conditions were maintained, including the use of a blanket of oxygen-free nitrogen, from which the last traces of oxygen had been removed by passage over a heated copper catalyst. Only traces of solvents were formed; the fermentation was essentially acidic in nature. Analysis of the fermentation products accounted for only 82 to 87% of sugar fermented, although additional products, such as ethanol, isopropanol acetoin, and lactate were not present in significant concentrations. These nonsolvent-producing cultures were also found to have lost their ability to sporulate when reincubated on either solid or in broth media. This loss of the ability to sporulate seemed to be irreversible.

A similar loss of solvent-producing ability occurred in nitrogen-limited chemostat cultures of strain NC1B8052 (Gottschal and Morris, 1981a) and of strain DSM1731 (Andersch *et al.*, 1982). In both cases acetone and butanol were produced in very small amounts; maximum concentration at pH 6.0 and dilution rate 0.127 hour^{-1} was 5.7 m*M* butanol (Bahl *et al.*, 1982a). On decreasing the pH of nitrogen-limited cultures stepwise, the proportion of solvents amongst the products increased at pH values below 5.4. However, the overall fermentation process slowed down, with less glucose fermented, and ammonium ions appeared in the effluent. It appeared that although nitrogen-limited cultures of *C. acetobutylicum* produced solvents at acid pH, the rate of product formation was adversely affected; thus, continuous cultivation seemed not to be useful for industrial scale solvent production.

Addition of butyric acid to the medium in glucose-limited cultures at pH 4.3 was reported to have a stimulatory effect on production of acetone and butanol. The stimulatory effect reached a plateau at an inlet concentration of 60 mM butyrate (4.62 g liter^{-1}) when the culture contained 11 mM butanol (0.814 g liter^{-1}) and 5 mM acetone (0.29 g liter^{-1}). Butyrate had no stimulatory effect above pH 5.0 and its effect could not be duplicated by acetate (Bahl et al., 1982a).

However, Gottschal and Morris (1982) noted that in turbidostat cultures of strain NC1B8052 at relatively low cell densities the fermentation was essentially acidogenic in nature, whereas with increasing culture density the fermentation gradually changed to an essentially solventogenic nature. Acetate and butyrate (9 mM each) present in the inflowing medium could stimulate low cell density cultures which were acidogenic to produce solvents. These solventogenic cultures were found to be asporogenous.

Similarly, Monot and Engasser (1983b) found that significant production of solvents (8.5 g liter^{-1} from 25.5 g liter^{-1} glucose fermented) could occur if glucose was always in excess and dilution rate was low. Maximum solvent production occurred with 20 g liter^{-1} of glucose unfermented (from the original 45.5 g liter^{-1}) at a pH of 5.0 and dilution rate of 0.038 hour^{-1}.

Using phosphate limitation at a pH of 4.3 it was possible to ferment 54 g liter^{-1} glucose to give 9.6 g liter^{-1} butanol and 4.3 g liter^{-1} acetone at dilution rate 0.025 hour^{-1}, with 20% glucose unfermented. A two-stage continuous system was utilized with a first stage operating at 37°C and dilution rate of 0.125 hour^{-1}, while the second operated at a dilution rate of 0.04 hour^{-1} and 33°C; 87.5% of the glucose supplied was converted to solvents, the rest going mostly to cells and minor amounts of acids. It was again noted that raising the pH stepwise from 4.3 decreased solvent production, while increasing acid production (Bahl et al., 1982b).

VIII. Solvent Production Using Immobilized Cells

Spores of C. butylicum immobilized on calcium alginate beads were used in a continuous solvent production system (Krouwel et al., 1981). The medium utilized was glucose (6% w/v), yeast extract (1% w/v) and $CaCl_2 \cdot 2H_2O$ (0.5% w/v). The operating temperature was kept at 37°C and a 10-cm-long conical column was run for 215 hours during which time column activity fell somewhat in correspondence with pH in the column. Mean productivity of the immobilized system was 1.0×10^{-3} kg butanol liter^{-1} hour^{-1} which is approximately four times that of the conventional batch fermentation. Yield of n-butanol and isopropanol on glucose was around 30%, final butanol concentration never rising above 5 g liter^{-1}, above which level the organism was very strongly inhibited. Throughout the experiment the cells were growing actively. Biomass concentrations in the effluent of 0.85 g liter^{-1} were achieved.

Haggstrøm and Molin (1980) immobilized a mixture of vegetative cells and spores of *C. acetobutylicum* ATCC824 (NC1B8052) on calcium alginate beads. A medium containing glucose plus various inorganic salts which did not encourage growth was used. Product formation using immobilized exponential phase cells was identical to a normal batch fermentation, except that yields were lower. Butanol yield on glucose was 17.6% and on acetone 3.9%.

Using immobilized stationary phase cells in a continuous process butanol was produced. Solvent production was most rapid with butyric acid in the feed, though yield was lower (butanol yield was only 11% on glucose). With glucose alone as substrate, production rate was halved but yield rose to 15.3%. When spores were immobilized, yields of butanol of 20.9% were achieved, using glucose as sole carbon source, and yields of 15.3% using glucose and butyric acid, although, in the latter case, specific productivity was higher.

Productivity of the system was 0.48 to 0.64 g butanol liter^{-1} hour^{-1}. The fermentation, using immobilized spores, was carried out over a period of 11 days during the first 2 to 4 days of which the rate of butanol production fell to 15% of the original rate.

IX. Future Prospects

To assess the future potential of the acetone/butanol fermentation a number of points must be considered: (1) Any fermentation process will have to compete with the existing petroleum-based processes. (2) It will be essential to utilize fully integrated plants using the most inexpensive sources of carbohydrates, e.g., waste products, cellulosics, and such crops as mycotoxin-contaminated grains, which are unusable in any food/feed context. (3) It would be advantageous if such future plants were flexible with respect to substrates used, thus allowing the exploitation of relatively short-term surpluses.

Bearing such factors in mind, it seems possible that the fermentation could become industrially significant again, especially if improvements could be made to the efficiency of product recovery. The process's inherent advantages, namely, the absence of the need to aerate or agitate vigorously, help to ensure that the fermentation is not too closely linked economically to rising energy prices.

REFERENCES

Andersch, W., Bahl, H., and Gottschalk, G. (1982). *Biotechnol. Lett.* **4**, 29–32.
Arroyo, R. (1938). U.S. Patent 2, 113, 471.
Bahl, H., Andersch, W., Braun, K., and Gottschalk, G. (1982a). *Eur. J. Appl. Microbiol. Biotechnol.* **14**, 17–20.
Bahl, H., Andersch, W., and Gottschalk, G. (1982b). *Eur. J. Appl. Microbiol. Biotechnol.* **15**, 201–205.
Barber, J. M., Robb, F. T., Webster, J. R., and Woods, D. R. (1979). *Appl. Environ. Microbiol.* **37**, 433–437.
Beesch, S. C. (1952). *Ind. Eng. Chem.* **44**, 1677.
Beesch, S. C. (1953). *Appl. Microbiol.* **1**, 85–95.
Buchanan, R. E., and Gibbons, N. E., eds. (1974). "Bergey's Manual of Determinative Bacteriology," 8th Edition. Williams & Wilkins, Baltimore.
Bu'Lock, J. D. (1975). *In* "Large Scale Fermentation of Organic Solvents." Octagon Papers II.
Calam, C. T. (1980). *Biotechnol. Lett.* **2**, 111–116.
Carnarius, E. H. (1940). U.S. Patent 2, 198, 104.
Davies, R. (1942). *Biochem. J.* **36**, 582–589.
Dellweg, H. (1977). *In* "Advances in Biotechnology" (M. Moo-Young and C. W. Robinson, eds.), Vol. II. Pergamon, London.
Doelle, H. W. (1975). "Bacterial Metabolism." Academic Press, London.
Dyr, J., Protiva, J., and Praus, R. (1958). Formation of neutral solvents in a continuous fermentation. *In* Symposium on Continuous Culture, Czech. Acad. Sci., Prague.
Finn, R. K., and Nowry, J. E. (1959). *Appl. Microbiol.* **7**, 29.
Frey, A., Gluck, H., and Ochme, H. (1939). U. S. Patent 2,166, 047.
Gavronsky, J. O. (1945). U.K. Patent 571,630.
George, H. A., Johnson, J. L., Moore, W. E. C., Holdeman, L. V., and Chen, J. S. (1983). *Appl. Environ. Microbiol.* **45**, 1160–1163.
Gibbs, D. F. (1983). *Trends Biotechnol.* **1**, 12–15.
Gottschal, J. C., and Morris, J. G. (1981a). *Biotechnol. Lett.* **3**, 525–530.
Gottschal, J. C., and Morris, J. G. (1981b). *FEMS Microbiol. Lett.* **12**, 385–389.
Gottschal, J. C., and Morris, J. G. (1982). *Biotechnol. Lett.* **4**, 477–482.
Haggstrøm, L., and Molin, N. (1980). *Biotechnol. Lett.* **2**, 241–246.
Hermann, M., Fayolle, F., Renard, J. M., Marchal, R., Vandecasteele, J. P., and Sebald, M. (1982). *Poster, XIII Int. Congr. Microbiol., Boston.*
Hongo, M., and Murata, A. (1965). *Agric. Biol. Chem.* **29**, 1135–1139.
Hongo, M., and Nagata, K. (1961). *Chem. Abstr.* **55**, 19126.
I. G. Farbenindustrie A. G. (1938). U.K. Patent 496, 428.
Jean, J. W. (1939). U.S. Patent 2, 182, 989.
Krouwel, P. G., van der Laan, W. F. M., and Kossen, N. W. F. (1981). *Biotechnol. Lett.* **2**, 253–258.
Kutzenok, A., and Aschner, M. (1952). *J. Bacteriol.* **64**, 829–836.
Lampen, J. O., and Peterson, W. H. (1943). *Arch. Biochem.* **2**, 443.
Langlykke, A. F., van Lanen, J. M., and Fraser, D. R. (1948). *Ind. Eng. Chem.* **40**, 1716.
Legg, D. A., and Stiles, H. R. (1938). U.S. Patent 2,122,884.
Leonard, R. H., and Peterson, W. H. (1947). *Ind. Eng. Chem.* **39**, 1443.
Lin, Y. L., and Blaschek, H. P. (1983). *Appl. Environ. Microbiol.* **45**, 966–973.
McCarthy, J. L. (1954). *In* "Industrial Fermentations" (L. A. Underkofter and R. J. Hickey, eds.). Chemical Publishing Co., New York.

McCoy, E., Fred, E. B., Peterson, W. H., and Hastings, E. G. (1926). *J. Infect. Dis.* **39**, 457–483.

McCutchan, W. N., and Hickey, R. J. (1954). *In* "Industrial Fermentations" (L. A. Underkofler and R. J. Hickey, eds.), pp. 347–352. Chemical Publishing Co., New York.

McNeil, B. (1984). Ph.D. thesis, University of Strathclyde, Glasgow.

Maddox, I. S. (1938). *In* "The Acetone-Butanol Fermentation and Related Topics" (J. C. Bu'Lock and A. J. Bu'Lock, eds.), pp. 95–97. Science and Technology Letters, Chameleon Press, London.

Martin, J. R., Petitdemange, H., Ballongue, J., and Gay, R. (1983). *Biotechnol. Lett.* **5**, 89–94.

Marvel, C. S., and Broderick, A. E. (1925). *J. Am. Chem. Soc.* **47**, 3045–3051.

Meade, G. P. (1963). "Cane Sugar Handbook," 9th Edition. Wiley, New York.

Meade, R. E., Pollard, H. L., and Rodgers, N. E. (1945). U.S. Patent 2,369,680.

Mes-Hartree, M., and Saddler, J. N. (1982). *Biotechnol. Lett.* **4**, 247–252.

Monot, F., and Engasser, J. M. (1983a). *In* "The Acetone-Butanol Fermentation and Related Topics" (J. D. Bu'Lock and A. J. Bu'Lock, eds.), pp. 117–119, Science and Technology Letters, Chameleon Press, London.

Monot, F., and Engasser, J. M. (1983b). *Biotechnol. Lett.* **5**, 213–218.

Monot, F., Martin, J. R., Petitdemange, H., and Gay, R. (1982). *Appl. Environ. Microbiol.* **44**, 1318–1324.

Muller, J. (1938). U.S. Patent 2,123,078.

Nakhmanovich, B. M., and Sheblikina, N. A. (1960). *Mikrobiologiya* **29**, 67–72.

O'Brien, R. W., and Morris, J. G. (1971). *J. Gen. Microbiol.* **68**, 307–318.

Oxford, A. E., Lampen, J. O., and Peterson, W. H. (1940). *Biochem. J.* **34**, 1588.

Peterson, W. H., and Fred, E. B. (1932). *Ind. Eng. Chem.* **24**, 237.

Petitdemange, H., Cherrier, G., Raval, G., and Gay, R. (1976). *Biochim. Biophys. Acta* **421**, 334–347.

Prescott, S. C., and Dunn, C. G. (1959). "Industrial Microbiology," 3rd Edition, pp. 250–283. McGraw-Hill, New York.

Reid, S. J., Allcock, E. R., Jones, D. T., and Woods, D. R. (1983). *Appl. Environ. Microbiol.* **45**, 305–307.

Reilly, J., Henley, F. R., and Thaysen, A. C. (1920). *Biochem. J.* **14**, 229–251.

Rose, A. H. (1961). "Industrial Microbiology," pp. 160–166. Butterworths, London.

Rosenfield, B., and Simon, E. (1950). *J. Biol. Chem.* **186**, 395–404.

Ross, D. (1961). *Prog. Ind. Microbiol.* **3**, 73–90.

Ryden, R. (1958). *In* "Biochemical Engineering" (R. Steel, ed.), pp. 125–129. Heywood, London.

Saddler, J. N., Yu, E. K. C., Mes-Hartree, M., Levitin, N., and Browell, H. H. (1983). *Appl. Environ. Microbiol.* **45**, 153–160.

Solomons, G. L. (1976). *Process Biochem.* **11**, 32–33.

Soni, B. K., Das, K., and Ghose, T. K. (1982). *Biotechnol. Lett.* **4**, 19–22.

Spivey, M. J. (1978). *Process Biochem.* **13**, 2–3.

Tong, G. E. (1978). *Chem. Eng. Prog.* **74**, 70–74.

Valentine, R. C., and Wolfe, R. S. (1960). *J. Biol. Chem.* **235**, 1948–1952.

van der Westhuizen, A., Jones, D. T., and Woods, D. R. (1982). *Appl. Environ. Microbiol.* **44**, 1277–1281.

Vierling, K. (1938). German Patent 659,389.

Walton, M. T., and Martin, J. L. (1979). *In* "Microbial Technology" (H. J. Peppler and D. Perlman, eds.), Vol. I. Academic Press, New York.

Webster, J. R., Reid, S. J., Jones, D. T., and Woods, D. R. (1981). *Appl. Environ. Microbiol.* **41**, 371–374.

Weizman, C. (1915). U.K. Patent 4845.

Weizmann, C. (1919). U.S. Patent 1,315,585.

Weizmann, C. (1945). U.S. Patent 2,377,197.

Weizmann, C., Bergmann, E., Sulzbacker, M., and Parisier, E. R. (1948). *J. Soc. Chem. Ind.* **67**, 225.

Wheeler, M. C., and Goodale, C. D. (1932). U.S. Patent 1,875,536.

Wiley, A. J., Johnson, M. J., McCoy, E., and Peterson, W. H. (1941). *Ind. Eng. Chem.* **33**, 606–610.

Wilson, P. W., Peterson, W. H., and Fred, E. B. (1927). *J. Biol. Chem.* **74**, 495–507.

Woodruff, J. C., Stiles, H. R., and Legg, D. A. (1937). U.S. Patent 2,089,522.

Yu, E. K. C., and Saddler, J. N. (1983). *FEMS Microbiol. Lett.* **18**, 103–107.

Zeikus, J. G. (1980). *Annu. Rev. Microbiol.* **34**, 423–464.

Zeikus, J. G. (1980). *Annu. Rev. Microbiol.* **34**, 423–464.

Survival of, and Genetic Transfer by, Genetically Engineered Bacteria in Natural Environments

G. Stotzky* AND H. Babich[†]

*Laboratory of Microbial Ecology, Department of Biology
New York University
New York, New York, and
†Laboratory Animal Research Center
Rockefeller University
New York, New York

I. Introduction

An estimated 10,000 laboratories are now conducting biotechnological research in public and private corporations, universities, and governmental agencies worldwide; more than 200 companies are marketing products of biotechnology (Saftlas, 1984). One aspect of biotechnology is genetic engineering, which involves introducing genes from viruses and prokaryotic and eukaryotic organisms into other organisms which do not normally exchange genetic information with these gene donors. This technology originated from studies that showed that DNA fragments from one organism (i.e., the donor system) could be enzymatically coupled *in vitro* to bacteriophage or plasmid DNA (i.e., the vehicle or vector system), and the recombinant DNA molecules so formed could then be introduced into bacteria (i.e., the host system), usually strains of *Escherichia coli* K12, in which they replicate and form chimeric microorganisms (Curtiss *et al.*, 1977). Considerable experimentation is being directed to the genetic engineering of microorganisms, and newly constructed genotypes are believed to be of considerable value not only for many areas of basic and applied research but also for economic exploitation.

There are many applications of this technology. For example, it has been possible to construct microbes capable of degrading recalcitrant pollutants,

ADVANCES IN APPLIED MICROBIOLOGY, VOLUME 31

such as mono- and dichlorophenyls and mono- and dichlorobenzoates, and persistent herbicides, such as 2,4,5-trichlorophenoxyacetic acid (2,4,5-T) (Reineke and Knackmuss, 1979; Chatterjee *et al.*, 1981; Pemberton and Don, 1981). Currently, there are 13 microbial pesticides (microbiological pest control agents: MPCA)—e.g., *Bacillus thuringiensis* var. *israelensis* used against mosquitoes—registered with the U.S. Environmental Protection Agency (EPA) for use in agriculture, forestry, and insect control, and recombinant DNA technology offers the prospect of greatly increasing the number and kinds of such microbial pesticides (Betz *et al.*, 1983). The drug industry has utilized recombinant DNA technology to construct bacteria capable of producing large quantities of human insulin and human growth hormone, as well as lymphotoxin, an anticancer agent (Saftlas, 1984). Recombinant DNA technology also holds the promise for the incorporation and expression of valuable new genetic traits in plants; an example is the use of the Ti plasmid of *Agrobacterium tumefaciens* as a vector for genes coding for nitrogen fixation (McDaniel, 1981; Shaw, 1986).

Much concern has been expressed about the possible biohazards resulting from the escape of genetically engineered microbes into natural environments (Curtiss, 1976; Sharples, 1983; Rissler, 1984). Considerable criticism has been focused on the use of *E. coli* in recombinant DNA research, as *E. coli* is a normal intestinal inhabitant of human beings, as well as of all warm-blooded animals, and one route of escape, with potential proliferation, is in the investigator. Furthermore, *E. coli* is "sexually promiscuous" and can transfer genetic information, especially that on plasmids, to representatives of over 40 gram-negative bacterial genera, including several pathogenic species. *E. coli* is also an "abnormal" occupant of rivers, streams, lakes, and estuarine waters and of soil in urban and agricultural areas (the result, in a large part, of the presence of human beings), thereby extending the possibility of genetic transfer to other bacteria in these environments. Because of the possibility of genetically engineered bacteria inadvertently escaping and subsequently establishing in the environment, debilitated strains of *E. coli* are commonly used as the host cells for engineered genes. For example, the host *E. coli* K12 strain χ1776 is sensitive to bile salts, thereby diminishing its potential survival specifically in the gastrointestinal tract; has a generation time that is two to four times greater than that of wild-type *E. coli*, thereby decreasing its ability to compete successfully with the indigenous microbiota; and requires diaminopimelic acid, an amino acid not commonly present in nature, and thymidine, the absence of which causes cell death and degradation of DNA (Curtiss *et al.*, 1977).

This article reviews the few studies that have evaluated the survival of bacterial hosts and cloning vectors (e.g., phages) and the transfer of genetic information, by the processes of conjugation, transduction, and transformation, in aquatic and terrestrial environments and on plants. Mention is also made of selected *in vivo* studies (i.e., in human beings and other animals), as the environmental factors that affect survival and genetic transfer *in vivo* have similarity to those that affect survival and genetic transfer *in situ* (i.e., in natural environments). For a more thorough review of studies on genetic recombination in bacteria *in vivo*, the following are suggested: Freter *et al.* (1979, 1983, 1984), Levy and Marshall (1979, 1981), Levine *et al.* (1983), and Duval-Iflah and Chappuis (1984).

II. General Concepts of Genetic Transfer in Bacteria

The transfer of genetic information in bacteria may involve extrachromosomal pieces of DNA, termed plasmids. Plasmids replicate independently of the chromosome of the host cell and, under most circumstances, are not essential for the growth and metabolism of the host cell. However, plasmids carry genes that may permit the host bacterium to survive better in adverse environments (Duval-Iflah and Chappuis, 1984). Plasmid-coded functions include resistance to antibiotics (e.g., penicillin, streptomycin, neomycin, chloramphenicol, erythromycin, kanamycin, tetracycline), heavy metals (e.g., arsenate, arsenite, bismuth, cadmium, lead, mercury, nickel), bacteriophages, bacteriocins, and ultraviolet light (plasmids that code for such resistances are termed R-plasmids); catabolism of sugars (e.g., lactose, raffinose, sucrose) and xenobiotics (e.g., toluene, xylenes, chlorobenzoates); functions associated with conjugation (e.g., formation of sex pili, transfer genes, DNA mobilization activity, surface exclusion properties); and interactions between bacteria and other organisms (e.g., crown gall tumor formation in plants by species of *Agrobacterium*, nitrogen fixation in legumes by species of *Rhizobium*) (Reanney, 1976; Beringer and Hirsch, 1984).

Plasmids occur as circular entities of supercoiled double-stranded DNA that range, in general, from 1 kilobase (kb; 1 kb contains 1000 nucleotide base pairs, which is enough to code for 1 or 2 genes; the average number of nucleotide base pairs in a gene is 900) to over 300 kb (Saunders, 1984). For example, a plasmid of 3.2 kb that coded for resistance to cadmium was identified in a species of *Staphylococcus aureus* (El Solh and Ehrlich, 1982); a plasmid with a mass of 150 megadaltons (MDa) (1 MDa $\simeq$ 1.5 kb) that specified both resistance to streptomycin and sulfonamides and degradation of toluene, *m*-xylene, and *p*-xylene was identified in a species of *Pseudomonas* isolated from a river (Yano and Nishi, 1980); whereas "megaplasmids," with a mass greater than 450 MDa, were identified in strains of *Rhizobium meliloti* and *Pseudomonas solanacearum* (Rosenberg *et al.*, 1982). Megaplasmids have also been identified in strains of *Alcaligenes eutrophus* (pHG1, 450 kb), *Alcaligenes hydrogenophilus* (pHG21a, 410 kb), *Pseudomo-*

nas facilis (pHG20, 350 kb), and *Paracoccus denitrificans* (pHG18, > 700 kb) (Hogrefe and Friedrich, 1984). Plasmids with a mass greater than 80 MDa are common in gram-negative bacteria, whereas those in gram-positive bacteria are usually much smaller. However, plasmids of 80 and 120 MDa were identified in actinomycetes isolated from the rhizosphere and root nodules of species of the genus *Alnus* (Dobritsa, 1984). As bacterial chromosomes are about 10^3 kb in size, a substantial portion of the total genetic material of some bacterial cells may occur as plasmid DNA (Saunders, 1984).

Plasmids may be maintained in multiple copies in the host bacterium, with small plasmids (< 35 kb) occurring at 10 to 40 copies per organism and large plasmids (> 35 kb) at 1 to 4 copies per organism (Saunders, 1984). For example, about 40 copies per cell of a 2.9-MDa plasmid that conferred resistance to tetracycline occurred in strains of *S. aureus* 649, whereas a 34.9-MDa plasmid that conferred resistance to arsenate, cadmium, and mercury occurred at 4 to 7 copies per cell (Chopra *et al.*, 1973). Cells of *Bacillus pumilus* strain NRS 576 contained 2 copies of a 28-MDa plasmid, whereas *B. pumilus* strain ATCC 7065 contained at least 10 copies per cell of a 6-MDa plasmid. Although the functions of these plasmids are not known (such plasmids are termed "cryptic" plasmids), the larger plasmid contained genes that coded for the inhibition of sporulation (Lovett, 1973). Furthermore, different plasmids may coexist in the same host cell. For example, a species of *Pseudomonas* isolated from soil contained two distinct plasmids: one plasmid, containing approximately 270 kb, coded for the catabolism of toluene, *m*-xylene, and *p*-xylene; the other, containing approximately 280 kb, coded for the ability to grow on phenylacetate and for resistance to mercury (Pickup *et al.*, 1983). A strain of *Pseudomonas stutzeri*, isolated from soil in a silver mine, contained three distinct plasmids, the largest (49.4 MDa) of which specified resistance to silver (Haefeli *et al.*, 1984).

A cell may lose its plasmids. The presence of a plasmid presumably creates an energy burden on the host. In the absence of the specific selection pressure (e.g., antibiotics, heavy metals) that favors the plasmid-containing cell, the plasmid may be lost or fragment, and plasmid-free cells accumulate in the nonselective environment (Freter, 1984; Saunders, 1984).

Plasmids often carry transposons and insertion sequences (IS), both of which are nucleotide sequences capable of transferring from one DNA molecule to another. Transposons carry a gene or genes that confer a selectable phenotype, whereas IS do not contain such markers (Williams, 1982). Unlike plasmids, transposons and IS cannot replicate independently and must be maintained as part of a functional replicon, such as a chromosome, a plasmid, or a bacteriophage. In addition to carrying genes for the process of transposition, transposons may carry resistance genes. For example, the transposon Tn1 (5.0 kb), codes for resistance to penicillin and occurs in many gram-negative bacteria, and Tn551 (5.2 kb) codes for resistance to erythromycin in *S. aureus* (Beringer and Hirsch, 1984; Saunders, 1984). Some transposons also possess genes for resistance to mercury (Radford *et al.*, 1981).

Both chromosomal and plasmid DNA can be transferred to other bacteria, of the same or different species, by several processes, including conjugation, transduction, transformation, and protoplast fusion. Conjugation requires cell-to-cell contact and involves the transfer of DNA from a donor to a recipient bacterium.

Conjugative plasmids (and, less commonly, conjugative transposons) in the donor cell both code for the production of sex pili, which are necessary for establishing cell-to-cell contact and through which the DNA molecules presumably move to the recipient cell, and contain transfer genes for the mobilization of the DNA molecules to the recipient cell. In addition to transferring their own plasmid DNA, conjugative plasmids may insert into and mobilize all or part of the host chromosomal DNA (Saunders, 1984). Such transfer of host chromosome occurs much less frequently than the transfer of plasmids (about 10^{-4} to 10^{-8} chromosomes per plasmid transferred) (Holloway, 1979). Conjugative plasmids are classified into incompatibility (Inc) groups. Incompatibility occurs when two different plasmids fail to coexist in a common host, with one plasmid being lost from the cell; this usually occurs between plasmids that have closely related systems for regulating their replication (Saunders, 1984).

Some conjugative plasmids have a limited host range, but other plasmids (so-called "promiscuous" or broad host range plasmids) are able to be transferred to recipient cells not of the same species or genus as the donor cell (Williams, 1982). For example, a TOL plasmid (i.e., a plasmid encoding for the degradation of toluene) in a species of *Pseudomonas* isolated from a river was transferred to strains of *Pseudomonas putida*, *Pseudomonas aeruginosa*, and *Pseudomonas ovalis* (Yano and Nishi, 1980). Plasmid pRD1, which specifies resistance to kanamycin, carbenicillin, and tetracycline, could be transferred from *E. coli* to *A. tumefaciens*, *Xanthomonas beticola*, and *Erwinia carotovora* (Kozyrovskaya *et al.*, 1984). Furthermore, *E. coli*, a chemoorganoheterotroph, can transfer plasmids (e.g., plasmids S-a, R388, and RP4 and its derivatives) to *Thiobacillus novellus*, a chemolithoautotroph (Davidson and Summers, 1983). In addition to transferring plasmid DNA to intergeneric recipient cells, certain promiscuous plasmids, notably R68.45, can mobilize host chromosomal DNA for transfer that extends beyond species and generic boundaries. However, because of the lack of homology, this chromosomal DNA cannot integrate into the chromosome of the new host, but it can be retained and inherited as an integrated, stable component of the plasmid, which now is housed in the new cell. Such plasmids are termed plasmid "primes" (Williams, 1982; Beringer and Hirsch, 1984).

Conjugation appears to be limited primarily to gram-negative bacteria, but a conjugation-like transfer of plasmid DNA has been observed in some gram-positive bacteria. For example, transfer of the plasmid, pAMβ1, which codes for resistance to erythromycin and lincomycin, occurred from a Lancefield group F streptococcus, strain DR1501, to *Streptococcus mutans*, *Streptococcus sanguis*, and *Streptococcus salivarius* (LeBlanc *et al.*, 1978) and from *Streptococcus faecalis* to *B. thuringiensis* (Lereclus *et al.*, 1983) and *Bacillus sphaericus* (Orzech and Burke, 1984). The transconjugants of *B. sphaericus* were able to transfer this plasmid to several isogenic strains of *B. sphaericus* (Orzech and Burke, 1984).

Many plasmids may integrate into the host chromosome, but this interaction is usually transient (Reanney *et al.*, 1983), although exceptions have been noted. For example, Kleeberger and Klingmüller (1980) mated cells of a nonnitrogen fixing, antibiotic-sensitive strain of *Enterobacter cloacae*, isolated from the rhizosphere of the grass, *Festuca heterophylla*, with *E. coli* K12 JC5466(pRD1). The pRD1 plasmid carried the genes for nitrogen fixation as well as for resistance to three antibiotics. Transconjugants were isolated that were able to fix nitrogen and were resistant to the antibiotics. However, subsequent subculturing in a minimal medium resulted in cells with different genotypes, and many of the cells either contained plasmids smaller than the original pRD1 or lacked plasmids. However, these plasmidless cells were able to fix nitrogen and were antibiotic resistant, suggesting that the pRD1 plasmid, with all its relevant genes, had become integrated into the chromosome of these cells.

Conjugative plasmids are generally assumed to constitute only a minority of the plasmids, and most plasmids are nonconjugative (i.e., they cannot themselves transfer their DNA) (Reanney *et al.*, 1983). For example, a survey of 433 isolates of enterobacteria obtained from 1917 through 1954 (i.e., the "preantibiotic" era) showed that 104 strains, or only approximately 24%, contained conjugative plasmids (Hughes and Datta, 1983). In general, conjugative plasmids are usually > 30 kb, in order to accommodate transfer and sex pili genes, whereas nonconjugative plasmids are smaller (Saunders, 1984). Both conjugative and nonconjugative plasmids may occur in the same cell. For example, *Neisseria gonorrhoeae* CDC67 has two plasmids, both conferring resistance to penicillin, with the 24.5-MDa plasmid being conjugative and the 4.4-MDa plasmid being nonconjugative (Roberts and Falkow, 1979).

Nonconjugative plasmids, however, may be mobilized to a recipient bacterium if a conjugative plasmid coresides in or is introduced into the donor cell (Saunders, 1984). For example, a strain of *P. stutzeri* contained the nonconjugative plasmid, pKK1, which confers resistance to silver ions. However, if the conjugative plasmid R68.45 was transferred from *P.*

aeruginosa PAO25 to the *P. stutzeri* strain, the resulting transconjugant was then able to transfer plasmid pKK1 to *P. putida* (Haefeli *et al.*, 1981). Such matings are termed "triparental matings."

To minimize the potential transfer of plasmids containing novel genes from a host bacterium to the microbiota indigenous to an environment (e.g., the gastrointestinal tract of human beings or aquatic and terrestrial environments), these genes are usually incorporated into nonconjugative plasmids that are poorly mobilizable by triparental mating (Curtiss, 1976). Such presumably safe plasmids include pBR322 and pBR325, which are generally hosted in debilitated strains of *E. coli* K12.

The insertion of foreign DNA into a nonconjugative plasmid appears to have little effect on the mobilizability of the plasmid. For example, Hamer (1977) constructed recombinants composed of various *E. coli* nonconjugative plasmids and fragments of fruit fly (*Drosophila melanogaster*) DNA. The recombinant plasmids remained nonconjugative, although they could be mobilized to a strain of *E. coli* by conjugative plasmids transferred from *Salmonella typhimurium* and *Salmonella panama*. Of the 47 recombinant plasmids studied, 46 were mobilized at approximately the same or slightly lower frequencies than the parental plasmids. For example, the mobilization frequencies (transconjugants/donor) of the nonconjugative plasmid, pSC101, using the mobilizing plasmids, Col IB and I, were 3.0×10^{-4} and 1.4×10^{-4}, respectively, and when a length of DNA from *D. melanogaster*, consisting of 2.9 kb, was inserted into pSC101, the respective mobilization frequencies were 1.8×10^{-4} and 1.2×10^{-4}. Only one recombinant plasmid [GM4(pDm2)] was mobilized 1000 times less frequently than its parental plasmid, presumably as the result of the specific piece of DNA from *D. melanogaster* that was inserted.

In the transfer of DNA by transduction, a temperate bacteriophage adsorbs to the surface of a recipient bacterium and injects its viral DNA, which is either inserted into the chromosome of the host bacterium or is maintained essentially extrachromosomally as self-replicating circular DNA (such as P1 phage) (Beringer and Hirsch, 1984). The phage DNA that is either integrated into the host chromosome or persists essentially as a plasmid is termed "prophage," and the host cell containing this phage DNA is termed "lysogenic." Temperate phages, as opposed to virulent phages, do not kill their hosts immediately after infection. However, exposure to chemical (e.g., mitomycin C, hydrogen peroxide) or physical (e.g., heat, ultraviolet light) agents that damage DNA can induce intracellular production of intact phage particles that then lyse the host cell. A small proportion of these new phage particles from the host cell may contain, in addition to their own viral DNA, chromosomal or plasmid DNA from the host cell. Such phages are termed "transducing phages" (Lacey, 1984; Saunders, 1984). Infection of other bacteria with these transducing phages may introduce novel bacterial DNA into the recipient cells. As the volume of the heads of bacteriophages is usually much smaller than the chromosomes of the host cells, it is unusual for even as much as 1% of the bacterial genome to be incorporated into the transducing phage (Beringer and Hirsch, 1984).

Transduction appears to be the major mechanism for the transmission of antibiotic resistance genes in lysogenic strains of S. aureus (Lacey, 1975) and Streptococcus pyogenes. For example, after treatment with mitomycin C, erythromycin-resistant lysogenic strains of S. pyogenes released transducing phages that were able to transfer chromosomally-borne erythromycin re-sistance to antibiotic-sensitive strains of S. pyogenes (Hyder and Streitfeld, 1978). Transduction of antibiotic resistance coded on a plasmid also had been noted; e.g., resistance to tetracycline, which was initially encoded on a nonconjugative plasmid (i.e., r_{ms7}tet) of S. aureus, became incorporated into the genetic constitution of staphylophage S1 (Inoue and Mitsuhashi, 1975).

Some temperate phages are species specific, but others can cross species and generic boundaries. For example, phage P1 can infect Klebsiella aero-genes, Enterobacter liquifaciens, Citrobacter freundii, and Erwinia amylovora, in addition to E. coli, and, therefore, can confer resistance to kanamycin to all these species when in the prophage state (Goldberg et al., 1974).

The frequency of prophage DNA in a host bacterium varies greatly, e.g., lysogeny is uncommon in P. putida, but it can attain 100% in P. aeruginosa; in E. coli, up to 30% of the total DNA may occur as prophage DNA. Lysogenization of a cell by phage may increase the size of the genome of the host by as much as 2% in a single genetic event. Furthermore, many species of bacteria are "polylysogenic" (especially species of Bacillus and Pseudomo-nas), in that they simultaneously contain several different prophages (Rean-ney, 1976; Reanney et al., 1983).

In the transfer of DNA by transformation, bacteria are able to take up "naked" DNA and integrate it into their genome. Although some bacteria, notably Bacillus subtilis and Streptococcus pneumoniae, can incorporate heterospecific DNA, others, such as species of Neisseria and Haemophilus, only incorporate homospecific DNA (Saunders, 1979). Competency, or the ability of a recipient cell to take up and incorporate naked DNA, is induci-ble, apparently occurring most often when the recipient cell is under ad-verse conditions (Reanney et al., 1983).

Bacteria may also transfer genes by protoplast fusion. This process, which is promoted by polyethylene glycol and calcium ions, relies on the fusion of protoplasts of gram-positive bacteria, followed by the regeneration of cell walls. Using this procedure, a plasmid that conferred resistance to chloram-phenicol was transferred from S. auerus to B. subtilis and from B. subtilis to

Bacillus megaterium, Bacillus licheniformis, and *Bacillus polymyxa* (Dancer, 1980).

III. *In Vivo* Studies

The survival and perpetuation of recombinant DNA in natural environments probably depend primarily on (1) the nature of the bacterial host and of the cloning vector, (2) the final ecological niches of the original and chimeric hosts, (3) the transmissibility of the recombinant DNA to other bacteria, and (4) the selective advantages or disadvantages conferred on the host by the presence of the foreign DNA (Curtiss *et al.*, 1977). One form of biological containment of recombinant DNA is based on the use of appropriate hosts and vectors. The most commonly used host–vector systems for biological containment currently employ strains of *E. coli* K12 and poorly mobilizable cloning vectors, such as plasmids pBR322 and pBR325. Several studies have shown that *E. coli* K12 does not colonize the human intestinal tract, even after the ingestion of billions of the bacteria by volunteers (Anderson, 1975a; Smith, 1975a; Cohen *et al.*, 1979; Levy *et al.*, 1980; Levy and Marshall, 1981). For example, elevated levels of E. coli K12 occurred in the feces of human volunteers only on the first day after ingestion, but thereafter, the numbers rapidly decreased, and no K12 strains were detected after 3 to 5 days (Anderson, 1975a; Smith, 1975a). Apparently, the indigenous intestinal microbiota has a selective advantage over introduced strains (Anderson, 1975b).

The intestines should presumably be a particularly favorable area for DNA transfer to occur via conjugation, as the consistently high population densities should provide maximal opportunities for cell-to-cell contact among bacteria (Reanney, 1977). However, for conjugation to occur efficiently in the intestinal tract, the human or other animal host must be colonized by large numbers of appropriate donor and recipient bacteria. It is difficult to colonize a host with nonenteropathic strains of Enterobacteriaceae unless the indigenous intestinal microbiota of the animals has been suppressed by antibiotics (e.g., Anderson, 1975b) or by starvation (e.g., Smith, 1976), or unless germ-free animals (e.g., Salzman and Klemm, 1968; LaFont *et al.*, 1984), new-born or very young animals in which the indigenous microbiota is either absent or not well established (Walton, 1966; Smith, 1970a), or normally germ-free organs (e.g., the bladder) of conventional animals (Richter *et al.*, 1973) are used. By the use of these techniques, especially the use of antibiotically suppressed animals, the conjugal transfer of R- and F-plasmids has been demonstrated in the intestines of humans (Smith, 1969; Farrar *et al.*, 1972; Anderson, *et al.*, 1973; Anderson, 1975a; Levine *et al.*, 1983), mice (Schneider *et al.*, 1961; Kasuya, 1964; Guinee, 1965; Salzman and Klemm, 1968), rats (Guinee, 1970), sheep (Smith, 1972, 1975b, 1976), pigs (Jarolmen and Kemp, 1968; Gyles *et al.*, 1977, 1978), cattle (Smith, 1970a), and turkeys (Nivas *et al.*, 1976) and other poultry (Walton, 1966; LaFont *et al.*, 1984), as well as in wounds (Hummel *et al.*, 1977), the urinary bladder (Richter *et al.*, 1973), and the respiratory tract (Gardner and Smith, 1969). However, at least two studies have noted plasmid transfer *in vivo* without suppression of the indigenous microbiota: R-plasmids were apparently transferred in the intestines of humans (Petrocheilov *et al.*, 1976) and sheep (Smith, 1975b) without pretreatment of the animals with antibiotics. The transfer of nonconjugative R-plasmids by triparental matings has also been noted in the intestines of human beings (Levine *et al.*, 1983).

Nevertheless, microbial competition may reduce genetic transfer by conjugation in animals. The frequency of transfer of a multiple antibiotic-resistance plasmid from *Salmonella typhosa* to *E. coli* in the bladder of healthy rabbits was as high as, and, in some instances, higher than, in *in vitro* systems containing either sterile urine or synthetic mating media. However, in the presence of other bacteria (exogens: i.e., *Proteus mirabilis* and nonconjugative *E. coli*), the frequency of transfer decreased significantly. This decrease was not the result of a physical (e.g., steric) interference of the exogens with the conjugation process, as polystyrene latex particles of the same size and at the same concentration as the exogens had essentially no effect on the frequency of plasmid transfer, suggesting that the exogens interfered chemically with conjugation (Richter *et al.*, 1973; Stotzky and Krasovsky, 1981).

Once R-plasmid-containing bacteria enter the gastrointestinal tract of mammals, they may (1) colonize the gut, (2) transfer their plasmids to bacteria that have already colonized, and/or (3) transfer their plasmids to sensitive pathogens with which the host may subsequently become infected, as shown, for example, by the *in vivo* transfer of multiple drug resistance from a resistant *E. coli* donor to sensitive *S. typhimurium* (Guinee, 1965), *Shigella flexneri*, and *Klebsiella pneumoniae* (Kasuya, 1964) recipients in the intestines of germ-free or antibiotic-fed mice. The transfer of multiple drug resistance among gram-negative bacteria has become a significant problem in nosocomial infections of humans (Gardner and Smith, 1969) and in the infections of other animals, primarily because antimicrobial prophylaxis has provided the selective pressure necessary to maintain bacteria containing R-plasmids as the etiological agents (Jukes, 1972).

The presence of foreign DNA may alter the survival of the cloning vectors or of the hosts that contain them. The survival and perpetuation of recombinant DNA presumably require that the foreign DNA confers some selective advantage, or at least no selective disadvantage, to the vectors or to the chimeric microbes, as the requirements for energy and precursors for the replication of the additional DNA in the absence of any derived benefit could be a selective disadvantage in natural environments. The synthesis of gene products, whether functional or nonfunctional, specified by the foreign DNA and which provide no benefit to the vector or host microbe could also impose an additional burden. Consequently, the presence of foreign DNA may be detrimental to the competitive survival of host vectors and cells (Curtiss, 1976; Helling *et al.*, 1981). For example, the survival in the intestines of humans of *E. coli* containing R-plasmids was significantly less than that of plasmid-free *E. coli* (Anderson, 1974), and strains of piliated *E. coli* K12 containing F-plasmids did not colonize the intestines of mice fed streptomycin to reduce their normal intestinal microbiota as well as did their plasmidless counterparts (Cohen *et al.*, 1979). Furthermore, competition studies between plasmidless and plasmid (pBR325, pJBK25, or F-amp)-containing strains of *E. coli* HS-4 showed that the plasmidless parental colonized the intestines of antibiotic-fed mice better than did the plasmid-containing derivatives (Laux *et al.*, 1982). In the absence of antibiotic pressure, plasmid-free strains of *E. coli* apparently had an ecological advantage in colonizing the intestines of gnotobiotic mice over strains harboring plasmids conferring antibiotic resistance (Duval-Iflah and Chappuis, 1984). Conversely, *E. coli* K12 strain χ1776, which was constructed to have poor survival in the human intestine, exhibited enhanced survival when containing plasmid pBR322 (Levy and Marshall, 1979): survival of *E. coli* χ1776 was extremely poor, with no detectable survival in human volunteers 1 day after ingestion; however, the same parental χ1776 strain that contained plasmid pBR322 was recovered from human volunteers for 4 days after ingestion. Similarly, survival of *E. coli* K12 strain χ1666 containing plasmid pBR322 was greater than that of the plasmidless parental (Levy *et al.*, 1980). However, the addition of two other plasmids, pLM2 and pSL222-4, to *E. coli* χ1666(pBR322) did not further enhance survival in the human intestine (Marshall *et al.*, 1981).

Genetic transfer *in vivo* by transformation or transduction has also been studied, although not as extensively as conjugation. Since Griffith's discovery of transformation in pneumococci in mice almost 60 years ago (Griffith, 1928) and the identification of the "transforming principle" as DNA (Avery *et al.*, 1944), numerous studies have evaluated transformation by purified DNA interacting with competent recipient cells. For example, genetic exchange by transformation was demonstrated during peritoneal infection of mice with pneumococci of different genotypes (Ottolenghi and MacLeod, 1963; Conant and Sawyer, 1967). Such transformations could be increased in frequency when the infected mice were treated with a drug that killed only one of the two infecting strains. Furthermore, interspecific transformation between pneumococci and streptococci was also demonstrated in mice infected interperitoneally or subcutaneously, with most of the transformants being streptococci that acquired pneumococcal DNA (Ottolenghi-Nightingale, 1969). Transformation of pneumococci by DNA from strains of the same species was also noted in the respiratory tract of mice (Conant and Sawyer, 1967) and human beings (Ottolenghi-Nightingale, 1972).

Transduction of extrachromosomal determinants of erythromycin resistance and of linked penicillin–erythromycin resistance occurred in the kidneys of mice infected with populations of lysogenic and nonlysogenic strains of S. aureus (Novick and Morse, 1967). Transduction in vivo of S. aureus to tetracycline resistance was demonstrated by the injection of staphylophage 80 (a transducing phage propagated in a hospital strain of S. aureus and which had been shown to transduce antibiotic resistance markers to sensitive recipients in vitro) into mice in which a kidney infection had been previously established with a tetracycline-sensitive strain of S. aureus (Jarolmen et al., 1965). When gnotobiotic mice harboring a defined microbiota of anaerobes and E. coli χ1666 (but no other E. coli) were fed E. coli K37 lysogenic for phage λ::Tn9, which codes for chloramphenicol resistance, four of the eight mice exposed contained E. coli χ1666 that was resistant to chloramphenicol throughout the 56 days of the study (Freter et al., 1979).

IV. *In Situ* Studies

The apparent transfer of antibiotic resistance via R-plasmids has been investigated in a variety of natural environments. Bacteria resistant to antibiotics have been isolated from hospital wastes (Fontaine and Hoadley, 1976), raw sewage (Linton et al., 1974; Fontaine and Hoadley, 1976; Bell et al., 1981; Corliss et al., 1981), waters receiving sewage effluents (Sturtevant and Feary, 1968; Garbow and Prozesky, 1973; Bell et al., 1981), sediments from an off-shore sewage dump site (Timoney et al., 1978), coastal sediments (Grabow et al., 1975; Goyal et al., 1979), fresh and marine recreational waters (Smith, 1970b, 1971; Feary et al., 1972; Smith et al., 1974; Goyal et al., 1979), estuaries (McNicol et al., 1980), rivers (Kelch and Lee, 1978), commercial fisheries (Watanabe et al., 1971; Aoki et al., 1973), abattoirs (Goyal and Hoadley, 1979), animal feed-lots (LaFont et al., 1981), plants (Talbot et al., 1980), soils (Cole and Elkan, 1979), and drinking waters (Armstrong et al., 1981).

Many plasmids that confer resistance to antibiotics also confer resistance to heavy metals, although the genes for these resistances may not necessarily be linked. The frequency of heavy metal resistance in clinical isolates of E. coli, K. pneumoniae, P. aeruginosa, and S. aureus was the same as, or higher than, that of antibiotic resistance. Many of the isolates (21.6% of E. coli; 27.2% of K. pneumoniae; 19.5% of P. aeruginosa; 8.0% of S. aureus) were resistant to heavy metals (arsenate, cadmium, lead, mercury) but were sensitive to antibiotics (streptomycin, tetracycline, chloramphenicol, kanamycin, gentamicin). The resistance to heavy metals was apparently plasmid-borne, as resistance to mercury, for example, could be transferred to recipient *E. coli* strains. Of the 117 strains of mercury-resistant *K. pneumoniae* studied, 110 carried plasmids that conferred resistance to mercury, and 102 of these strains also carried plasmids that coded for resistance to arsenate. Of the 231 *E. coli* strains that were resistant to mercury, 207 carried these resistance genes on plasmids, and 195 strains also showed resistance to arsenate (Nakahara *et al.*, 1977).

Plasmid-borne antibiotic resistance can be transferred from environmental isolates to various laboratory and clinical recipients *in vitro*. For example, strains of *E. coli* resistant to heavy metals and antibiotics isolated from polluted rivers and lakes, effluents from pig and cattle breeding stations, and urine and feces from humans treated with antibiotics transferred their resistance to heavy metals and, to a lesser extent, antibiotics to recipient *E. coli* cells *in vitro* (Cenci *et al.*, 1982). *E. coli, K. pneumoniae, Enterobacter* sp., and *Salmonella* sp. isolated from municipal waste waters contained R-plasmids that conferred resistance to antibiotics, and this resistance could be transferred *in vitro* to *E. coli* K12 (Kralikova *et al.*, 1983). Strains of *E. coli*, isolated from estuarine waters, that were resistant to multiple antibiotics were able to transfer this drug resistance to *E. coli* K12 *in vitro* (Shaw and Cabelli, 1980), and the antibiotic resistance of *E. coli* strains isolated from rivers receiving sewage was transferable *in vitro* to *Salmonella typhi* and *S. typhimurium* (Smith, 1970b). Of the 423 total coliform, 300 fecal coliform, and 100 salmonella isolates from the aqueous phase and sediments of coastal marine canal communities used for recreational activities, 74.9% of the total coliforms, 61.3% of the fecal coliforms, and 71.0% of the salmonellae were resistant to one or more of the 12 antibiotics tested. When 178 total coliform, 137 fecal coliform, and 51 salmonella isolates were tested *in vitro* for their ability to transfer antibiotic resistance, 53.9, 53.3, and 56.9%, respectively, were capable of transferring resistance to *E. coli* and/or *Salmonella choleraesuis* (Goyal *et al.*, 1979).

A. CONJUGATION

1. Aquatic Environments

Plasmids appear to be ubiquitous in bacterial populations in aquatic as well as in other natural environments. For example, 31% of 155 psychrophilic and psychrotrophic bacteria isolated from sea ice, sea water, sediments, and benthic or ice-associated animals in Antarctica contained at least one plasmid, with several isolates containing multiple plasmids. The majority of the plasmids had a mass of 10 MDa or less (Kobori *et al.*, 1984). Of 58 bioluminescent marine bacteria, 43% contained plasmids ranging from 5 to 120 MDa (Simon *et al.*, 1982).

Higher numbers of plasmid-containing bacteria have been isolated from polluted than from nonpolluted sites. When 440 planktonic marine species of *Vibrio* from an operational offshore oil field in the Gulf of Mexico and from a coastal area approximately 8 km from the production area were examined, there was a higher incidence of plasmids in isolates from the oil field site (35%) than from the control site (23%) (Hada and Sizemore, 1981). Similarly, more plasmid-containing *Pseudomonas*-like isolates were detected in a polluted site in a river in South Wales than in a nearby nonpolluted site (Burton *et al.*, 1982). Glassman and McNicol (1981) isolated 218 aerobic heterotrophs in Chesapeake Bay and noted that, in general, the bacteria isolated from the cleaner station contained predominantly small plasmids (ca. 3 MDa), whereas bacteria from Baltimore's inner harbor tended to contain plasmids with a mass greater than 30 MDa and, thus, were large enough to contain genes to enable conjugal transfer. Portions of the fecal coliform and salmonella populations in the Red River of Manitoba, Canada, near urbanized areas were resistant to as many as 12 antibiotics; 52.9% of the fecal coliforms resistant to one or more of the antibiotics was able to transfer resistance to *S. typhimurium* and 40.7% to *E. coli* K12. Of the resistant salmonellae, 57% was able to transfer one or two determinants to the *Salmonella* recipient and 39% to the *E. coli* recipient (Bell *et al.*, 1980). In comparison, only 7.1% of the fecal coliform population in the remote Slave River in northern Canada was resistant to antibiotics (Bell *et al.*, 1983).

Gowland and Slater (1984) studied the transfer of compatible (R1 and TP120) and incompatible (TP113 and TP125) plasmids between *E. coli* K12 strains in a continuous flow culture system (i.e., a water-jacketed fixed-bed column fermentor) and in pond water. Transfer of plasmids between *E. coli* K12(R1) and *E. coli* K12(TP120) occurred in the fermentor, both in the absence and presence of any selection pressure (i.e., antibiotics); e.g., after 3 hours of incubation in the absence of any selection pressure and at a flow rate of 50 ml/hour, 100% of the population was composed of transconjugants that contained the antibiotic markers of both TP120 and R1. Tests for auxotrophy of the recombinant isolates showed that plasmid R1 (which coded for resistance to ampicillin, streptomycin, sulfanilamide, chloramphenicol, and kanamycin) was transferred to the *E. coli* strain containing plasmid TP120 (which coded for resistance to ampicillin, streptomycin, sulfanilamide, and tetracycline). Furthermore, *in vitro* matings of a recombinant isolate with a nalidixic acid-resistant strain of *E. coli* showed that unless both plasmids were specifically selected for, the R1 plasmid would preferentially transfer into the nalidixic acid-resistant *E. coli* strain. With strains containing the incompatible plasmids, recombinants accounted for only 2% of the population in the absence of an antibiotic-selection pressure. However, in the presence of a suitable antibiotic-selection pressure (i.e., in the presence of kanamycin and streptomycin) and after an additional 300 hours of growth, 100% of the population was composed of transconjugants.

When *E. coli* strains containing either plasmids R1 or TP120 or strains carrying plasmids TP113 or TP125 (which coded for resistance to kanamycin and to streptomycin, sulfanilamide, chloramphenicol, and tetracycline, respectively) were inoculated into dialysis bags containing sterile pond water and the bags submerged in pond water, transconjugants at very low frequencies (i.e., 4.4×10^{-7} after 192 hours and 4.7×10^{-8} transconjugants/donor plus recipient after 360 hours of incubation) were obtained only from strains carrying R1 or TP120 and then only when high population densities ($> 10^{10}$ cells/ml) were used. This low transfer frequency, when compared to an *in vitro* transfer frequency of 2.6×10^{-2}, was suggested to be a reflection of the poorer nutritional status and lower temperature of the natural environment. Strains of *E. coli* containing both the TP120 and R1 plasmids were stable over 5 months of continuous cultivation under carbon-limited conditions in the presence of the antibiotics that selected for the maintenance of both plasmids. However, there was no formation of a single cointegrate plasmid from the two separate plasmids (Gowland and Slater, 1984).

Stewart and Koditschek (1980) examined the survival and genetic recombination at 10°C of *E. coli* inoculated into glass vessels containing either sterile sediment (obtained from a sewage dump site area in the New York Bight) and sterile overlying sea water or sterile sea water alone (the sea water was obtained near Sandy Hook, NJ). In the system containing sediment plus sea water, survival in the sediment of donor (a environmental *E. coli* isolate, designated W510, resistant to nalidixic acid, streptomycin, and tetracycline) and recipient (an F^- strain of *E. coli* K12) cells decreased from approximately 10^7 cells/ml wet sediment to a constant population of 2×10^2 to 3×10^2 cells/ml wet sediment after 14 to 28 days, and neither donor nor recipient cells were detected in the overlying sea water after 14 days. In the sediment-free vessels, neither donor nor recipient cells were detected after 21 and 28 days, respectively, indicating that the survival of the donor and recipient strains was greater in the sediment than in the water column. Transconjugants were detected in the water phase of both the sea water–sediment and the sediment-free vessels for only as long as 1 day after inoculation of the parental strains. In contrast, transconjugants in the sediment were detected for 28 days and, in one experiment, even after 46 days of incubation. It was suggested that the continued presence of transconjugants may have reflected either their better survival than that of the parentals or continued conjugation despite the low population densities of the parentals, or both. Krasovsky and Stotzky (1986) also noted that survival of transconjugants in soil was generally better than that of the parentals, especially in soils amended with the clay mineral, montmorillonite.

Bacteria containing plasmids that code for the degradation of xenobiotics have been isolated from sediments. For example, Kamp and Chakrabarty (1979) isolated a polychlorinated biphenyl (PCB)-degrading bacterium, identified as *K. pneumoniae*, from PCB-contaminated sediment of the Hudson River. The *K. pneumoniae* isolate contained three plasmids, the largest of which was 65 MDa and not only had genes for the degradation of PCB, but it could be transferred to a plasmid-cured strain of the *K. pneumoniae* isolate.

It must be stressed that reports of the presence of R-plasmids in bacteria isolated from natural environments and of the transfer of R-plasmids from such environmental isolates to laboratory strains are only suggestive of the occurrence of genetic transfer *in situ*. Relatively few studies on the transfer of plasmids or of chromosomal DNA have been conducted in samples of aquatic and terrestrial environments, and most of these studies have focused on the possible transfer in sewage treatment plant facilities.

2. Sewage

Grabow *et al.* (1973) studied the population dynamics of multiple drug-resistant coliforms in a series of sewage maturation ponds, which function in reducing coliform counts and are used for the tertiary treatment of sewage. The average reduction through the ponds of coliforms with multiple antibiotic resistance [including both transferable (R^+) and nontransferable (R^-) resistance to multiple antibiotics] was 6.8% less than that of antibiotic-sensitive coliforms. This difference was correlated with a threefold increase in R^+ coliforms in the ponds, i.e., from 0.86 to 2.45%. Transferability of R^+ resistance was determined by *in vitro* matings with drug-sensitive *E. coli* E25 and *S. typhi*. It was postulated that either the R^+ coliforms were more tolerant of the adverse conditions in the maturation ponds (i.e., plasmids conferring resistance to the antibiotics perhaps also conferred resistance to other antimicrobial effects of the ponds) or the increase in numbers was the result of *in situ* transfer of the R-plasmids to sensitive recipients. The investigators also followed the incidence of coliform bacteria capable of producing indole in tryptophan broth at 44.5°C (referred to as *E. coli* I and which were the majority of the R^+ coliforms in the influent) and noted that in passage through the ponds, the average incidence of *E. coli* I cells carrying R^+ resistance decreased from 66 to 62%. This reduction in the effluent of the major donor among the R^+ coliforms supported the possibility that genetic transfer had occurred.

Grabow *et al.* (1975) also studied the population dynamics of R^+ and R^-

coliforms in waters of a river and of a dammed impoundment into which it flowed; the effluent from a sewage treatment plant was discharged into the river about 10 km upstream of the dam. The ratio of antibiotic-resistant coliforms (both R^+ and R^-) to total coliforms decreased between the sewage treatment plant and the dam, but it increased between the influent and effluent of the dam, and the ratio in the dam effluent was higher than in the river upstream of the sewage plant. To determine whether these differences in ratio were the result of the differential survival of antibiotic-resistant and antibiotic-sensitive coliforms, dialysis bags were inoculated with known R^+, R^-, and antibiotic-sensitive strains of E. coli, submerged in river and dam waters, and viability was determined during a 55-day period. No significant differences in survival were noted among these strains. To evaluate the possibility that the observed differences in ratio resulted from genetic transfer, 10 R^+ strains isolated from the effluent of the dam were used in mating studies with recipient E. coli E25. Mating pairs were inoculated into dialysis bags, which were submerged in the waters in situ or in water samples maintained in the laboratory. Three of the 10 donor strains transferred antibiotic resistance to E. coli E25, but transfer occurred at a low frequency. It was postulated that the increased number of antibiotic-resistant bacteria in the dammed impoundment resulted from the transfer of R-factors to sensitive coliforms, whereas the turbulence of the river reduced the transfer of R-factors and contributed to the decrease in percentages of antibiotic-resistant coliforms in the river, in contrast to the increases noted in the dammed impoundment (Grabow et al., 1975) and in the maturation ponds (Grabow et al., 1973).

The survival of R^+ and R^- coliforms was also studied in settled sewage and in effluent waters after the treatment processes of biofiltration, secondary sedimentation, chlorination, and sand filtration. The ratio of R^+ to R^- antibiotic-resistant coliforms increased during secondary sedimentation and chlorination, indicating that either the R^+ coliforms were resistant to these specific purification processes or R-plasmids were transferred to sensitive coliforms during these processes. To explain the lack of a similar increase in the numbers of R^+ coliforms during the other purification processes, it was suggested that the rapid passage of the water over the stony surfaces in the biological and sand filters was not conducive to conjugation and may, in fact, have damaged the sex pili. In contrast, R-plasmid transfer could be expected to occur under relatively stagnant conditions, such as those in the sedimentation and chlorination ponds (Grabow et al., 1976).

In these studies by Grabow et al. (1975), the presence of plasmids, even in the absence of selection pressure (i.e., antibiotics), was apparently not detrimental to the survival of the host cells, as there were no significant differences in survival of E. coli E25 and E1N with and without plasmids when inoculated into dialysis bags submerged in natural waters. Sturtevant and Feary (1969) also observed that the survival of coliforms containing R-plasmids in conventional sewage purification systems was similar to that of antibiotic-sensitive coliforms, and the survival of *E. coli* strains in sea water was unaffected by the presence of R-plasmids (Smith *et al.*, 1974).

The relative survival of plasmid-containing and plasmid-free bacteria has been studied in chemostats under a variety of nutrient-limiting conditions, and most studies have shown that when plasmid-free cells are "more fit" (i.e., as noted by an increased growth rate) under a given set of growth conditions, the plasmid-containing cells will be displaced rapidly from the population (Bialey, 1984). However, *E. coli* lysogenic for λ phage grew more rapidly than nonlysogens; a 1.8-kb segment of the λ phage DNA, i.e., the *cos* fragment, was apparently involved in enhancing the fitness of the bacterial host. When grown together in a glucose-limited chemostat, *E. coli* RR1 containing plasmid pSa151 grew at a reduced growth rate when compared to that of its plasmid-free counterpart. However, when the *cos* fragment was inserted into pSa151, now termed pSa747, both the plasmid-containing and the plasmid-free strains grew equally well when inoculated together into a glucose-limited chemostat. Apparently, some property of the *cos* nucleotide sequence influenced positively bacterial fitness. The growth rate in a glucose-limited chemostat of bacteria carrying the *cos* fragment on the plasmid was 20% higher than the growth rate of plasmid-containing bacteria lacking this DNA fragment (Edlin *et al.*, 1975, 1984).

Fontaine and Hoadley (1976) studied transferable antibiotic resistance in hospital sewage. Raw sewage containing, as part of its indigenous biota, tetracycline-resistant fecal coliforms capable of transferring drug resistance (as shown *in vitro*) was amended with sterilized hospital sewage previously inoculated with *S. choleraesuis* recipients and incubated at 23 and 30°C. Tetracycline-resistant cells of *S. choleraesuis* were detected after 1 hour at 30°C and after 3.5 hours at 23°C, and their numbers increased after 12 hours to $10^{2.8}$ and $10^{3.4}$ cells/ml at 23 and 30°C, respectively.

The survival of *E. coli* host–vector systems was studied in a bench-scale model of an activated sludge domestic sewage treatment plant that utilized primary and secondary processes (Sagik and Sorber, 1979; Sagik *et al.*, 1981). There were three additional processes ancillary to this system: lagooning of both primary and secondary effluents and anaerobic digestion of wasted sludge. *E. coli* GF215 (a genetically-marked sewage isolate), *E. coli* K12 strain χ2656 (which contained plasmid pBR322), *E. coli* K12 strain GF2174 (which contained plasmid pBR325), and the λ coliphage, Charon 4A, were seeded into raw waste water, and survival was determined in the raw wastewater reservoir (0 to 48 hours) and in the primary and secondary lagoons (48 to 120 hours). The decay constants for survival in the raw wastewater reservoir were 2.2, 1.2, 2.7, and 5.0 for *E. coli* GF215, χ2656(pBR322), GF2174(pBR325), and Charon 4, respectively, indicating that the phage was the most labile. In the primary lagoon, the respective decay constants were 1.9, 1.9, 1.9, and 1.5, and in the secondary lagoon, they were 0.96, 1.5, 0.67, and 1.5. *E. coli* χ2656(pBR322) was more stable in the anaerobic digester than was GF2174(pBR325), with a 90% reduction in χ2656(pBR322) occurring within 30 hours, whereas viable GF2174(pBR325) cells were only sporadically recovered 20 hours after seeded sludge was inoculated into the digester. Similar studies with *E. coli* strains DP50SupF and χ1776 showed that χ1776 was inactivated more rapidly in raw waste water and primary effluent (with decay constants of 3.6) than was DP50SupF (with decay constants of approximately 2.0). There was also a more rapid disappearance of χ1776 in the anaerobic digester, with no viable cells being detected after 20 hours, whereas there was a 10% survival of DP50SupF after 20 hours.

These studies were extended to evaluate plasmid transfer in sewage sludge (Sagik *et al.*, 1981). When *E. coli* χ2656(pBR322) was added to raw sewage or primary sludge, there was a rapid disappearance of this strain with a concomitant increase in indigenous coliforms showing resistance to tetracycline and carbenicillin (resistances to these antibiotics were coded on pBR322). However, this increased antibiotic resistance of the indigenous sewage bacteria was attributed to the "test" conditions, which apparently selectively promoted growth of this antibiotic-resistant population, as no transfer of pBR322 could be demonstrated. Studies conducted in laboratory medium showed that the transfer of either pBR322 or pBR325 to the indigenous sewage bacteria occurred only at a low frequency and that the addition of the mobilizer strain, *E. coli* χ1784, increased the frequency of the transfer of the plasmids. When the transfer of plasmid pBR325 from *E. coli* GF2174 (pBR325) to the indigenous bacteria in the raw waste water reservoir—both in the absence and presence of *E. coli* χ1784—was evaluated, the numbers of antibiotic-resistant sewage bacteria increased by a factor of 8.1 after 24 hours when only *E. coli* GF2174(pBR325) was introduced, but when both GF2174(pBR325) and χ1784 were introduced, there was a 25-fold increase in the numbers of antibiotic-resistant bacteria. In the absence of either GF2174(pBR325) or χ1784, the numbers of indigenous bacteria resistant to antibiotics increased by a factor of 4.7 after 24 hours. The investigators concluded that "such observations are suggestive of plasmid transfer" (Sagik *et al.*, 1981).

Altherr and Kasweck (1982) studied the transfer of antibiotic resistance in environmental isolates of *E. coli* in raw sewage or in waters receiving sewage effluent. The donor strain, MA527, isolated from a residential sewage and containing a 60-MDa plasmid that conferred resistance to streptomycin and tetracycline, and the recipient strain, MA728, isolated from a creek, were inoculated into membrane diffusion chambers that were submerged in the degritter tank of a sewage treatment facility and in effluent-receiving waters 500 m from the treatment facility. The *in situ* transfer frequency in the raw sewage (i.e., in the degritter tank) ranged from 3.2×10^{-5} (at an ambient temperature of 22.5°C) to 1.0×10^{-6} transconjugants/donor (at 29.5°C) whereas the *in vitro* transfer frequency was 1.6×10^{-4} at 20°C and 4.4×10^{-5} at 30°C. No transfer of plasmids apparently occurred in the diffusion chambers submerged in the waters receiving effluent. It was postulated that these receiving waters constituted a stressful environment (i.e., the concentration of nutrients in the effluent-receiving waters was less than in the raw sewage, and the salinity of the receiving waters was 10 times higher) and that the maintenance or expression of an R-plasmid under such detrimental conditions did not confer any selective advantage to the host but, instead, may have been a burden.

The ability of *Salmonella enteritidis*, *P. mirabilis*, and *E. coli*—isolated from clinical specimens and primary sewage effluents and containing plasmids for multiple resistance to antibiotics—to transfer resistance to sensitive *E. coli* strains or *Shigella sonnei* in a wastewater treatment plant was studied by Mach and Grimes (1982). Studies were performed *in vitro*, by inoculating mating pairs into broth or sterile sewage, and *in situ*, by inoculating mating pairs into membrane-filter diffusion chambers containing sterile sewage that were submerged in primary and secondary clarifiers. The *in vitro* frequencies of R-plasmid transfer were similar, in general, in broth and sterile sewage (approximately 2.1×10^{-3} transconjugants/donor), whereas the frequency of transfer *in situ* was approximately 5.9×10^{-5}. It was suggested that the lower frequencies in the *in situ* matings may have resulted, in part, from the lower water temperatures, which averaged 10.6°C as compared to 20°C at which the *in vitro* matings were conducted. The frequency of transfer *in situ* was greater in the secondary (7.5×10^{-5}) than in the primary clarifier (4.9×10^{-5}), with differences in the physicochemical characteristics (e.g., pH, BOD) of the two clarifiers presumably being responsible for the difference in frequency.

3. Soil

Schilf and Klingmüller (1983) studied the transfer of the wide-host-range plasmid, pRD1, carried by *E. coli* strain JC5466, to the indigenous bacterial populations of soil and freshwater. Studies *in vitro* with soil and aquatic isolates showed that 1.3% of the soil bacteria and 17.3% of the freshwater bacteria could serve as recipients for pRD1 as well as for another wide-host-range plasmid, RP4, carried by *E. coli* strain J5. The *in vitro* transfer rate of pRD1 to mixed populations of bacteria obtained from soil ranged from $7.1 \times$

10^{-8} to 4.5×10^{-6} transferants/recipient and that to mixed populations of bacteria obtained from fresh water ranged from 1×10^{-6} to 5×10^{-6}. Studies *in situ*, however, yielded much lower transfer rates: when *E. coli* JC5466(pRD1) cells were mixed with an agricultural soil, the rate of transfer of pRD1 to the indigenous bacterial population was only 1×10^{-9}, and the *in situ* transfer rate in a pond water was 3.3×10^{-8}.

Schilf and Klingmüller (1983) also studied the persistence in natural terrestrial and aquatic environments of *E. coli* strains J5(RP4) and JC5466(pRD1) and of the transconjugants obtained from matings *in vitro:* i.e., Enterobacteriaceae strain 1(pRD1), Enterobacteriaceae strain 2(RP4), and *Pseudomonas fluorescens*(pRD1). These plasmid-containing bacteria were inoculated into soil—maintained either at 4 or 20°C with a constant water content of 16% or with drying to a water content of 4% at 20°C—and into pond water—maintained at either 4 or 20°C. In both environments and with all treatments, there was a decrease with time in the survival of the plasmid-containing bacteria, but the titer of the indigenous bacteria remained constant during the study. The researchers postulated that the apparent decrease in survival of the plasmid-containing strains may have resulted either from a selective disadvantage to the plasmid-containing bacteria or from a loss of the plasmids when the bacteria were cultured under nonselective conditions.

Weinberg and Stotzky (1972) studied the growth and conjugation of prototrophic and auxotrophic strains of *E. coli* in sterile soils. When a soil from Ossining, NY, that did not naturally contain the clay mineral, montmorillonite, and with a pH of 5.3 was amended with 20% (v/v) montmorillonite (referred to as KM20 soil; resultant pH of 6.6), the growth rate [i.e., spread through soil as determined by the soil replica plate method (Stotzky, 1965, 1973)] was approximately 0.5, 0.3, 0.2, 0.15, and 0.2 mm/day over a 30-day period for the prototrophic donor strains, $\chi 503$ (Hfr) and $\chi 209$ (F$^+$), for the auxotrophic recipient strain, $\chi 705$ (F$^-$), and for auxotrophic strains 1252 and 1230 (sex unknown), respectively. The respective growth rates in a soil from La Lima, Honduras, with a pH of 5.4 and which also did not naturally contain montmorillonite but was amended with 10% montmorillonite (referred to as SAM10 soil; resultant pH of 6.3) were approximately 0.1, 0.2, 0.15, 0.1, and 0.1 mm/day, respectively. It was suggested that the slightly higher pH and the greater amounts of organic matter, inorganic nutrients, and montmorillonite contributed to the better growth of all the strains in the KM20 soil. In soils not amended with clay minerals or amended with comparable concentrations of kaolinite (which was naturally present in these soils), growth of the bacteria was considerably slower and most strains died out after 5 to 12 days.

When donor and recipient strains were placed either into the same or separate sites in the soils, conjugation was evident both in the soil sites into which the parentals had been inoculated together and in the areas between the sites where the parentals had been inoculated individually, indicating that the donors and recipients grew toward each other sufficiently for effective pairing during incubation of the soils. The relative frequency of conjugation—using the *E. coli* donor strains, χ503 and χ209, and the *E. coli* recipient strains, χ705 and χ696 (F⁻)—was higher in the KM20 soil than in the SAM10 soil. In both soil types, the frequency of conjugation was higher with the χ503 Hfr donor than with the χ209 (F⁺) donor. However, only partial recombinants were obtained, regardless of the soil system and the mating pairs used. The relative frequencies of conjugation in soils without montmorillonite were, in general, lower than in soils to which the clay mineral had been added. Montmorillonite apparently enhanced the growth of the parentals and, thereby, increased the number of contacts between donors and recipients, as well as increasing the pH of the soils to levels more conducive to conjugation in these laboratory strains (see Stotzky, 1986).

The influence of pH on the survival and genetic recombination of *E. coli* K12 strains in soil was studied further (Stotzky and Krasovsky, 1981; Krasovsky and Stotzky, 1986). When strains χ503 and χ696 were inoculated into sterile soil amended with either kaolinite or montmorillonite—which raised the natural bulk pH of the soil from 4.7 to 4.9 when amended with 9% kaolinite and to 5.5 when amended with 9% montmorillonite—no transconjugants were detected. However, survival of both donor and recipient cells increased as the pH was increased to 5.5. When the pH of the natural and clay-amended soils was progressively increased to 5.7–6.0 and to 6.2–6.6 with $CaCO_3$, survival of the parentals was increased, but transconjugants were still not detected. However, when the pH of the soils was increased to 6.8–6.9, transconjugants were isolated from the soils, and the survival rates of the parentals were enhanced. Increasing the pH of the soils to 7.0 did not further enhance the survival of the parentals but reduced the numbers of transconjugants detected.

Krasovsky and Stotzky (1986) also studied the influence of exogenous microbial populations on the conjugation of *E. coli* in soil. *E. coli* K12 strain χ493 (Hfr) and the recipient strain, χ696, were inoculated either into sterile soil (unamended or amended with 9% kaolinite or montmorillonite and adjusted to pH 6.8–6.9 with $CaCO_3$) in the absence and presence of *P. fluorescens* (5×10^9 cells/g soil) or *Rhodotorula rubra* (3×10^7 cells/g soil) or into nonsterile soil (approximately 10^8 bacteria and approximately 10^5 fungi/g soil), and survival and frequency of recombination were evaluated after 4 days. The frequency of recombination (which ranged from 10^{-2} to 10^{-3}) was generally higher in the clay-amended sterile soils than in the sterile soil not amended with clays. In the presence of *P. fluorescens*, the frequency of recombination in the unamended and kaolinite-amended soils was equivalent to that in the comparable sterile soils, but it was enhanced in the montmorillonite-amended soil. In the presence of R. *rubra*, the frequency of recombination was essentially the same as in the unamended or clay-amended sterile soils. The frequency of recombination was lower in the nonsterile than in the sterile soil, and the addition of the clays to the nonsterile soil did not affect the frequency of recombination. When the transconjugants were isolated from soil, grown in laboratory media, and reinoculated into the nonsterile soil systems, their survival after 4 days was better than that of the parentals, especially in the montmorillonite-amended soil, wherein growth of the transconjugants occurred.

Kelly and Reanney (1984) studied the prevalence of plasmids conferring resistance to mercury in bacteria isolated from soils not polluted with mercury or with other metals. Of the 504 soil bacteria isolated, 97 were resistant to mercury (Enterobacteriaceae, 30; *Pseudomonas*, 27; *Bacillus*, 24; *Mycobacterium–Nocardia*, 6; *Flavobacterium–Cytophaga*, 5; gram-positive cocci, 3; *Alcaligenes*, 2). Many of the mercury-resistant isolates were also resistant to antibiotics, especially to carbenicillin. When 76 of the 97 mercury-resistant isolates were evaluated for their ability to transfer mercury resistance to recipient cells (e.g., *E. coli*, *P. aeruginosa*, *Erwinia herbicola*, *P. mirabilis*), only 4 were able to transfer this resistance. A *Citrobacter* sp., designated LC1, and an *Enterobacter* sp., designated LC4, were able to transfer linked resistance to mercury and three antibiotics (i.e., streptomycin, sulfanilamide, and tetracycline) at frequencies of 10^{-2} to 10^{-3} transferants/donor. The stability of this linkage indicated that the resistance determinants were located on common plasmids, designated pWK1 and pWK2, respectively. However, the majority of the mercury-resistant isolates did not apparently have the genes for mercury resistance on conjugative plasmids, although some of these mercury resistance genes appeared to be on nonconjugative plasmids. When plasmid RP1 was transferred, using *E. coli* and *P. aeruginosa* strains as donors, to 35 of 42 of the gram-negative mercury-resistant isolates, 9 of these isolates were then able to transfer both mercury resistance and the RP1 markers to an *E. coli* K12 recipient. Transduction studies with phage P1 further showed that 7 of these transfers were the result, at least in part, of the formation of a RP1-mercury resistance cointegrate plasmid. Radford *et al.* (1981) had previously studied these 42 gram-negative mercury-resistant isolates and noted that 3 (a *Klebsiella* sp., a *Citrobacter* sp., and *P. fluorescens*) contained transposons conferring resistance to inorganic mercury and, in the case of *P. fluorescens*, also to phenylmercuric ions.

Bacteria containing plasmids that code for the degradation of xenobiotics have also been isolated from soil. For example, Williams and Worsey (1976) isolated 13 strains of *Pseudomonas* from 9 soils by enrichment culture on *m*-toluate as the sole source of carbon. These bacteria carried the TOL plasmid, which codes for the metabolism of toluene and xylenes. Eight of these strains transferred the TOL plasmid to their own plasmid-cured strains, and 5 strains transferred the plasmid to a plasmid-cured derivative of *P. putida* mt-2. In addition, some of these isolates contained multiple plasmids (Duggleby *et al.*, 1977). Don and Pemberton (1981) isolated 6 plasmids from *Alcaligenes paradoxus* and *A. eutrophus* that coded for the degradation of the herbicides, 2,4-dichlorophenoxyacetic acid and 4-chloro-2-methylphenoxyacetic acid. Four of the plasmids, designated pJP3, pJP4, pJP5, and pJP7, could be transferred to strains of *E. coli*, *Rhodopseudomonas sphaeroides*, *Rhizobium* sp., *A. tumefaciens*, *P. fluorescens*, and *Acinetobacter calcoaceticus*. A plasmid of 37 kb that codes for the degradation of styrene was identified in *P. fluorescens* (Bestetti *et al.*, 1984).

Pertsova *et al.* (1984) studied the degradation of 3-chlorobenzoate in soil columns that were uninoculated or inoculated either with a soil isolate of *P. putida*, which contained a plasmid coding for the degradation of 3-chlorobenzoate, or with *P. aeruginosa*, which was incapable of growth on 3-chlorobenzoate but which contained the inserted plasmid, pBS 2, which codes for ortho-cleavage of the aromatic ring of organic compounds. Degradation of 3-chlorobenzoate was not detected in the uninoculated soils, indicating that the indigenous microbiota was not capable of degrading 3-chlorobenzoate, but degradation was detected in the soils inoculated with either species of *Pseudomonas*. Furthermore, pseudomonads capable of utilizing 3-chlorobenzoate as the sole source of energy and carbon were isolated from the inoculated soils. However, based on taxonomic characteristics, the bulk of the isolated 3-chlorobenzoate-degrading pseudomonads were not similar to *P. aeruginosa* or *P. putida*, suggesting that there had been an exchange of genetic information between the indigenous microbiota and the plasmid-carrying strains of *P. aeruginosa* and *P. putida*.

Several strains of *E. coli*, containing plasmids that ranged in size from 3.9 to 96 kb, were inoculated into soil at 10^4 to 10^5 cells/g soil (Devanas *et al.*, 1986). The initial total bacterial population was approximately 10^7 colony-forming units (CFU)/g soil, and the gram-negative bacterial population ranged from 10^5 to 10^6 CFU/g soil. Changes in various populations were followed with a dilution-drop plating technique and various selective media, which enabled the detection of as few as 1 to 20 introduced CFU/g soil. The introduced strains either remained at approximately 10^4 CFU/g soil or decreased to undetectable levels during a 28-day incubation, depending on the host strain but not on the type or size of the plasmid. The addition of nutrients, initially or during the incubation, resulted in an increase in all bacteria and decreased the rate of decline of the introduced strains. There appeared to be no loss of the plasmids, and the rapid decline of some host/vector systems was a function of the host strain and not of the plasmid that it contained. These results indicated that the survival of some genetically engineered bacteria in soil is primarily a function of the bacterial strain and not of the contained plasmid and is influenced by the nutritional status of the soil. Furthermore, these studies indicated that the survival in soil of some genetically engineered bacteria that are not indigenous to soil can be long enough (i.e., at least 28 days) to suggest that they could transfer their plasmids to indigenous bacteria. However, no such transfer was observed.

Devanas and Stotzky (1986) also studied the fate in soil of a recombinant plasmid, C357, composed of pBR322 containing cDNA that codes for an egg yolk protein from *Drosophila grimshawi*, in *E. coli* strain HB101. The HB101 host, HB101(pBR322), and HB101(C357) were inoculated into sterile and nonsterile soil at 10^4 to 10^5 cells/g soil. The initial total natural population was approximately 10^7 CFU/g soil, and the gram-negative bacterial population ranged from 10^5 to 10^6 CFU/g soil. Changes in the various populations were followed with the dilution-drop plating technique, various selective media, and a ^{32}P-labeled DNA probe for C357. The survival of the host/plasmid systems was similar in sterile and nonsterile soils, but HB101(C357) declined more rapidly than HB101 and HB101(pBR322) in both soils and was not detected after 28 days of incubation, even though the techniques enabled the detection of as few as 1 to 20 CFU/g soil. There was no loss of plasmids, and the more rapid decline of HB101(C357) appeared to be a function of the host/vector combination. The addition of nutrients, initially or during the incubation, resulted in an increase in all bacteria and decreased somewhat the rate of decline of HB101(C357). These results indicated that heterologous DNA in novel bacteria can survive and replicate in soil. This survival and replication—and possible subsequent transfer of the novel genes to indigenous bacterial populations—must be considered before the release of genetically engineered bacteria to the environment.

4. Plants

The ability of *A. tumefaciens* to induce crown gall formation in dicotyledonous plants is encoded on a large plasmid [the tumor-inducing (Ti) plasmid; 120 MDa]. Watson *et al.* (1975) inoculated a tomato seedling with *A. tumefaciens* C-58 (a virulent, antibiotic-sensitive strain that contained the Ti plasmid) and 6 days later inoculated the plant with *A. tumefaciens* A136 (an avirulent, antibiotic-resistant strain). After 5 weeks of incubation, tumor tissue was excised and transconjugants that exhibited both virulence and antibiotic resistance were isolated. The transconjugants contained a large plasmid, which exhibited DNA homology in hybridization studies with the Ti plasmid of the donor cell. Kerr *et al.* (1977) mated, *in planta*, a donor strain of *Agrobacterium* (strain NCPPB 1001, which was pathogenic and sensitive to antibiotics) with a recipient strain of *Agrobacterium* (strain K57, which was nonpathogenic and resistant to antibotics). Both donor and recipient strains were inoculated onto tomato seedlings, and transconjugants that were both virulent and antibiotic-resistant were detected after 3, 4, and 5 weeks of incubation at a frequency of 4.0×10^{-4}, 4.8×10^{-3}, and 3.7×10^{-2}, respectively.

van Larebeke *et al.* (1975) inoculated the plant, *Kalanchoe daigremontiana*, with the virulent, plasmid-containing, antibiotic-sensitive *Agrobacterium rhizogenes* strain 223 and the avirulent, antibiotic-resistant *A. tumefaciens* strain C58-C1 (which had been cured of its tumor-inducing plasmid). Eight weeks after inoculation, 50% of the antibiotic-resistant bacteria that were isolated were able to induce crown gall tumor formation on *K. daigremontiana* as well as on seedlings of sunflower and pea. These virulent, antibiotic-resistant strains were demonstrated to be *A. tumefaciens* transconjugants, as they contained a plasmid of a size similar to that occurring in *A. rhizongenes* strain 223. Further studies involved the inoculation of sterile pea seedlings with the antibiotic-sensitive, crown gall-inducing *A. tumefaciens* strain Kerr 14, and 12 days later, with the antibiotic-resistant, avirulent *Agrobacterium radiobacter* strain S1005. Transconjugants of *A. radiobacter* were later detected that were antibiotic resistant, virulent, and contained the large plasmid.

In planta transfer of plasmid RP1 was demonstrated by Lacy and Leary (1975). Immature pods of lima beans (*Phaseolus lunatus*) were inoculated with donor and recipient bacteria, or the donors and recipients were inoculated either into the leaflets of trifoliate leaves by water congestion of the abaxial surface or onto the surface of leaflets by misting. RP1 was transferred in the pods from donor *E. coli* χ705 to recipient *Pseudomonas glycinea* PG9 and *Pseudomonas phaseolicola* HB10Y (no comparative transfer frequencies were presented). The transfer frequencies of RP1 from *E. coli* χ705 to *P. glycinea* PG9 were 2.8×10^{-3} and 1.6×10^{-2} *in vitro* and in congested leaves, respectively. No changes in host range or pathogenicity were evident in *P. glycinea* PG9 as the result of the presence of the RP1 plasmid, as the host reactions to *P. glycinea* PG9 and *P. glycinea* PG9(RP1) on four cultivars of soybean (*Glycine max*) and on one cultivar each of lima bean and pole bean (*Phaseolus vulgaris*) were comparable. RP1 was also transferred from *P. glycinea* PG9(RP1) to *P. phaseolicola* HB36: the *in vitro* transfer frequencies ranged from 4.9×10^{-3} to 5.8×10^{-2}, and the *in planta* transfer frequencies were 5.5×10^{-1} (in pods), 8.2×10^{-2} (in congested leaves), and 3.1×10^{-1} (in misted leaves).

Lacey *et al.* (1984) also demonstrated the *in planta* transfer in pear blossoms of plasmid RP1, that mediated resistance to oxytetracycline, from *Erwinia herbicola* and *Pseudomonas syringae* pv. *syringae* to *Erwinia amylovora*. The plasmid conferred resistance to high concentrations (> 1624 μg/ml) of the antibiotic and did not prevent the pathogenesis of *E. amylovora*. However, the pathogen was cured during infection of plant tissues, and only about 0.03% of the bacteria recovered maintained the plasmid phenotype. Although these results suggested that plasmid-borne resistance could interfere with the efficacy of oxytetracycline in the control of fire blight, no transconjugants of *E. amylovora* showing resistance to the antibiotic have been reported to occur *in situ*.

Bacteria of the genus, *Rhizobium*, fix nitrogen in symbiotic association with leguminous plants. In *Rhizobium trifolii*, which nodulates clover, the genes for symbiotic nitrogen fixation are located on plasmid pRtr5a. The *in vitro* transfer of pRtr5a from *R. trifolii* to *Rhizobium leguminosarum* resulted in transconjugants capable of nodulating and fixing nitrogen not only in pea and vetch, the normal symbionts of *R. leguminosarum*, but also in clover. Furthermore, transfer of pRtr5a from *R. trifolii* to mutants of *R. leguminosarum* that were not capable of nodulating their normal symbionts resulted in transconjugants that could nodulate and fix nitrogen in clover but still could not nodulate pea and vetch, and transfer of pRtr5a from *R. trifolii* to mutants of *R. leguminosarum* that could nodulate but not fix nitrogen in pea and vetch resulted in transconjugants that could nodulate and fix nitrogen in clover. *R. trifolii* was also capable of transferring, *in vitro*, its pRtr5a plasmid to a strain of *A. tumefaciens* in which the Ti-plasmid had been cured, and the transconjugants were capable of nodulation but not of nitrogen fixation in clover (Hooykaas *et al.*, 1981).

Talbot *et al.* (1980) studied the transfer of antibiotic resistance among *Klebsiella* strains inoculated into "botanical environments." Radish seeds were soaked in a suspension of recipient antibiotic-sensitive *Klebsiella* cells, loosely packed in a nonsterile sandy loam soil, inoculated over their exposed surfaces with donor antibiotic-resistant *Klebsiella* cells, and then completely covered with soil. At weekly intervals, the radish seedlings were removed, homogenized, and transconjugants were enumerated on selective media. Transfer of antibiotic resistance was detected in 5 of 21 donor-recipient mating combinations, with transfer frequencies of 10^{-6} to 10^{-7} during the first week of sampling, but no transconjugants were detected thereafter. When donor and recipient strains were inoculated into an aqueous suspension of redwood sawdust, transfer of antibiotic resistance was detected after 3 days of incubation in 6 of 21 donor–recipient combinations at frequencies of 10^{-5} to 10^{-8}, whereas in broth, transfer of antibiotic resistance was detected in 12 of 21 donor–recipient combinations at frequencies of 10^{-3} to 10^{-6}.

The role of plasmids in interactions between microbes and plants has recently been reviewed (Shaw, 1986).

B. Transduction

Transduction of *P. aeruginosa* strain RM2054 to streptomycin resistance by the generalized transducing phage, F116, occurred in flow-through environmental test chambers submerged in a freshwater reservoir. The environmental chambers were inoculated with *P. aeruginosa* RM2054 and either the lysates from or intact cells of the lysogenic strain, *P. aeruginosa* RM2060 (which was resistant to streptomycin), and submerged in the reservoir at a site approximately 1 km downstream from an outfall from a secondary sewage treatment plant. Transduction of strain RM2054 by lysates containing phage F116 occurred at frequencies of 5×10^{-6}, 1.2×10^{-2}, and 9.5×10^{-1} after 1 hour, 4 days, and 10 days, respectively. The transduced cells increased in number during the first 4 days and had survival characteristics similar to those of strain RM2054. In the chamber containing the lysogenic strain, the transduction frequencies were 1.4×10^{-5}, 5.9×10^{-3}, and 8.3×10^{-2} after 1 hour, 4 days, and 10 days, respectively, and the transduced cells attained population densities comparable to those of the lysogenic donor (Morrison *et al.*, 1978).

Baross *et al.* (1974) demonstrated transduction, in oysters, of the ability of *Vibrio parahaemolyticus* to degrade agar ("agarase characteristics"). Oysters were allowed to take up a streptomycin-resistant strain of *V. parahaemolyticus* in marine aquaria and were then placed in aquaria containing sterilized sea water and a phage (P4) of a psychrophilic *Vibrio* sp. that contained genes for the degradation of agar. After 154 hours, streptomycin-resistant agar-degrading vibrios were detected in samples of the oysters. Transduction has apparently not been studied nor demonstrated in soil, whether sterile or natural.

C. Transformation

Transformation in *B. subtilis* in sterile soil was demonstrated by Graham and Istock (1978, 1979). Strains of *B. subtilis*, each with three different chromosomally labeled and linked markers, were inoculated into soil, and after 8 days of incubation, 79% of the randomly selected colony-forming units exhibited a phenotype containing markers from both parents. The parental strains, however, were not detected after the first day of incubation. The observed genetic exchange was presumed to have occurred by transformation, as *B. subtilis* is known to release transforming DNA in liquid culture, and there were no detectable plasmids or generalized transducing phages in the strains employed. The frequencies of triple transformants obtained in soil were relatively low (e.g., 1.8×10^{-4} 1 day after inoculation) and did not increase in concert with increases in the mixed population (i.e.,

transformants with a phenotype containing some markers from both parentals), presumably because of "natural selection against one or more of the 3
genetic markers." However, when the strains were grown individually in
soil in the presence of DNA from the other strain, the frequencies of triple
transformants increased to 17%, and the frequencies increased in concert
with the increase in numbers of the individual populations. The addition of
DNase to the soil did not affect the number of triple transformants isolated.
The investigators suggested that either the DNase absorbed on the soil
particulates and was inactivated or the released bacterial DNA may have
been complexed with protein or bits of bacterial membrane, thereby making
it unavailable for degradation by DNase.

Aardema *et al.* (1983) found that DNase reduced the transformation frequencies of "free" DNA (i.e., DNA in solution) from *B. subtilis* to a greater
extent than that of DNA adsorbed on "marine sediment" (i.e., sea sand
consisting primarily of quartz). Lorenz *et al.* (1981) also noted that calf
thymus DNA adsorbed on this marine sediment was protected against degradation by DNase. This presumed protection of adsorbed DNA against
enzymatic degradation was used to explain the higher frequency of transformation of *B. subtilis* that occurred in the presence of the marine sediment.
However, in the absence of DNase, the transformation frequency of the
adsorbed DNA was only about 60% of that of "free" DNA. It was not
determined whether *B. subtilis* was transformed by DNA that was released
from the sediment or by adsorbed DNA (Aardema *et al.*, 1983).

The transformation studies of Graham and Istock (1978, 1979) and of
Aardema *et al.* (1983) were conducted in sterile systems and, thus, may not
accurately reflect the interactions between "naked" DNA and the microbial
populations in natural systems. For example, Greaves and Wilson (1970)
showed that DNA or RNA added to nonsterile soil was readily degraded and
resulted in concomitant increases in the indigenous microbial populations.
Furthermore, although DNA or RNA is apparently strongly adsorbed on soil
particulates, such as montmorillonite (Blanton and Barnett, 1969; Greaves
and Wilson, 1969), the addition of montmorillonite and DNA or RNA to a
coarse sand did not prevent microbial degradation of the nucleic acids
(Greaves and Wilson, 1970). Transformation has apparently not been studied
nor demonstrated in nonsterile soils.

V. Effects of Biotic and Abiotic Environmental Factors on Survival of, and Genetic Transfer by, Engineered Bacteria

Despite the remarkable advances in the isolation, analysis, construction,
and methods of introducing new genes into organisms, the ultimate fate of

natural and manipulated genetic material is dependent on the survival, establishment, and growth of the microbial hosts, which house the genetic material, in the natural habitats into which the hosts are introduced. Survival, establishment, and growth are, in turn, dependent on the genetic constitution of the hosts and on the physical (e.g., temperature, pressure, electromagnetic radiation, surfaces, spatial relations), chemical (e.g., carbonaceous substrates, inorganic nutrients, growth factors, ionic composition, available water, pH, oxidation-reduction potential, gaseous composition, toxicants), and biological (e.g., characteristics of and positive and negative interactions between microbes) factors of the various habitats (Stotzky, 1974; Stotzky and Krasovsky, 1981).

For example, to study the transfer of genes in *E. coli* K12 strains *in vivo*, *in vitro*, and *in situ*, the competition between indigenous microbial populations and the introduced *E. coli* necessitated the use of antibiotic-fed, starved, or germ-free animals *in vivo*, of sterile soils *in vitro*, and of environmental chambers in aquatic systems *in situ*, including those of sewage treatment facilities, wherein the microbes of interest were physically separated from the indigenous populations. There is little information on the potential survival, establishment, and growth of, and subsequent genetic transfer from, *E. coli* K12 host–vector systems—or of any other host–vector systems—in natural environments that retain their indigenous microbial populations as potential competitors against the introduced host–vector systems. Differences in the potential survival and establishment of non-genetically engineered microorganisms introduced into sterile and nonsterile environments in which many were not natural residents were observed by Liang *et al.* (1982). When inoculated into sterile sewage, *R. meliloti* and, to a lesser extent, *A. tumefaciens*, *K. pneumoniae*, and *S. typhimurium* grew and persisted at elevated levels; however, when inoculated into nonsterile raw sewage, the numbers of *R. meliloti* and *A. tumefaciens* declined slowly, whereas those of *S. typhimurium* and *K. pneumoniae* declined rapidly. Differences in viability were also noted in sterile and natural lake water and soil. In sterile lake water, *R. meliloti* grew slightly, *A. tumefaciens* and *S. typhimurium* declined slowly to a stable population size, and *K. pneumoniae* was rapidly killed. In natural lake water, however, the numbers of *R. meliloti* and *A. tumefaciens* declined slowly and reached a stable population size, whereas the numbers of *K. pneumoniae* and *S. typhimurium* declined rapidly. In sterile soil, there was a 1 to 2 logarithmic increase in numbers of *S. typhimurium*, *A. tumefaciens*, and *K. pneumoniae* (*R. meliloti* was not studied), whereas in natural soil, there was a gradual reduction in numbers of all bacteria. Krasovsky and Stotzky (1986) observed that survival and genetic recombination of *E. coli* χ493 and χ696 was reduced to a greater extent in nonsterile than in sterile soil.

The ability of an organism to survive, grow, and colonize new habitats will influence its ability to transfer successfully genetic information. This is especially true for conjugation, which requires large cell populations to ensure cell-to-cell contact between donor and recipient cells. The yield of plasmid-containing transconjugants is decreased by approximately the square of the dilution of the parental densities below 10^8 cells/ml. Thus, the number of transconjugants that inherited plasmids was reduced by three to four orders of magnitude by reducing the mating titers from 10^8 to 10^6 cells/ml (Curtiss, 1976).

The rate of reproduction of the host organism can be influenced, in part, by its plasmid, which could ultimately affect the persistence of the host cells in the environment. In competition studies between a strain of *E. coli* K12, that contained a plasmid (31.6 MDa) conferring resistance to ampicillin, streptomycin, sulfonamide, and tetracycline, and its mutant strain, containing the same plasmid (now, 20.0 MDa) but lacking the gene for resistance to tetracycline, the mutant strain with the smaller plasmid had a faster growth rate, which gave it a considerable advantage over the parental strain when grown in mixed culture (Godwin and Slater, 1979). Zund and Lebek (1980) evaluated the influence of plasmid size on the generation time of host cells. Approximately 100 plasmids conferring resistance to antibiotics were identified in clinical strains of *E. coli*, *Salmonella*, *Enterobacter*, and *Klebsiella* and conjugally transferred *in vitro* to an *E. coli* K12 recipient. In the absence of any added plasmid, the generation time of the *E. coli* recipient was 30 minutes, but 27% of the R-plasmids tested increased the generation time of the transconjugants by more than 15% (up to over 35 minutes), and approximately 50% of the plasmids had lesser effects on prolonging the generation time. However, the majority of the R-plasmids that increased the generation time of the host by more than 15% contained more than 80 kb, whereas the majority of plasmids containing less than 80 kb had no such adverse effect.

Insufficient research has evaluated the subtle alterations in the physiology of transconjugants that could occur as a result of the acquisition of plasmids. Lacy and Leary (1975) showed that the presence of plasmid RP1 in *P. glycinea* did not change its host range or its pathogenicity. However, Kozyrovskaya *et al.* (1984) reported that the transfer of plasmid pRD1, which carried genes for nitrogen fixation and for resistance to kanamycin, tetracycline, and carbenicillin, from *E. coli* K12 J62-1 to the phytopathogenic bacteria, *A. tumefaciens, X. beticola,* and *E. carotovora* subsp. *carotovora,* caused several unexpected alterations in the physiology and biochemistry of the transconjugants. Expression of antibiotic resistance in the transconjugants exceeded that of the donor, e.g., the minimal inhibitory concentrations (MIC) of kanamycin and tetracycline were 1000 and 400 μg/ml, respectively, for the donor and 2 μg/ml each for the three recipients, but the MIC were 1100 and 700 µg/ml, respectively, for the transconjugants of X. be-ticola, 1400 and 800 µg/ml, respectively, for the transconjugants of E. car-otovora subsp. carotovora, and 1200 and 800 µg/ml, respectively, for the transconjugants of A. tumefaciens. Although the data were not presented, similar differences were presumably observed for resistance to carbenicillin. Furthermore, transconjugants of X. beticola also acquired resistance to chloramphenicol, with transconjugants that contained genes for nitrogen fixation being more resistant to chloramphenicol than transconjugants that did not contain these genes: the MIC for the donor, the X. beticola recip-ient, the X. beticola(pRD1) transconjugant not containing genes for nitrogen fixation, and the X. beticola(pRD1) transconjugant with genes for nitrogen fixation were 2, 2, 50, and 100 µg/ml chloramphenicol, respectively. More-over, the chromosomal marker for resistance to nalidixic acid that was ex-pressed in the donor was also expressed in the X. beticola(pRD1) transcon-jugants containing genes for nitrogen fixation but not in the transconjugants lacking these genes or in the transconjugants of the other two recipients.

Differences were also reported in the virulence of the plasmidless parental strains and the transconjugants: crown galls appeared on carrot slices inocu-lated with the parent and transconjugant strains of A. tumefaciens after 10 and 7 days, respectively, and on tobacco cultivars after 7 and 5 days, respec-tively. On beet slices inoculated with parent and transconjugant strains of X. beticola, excrescences developed after 7 and 4 days, respectively, with the excrescences formed by the transconjugant being larger than those formed by the parent. In contrast, the pathogenicity of the E. carotovora subsp. carotovora transconjugant was less than that of its parental strain, although the transconjugants had a greater capability to macerate plant tissue (Kozy-rovskaya et al., 1984).

The biochemistry of the transconjugants was also apparently different from that of the parental strains. When 44 cultural and biochemical charac-teristics—which were not encoded on the plasmid pRD1 (e.g., gas and acid production from the catabolism of glucose, lactose, maltose, sucrose, man-nitol, salicyn, rhamnose, dulcitol, and sorbitol)—of the donor, the X. be-ticola recipient, and several transconjugants of X. beticola(pRD1) were stud-ied, 45.5% of the biochemical characteristics expressed by a nonnitrogen-fixing transconjugant were shared by both parental strains, 36.4% were characteristics expressed only by the X. beticola parent, 12.1% were charac-teristics expressed only by the E. coli donor, and 6% of the characteristics were novel and not expressed by either the donor or the recipient. Trans-conjugants of E. carotovora subsp. carotovora(pRD1) also differed from the parental strains in their utilization of amino and other organic acids (Kozyrovskaya et al., 1984). These subtle differences in antibiotic resistance, virulence, and biochemistry between donor, recipient, and transconjugants would probably influence the ecology and population dynamics of these competing cells in natural environments.

Curtiss (1976), Stotzky and Krasovsky (1981), Freter (1984), and Stotzky (1986) have discussed some of the physicochemical factors that may affect the survival, establishment, growth, and genetic recombination of microbes in natural habitats. Temperature exerts an influence on conjugal plasmid transfer; e.g., the optimum temperature for the transfer of the F, I, and N incompatibility plasmid groups commonly found in enterics is near 37°C, and essentially no transfer occurs below 28°C, as donors are apparently unable to express the donor phenotype at the lower temperatures (Curtiss, 1976). The maximum numbers of F-pili per cell were formed by an F^+ strain of *E. coli* B/r at temperatures from 37 to 42°C; below 37°C, the numbers of pili decreased, and no pili were formed at 25°C (Novotny and Lavin, 1971). No mating pair formation occurred between the donor, *E. coli* HfrH, and the recipient, *E. coli* PA309, below 24°C; increasing the temperature from 30 to 41°C progressively increased pair formation, but a sharp decline in mating pair formation occurred at 41 to 45°C (Walmsley, 1976). Curtiss (1976) suggested that the *in situ* transfer of plasmids in enteric bacteria probably occurs in intestinal tracts and in fermenting excreta (e.g., manure piles, sewage treatment plants) but that the conjugative plasmids commonly present in bacteria indigenous to soil and aquatic systems are more likely to be transferred at higher efficiencies in the temperature range of 20 to 30°C.

Maximum conjugal transfer of the R-plasmid, R1drd-19, in strains of *E. coli* occurred at 37°C, and transfer progressively decreased when the temperature was sequentially decreased to 17°C; no transconjugants were detected in matings maintained at 15°C (Singleton and Anson, 1981). The frequency of transfer of a plasmid conferring resistance to kanamycin from a strain of *Proteus vulgaris*, isolated from the urinary tract of a patient with postoperative pyelonephritis, to *E. coli* 2050 recipient cells was about 10^5 times higher at 25 than at 37°C (Terawaki *et al.*, 1967). Smith *et al.* (1978) differentiated antibiotic-resistant enterobacteria isolated from human beings, animals, polluted rivers, and sewage treatment plants on the basis of the thermosensitivity of their plasmids: "thermotolerant" plasmids were designated as those that were transferred equally well at 22 to 37°C, and "thermosensitive" plasmids were those that were transferred at high rates at 22 or 28°C but at very low rates at 37°C. Of the 775 conjugative plasmids conferring antibiotic resistance, only 24 (3.1%) were thermosensitive and occurred most often in *K. pneumoniae*. Many of the thermosensitive plasmids also conferred resistance to mercury, arsenate, and tellurite, and they could also be transferred at 15°C, although less efficiently than at 22 or 28°C.

Kelly and Reanney (1984) evaluated the influence of temperature on the frequency of conjugation in isolates from soil and in laboratory strains. A

Citrobacter sp. and an *Enterobacter* sp. isolated from nonpolluted soil contained plasmids, designated pWK1 and pWK2, respectively, that conferred resistance to mercury and to antibiotics. When an *E. coli* strain was used as the recipient, pWK1 and pWK2 were transferred optimally at 28°C, and the frequencies were greatly reduced at 15 and 37°C. In contrast, pRD1, a laboratory-maintained plasmid isolated from a clinical strain of *Pseudomonas*, was transferred optimally at 37°C from an *E. coli* K12 donor to the *E. coli* recipient, and the frequencies progressively decreased as the temperature was reduced to 15°C. When *E. herbicola* was the recipient, transfer of pWK1 occurred even at 12.5°C.

When a donor strain of *E. coli* isolated from a residential sewage and containing a plasmid conferring resistance to streptomycin and tetracycline was mated with a recipient strain of *E. coli* isolated from a creek, the *in vitro* frequency of transfer of antibiotic resistance was greatest at 25°C (1.4×10^{-4}) and lowest at 35°C (1.5×10^{-7}). The frequencies for matings performed at 15, 20, and 30°C were 4.3×10^{-6}, 1.6×10^{-4}, and 4.4×10^{-5}, respectively. Similarly, the frequency of transfer in raw sewage *in situ* was 3.2×10^{-5} at 22.5°C and 1.0×10^{-6} at 29.5°C (Altherr and Kasweck, 1982).

In addition to the effect of temperature on the transfer of plasmids, the maintenance of some plasmids by their host cells is also temperature dependent. For example, a plasmid (Rts1) coding for resistance to kanamycin was lost spontaneously in cells of a strain of *P. vulgaris* grown at 42°C, whereas no elimination of the plasmid occurred in cells grown at 25°C. Growth at 42°C also eliminated this plasmid from *E. coli* and *S. typhimurium* transconjugants that had obtained the plasmid from the *P. vulgaris* donor (Terawaki *et al.*, 1967). A strain of *S. aureus* that contained genes for tetracycline resistance and penicillinase production on two distinct plasmids, lost both plasmids when grown at 44°C (May *et al.*, 1964). Avirulent derivatives of *A. tumefaciens* C-58 were obtained when the cells were grown at 37°C, with the loss of tumor-inducing ability being correlated with the loss of the Ti plasmid (Watson *et al.*, 1975). The nodulation ability of *R. trifolii* strains 24 and T12, which is coded by plasmid pW22, was eliminated when the host cells were grown at 37°C, and the loss of the nodulation capacity was correlated with the loss of the plasmid (Zurkowski and Lorkiewicz, 1979). The survival of *E. coli* J5, containing plasmid RP4, and of *E. coli* JC5466, containing pRD1, was reduced to a greater extent in soil maintained at 20 than at 4°C (Schilf and Klingmüller, 1983).

Another physicochemical factor that affects the transfer of genetic material is pH. Conjugal plasmid transfer of the F, I, and N incompatible plasmid groups of *E. coli* seemed to be restricted to pH levels between 6 and 8.5 (Curtiss, 1976). Genetic recombination between *E. coli* strains χ503 and χ696 did not occur in soils with a bulk pH of 6.6 or below; in soils adjusted to pH 6.8–6.9, conjugation occurred, and increasing the soil pH to 7.0 reduced the frequency of conjugation without affecting the survival of the parentals (Stotzky and Krasovsky, 1981; Krasovsky and Stotzky, 1986). Weinberg and Stotzky (1972) observed that *E. coli* strains χ503, χ209, χ705, and 1230 grew better in a soil with a pH of 6.6 than in a soil with a pH of 6.3.

A dual stress by adverse pH and temperature may interact synergistically and exert a greater influence on genetic transfer than a stress by each factor individually. For example, the inhibition of *in vitro* conjugal transfer of plasmid R1drd-19 in *E. coli* by deviations in pH from the optimum of 6.9 (i.e., at pH 6.3 or 7.8) was more pronounced as the temperature was decreased from the optimum of 37 to 17°C (Singleton and Anson, 1983).

The water content of soil influenced the survival of the plasmid-containing donor strains, *E. coli* J5(RP4) and *E. coli* JC5466(pRD1), and of the transconjugants, Enterobacteriaceae strain 1(pRD1) and *P. fluorescens*(pRD1). A lower reduction in numbers occurred in soil maintained at a 16% water content than when the soil was dried to a 4% water content within 14 days (Schilf and Klingmüller, 1983).

Conjugal transfer of plasmids may also be influenced by the oxygen tension of the environment. The frequency of conjugal transfer between the donor, *E. coli* χ895 Hfr OR75, and the recipient, *E. coli* χ160 F⁻, was equivalent under aerobic and anaerobic conditions (Stallions and Curtiss, 1972). However, the frequency of the conjugal transfer of R-plasmids by two donor strains of *E. coli* isolated from human feces was reduced by a factor of 10 to 1000 under anaerobic conditions (Moodie and Woods, 1973). The expression of antibiotic resistance by 45 different plasmids in strains of *E. coli* K12 was the same under conditions of anaerobic and aerobic growth, but anaerobiosis adversely affected the formation of sex pili in some strains (Burman, 1977).

Surfaces, such as those provided in many natural habitats by clay minerals and other inorganic as well as organic particulates, appear to affect conjugation as, according to Freter (1984), "it is well known to plasmid geneticists that filter matings, i.e., the aggregation of donor and recipient bacteria on a filter surface, significantly promote plasmid transfer rates over those which take effect in bacterial suspensions." The addition of montmorillonite, which has a large specific surface area (approximately 800 m²/g), to soil increased the frequency of conjugation between strains of *E. coli* inoculated into soil, as well as enhancing the survival of the transconjugants. The addition of comparable amounts of kaolinite, which has a relatively low specific surface area (approximately 16 m²/g), had no effect on conjugation or survival (Weinberg and Stotzky, 1972; Stotzky and Krasovsky, 1981; Krasovsky and Stotzky, 1986). However, this differential effect between montmorillonite and kaolinite may have been the result of differences in their cation exchange capacities (CEC; the CEC of the montmorillonite used was approximately 98 mEq/100 g clay, whereas that of the kaolinite was approximately 6 mEq/100 g clay) rather than in their specific surface areas (Stotzky, 1986). Conversely, studies performed *in vitro* showed that colloidal montmorillonite reduced the frequency of transfer of plasmid R1drd-19 in *E. coli* from 2.2×10^{-2} in the absence of clay to 1.2×10^{-4} (Singleton, 1983). Whether these effects of clay minerals were a function of their specific surface areas or their CEC—or their other physicochemical characteristics (Stotzky, 1980, 1986)—has not been clarified.

Clay minerals also bind soluble organics and viral particles (Stotzky, 1980, 1986; Stotzky *et al.*, 1981) and, thereby, may also influence transformation and transduction, respectively, in natural environments. Aardema *et al.* (1983) noted that although DNA adsorbed on sea sand (consisting primarily of quartz) and was, thereby, presumably protected against degradation by DNase, the adsorbed DNA was able to transform *B. subtilis*, albeit at a lower frequency than "free" DNA. Greaves and Wilson (1970) showed that although DNA adsorbed on clay minerals (Blanton and Barnett, 1969; Greaves and Wilson, 1969), most of the adsorbed DNA was readily degraded in nonsterile soil by the indigenous soil microbiota. The DNA adsorbed on montmorillonite was degraded in nonsterile soil but not in γ-irradiated soil, indicating that nucleases from living cells, rather than cell-free nucleases, were responsible for the degradation. Ivarson *et al.* (1982) showed that nucleic acid bases (i.e., guanine, adenine, cytosine, thymine, and uracil) adsorbed on montmorillonite, illite, kaolinite, gibbsite, goethite, and a fulvic acid–montmorillonite complex were degraded by mixed populations of soil microbes, but only after an initial lag period, and the rate of biodegradation varied with the type of adsorbent.

Although the influence of particulates on transduction *in situ* has not been studied, the adsorption of animal, plant, and bacterial viruses on clay particles and the effects of such sorption on the infectivity of the viruses has been studied (see Stotzky *et al.*, 1981; Lipson and Stotzky, 1984). Clay minerals appear to enhance the persistence of viruses in natural environments, such as lake waters (Babich and Stotzky, 1980) and soils (Duboise *et al.*, 1979), as viruses adsorbed on clays are apparently protected against biological and abiological inactivation.

UV radiation usually reduces the survival of microbes in many natural habitats, but some plasmids appear to confer resistance to UV radiation on their hosts (Marsh and Smith, 1969). For example, in a long-term retention lagoon of a sewage treatment plant at Fort Smith in the Northwest Territories, Canada, where the long hours of sunlight result in a high amount of incident UV radiation, the percentage of fecal coliforms with plasmids conferring multiple resistance to antibiotics was 21.5% higher in the final effluent than in the raw sewage (Bell *et al.*, 1983).

The ionic composition of an environment can also influence the survival of and genetic transfer in bacteria. Smith *et al.* (1974) reported that the survival of R^+ strains of *E. coli* in sea water was unaffected by the presence of plasmids, whereas the survival of antibiotic-sensitive fecal coliforms was reduced. Singleton (1983) indicated that transfer of the R1drd-19 plasmid in *E. coli* was stimulated by concentrations of NaCl that are present in estuaries.

The survival of *E. coli* host-Charon 4A phage systems in liquid aerosols was affected by the relative humidity (RH): the detrimental effects of RH on airborne cells of *E. coli* strain DP50SupF followed the sequence of 50% > 70% > 30%; for strains χ1666 and χ1776, the sequence was 30% > 50% > 70%; and lysates of Charon 4A maintained at 40 to 60% RH lost approximately 98% of their ability to produce plaques within the first minute of aerosolization (Chatigny *et al.*, 1979).

In environments lacking the appropriate selection pressure encoded on the plasmid, the plasmid could be a burden to the host cell, and all or part of the plasmid may be eliminated (Curtiss, 1976). Although heavy metal pollution of natural environments may promote the maintenance in host cells of R-plasmids conferring resistance to heavy metals (Timoney *et al.*, 1978; Olson *et al.*, 1979), the same metal pollutants may hinder *in situ* transformation. Treatment of competent *B. subtilis* cells with mercury or cadmium inhibited transformation without decreasing cell viability, apparently by blocking the uptake of DNA (Groves *et al.*, 1974), and the treatment of competent *B. subtilis* cells with low concentrations of copper, nickel, or zinc inhibited the binding of and transformation by DNA (Young and Spizizen, 1963).

Even these few studies that have been conducted on the mediating effects of environmental factors on gene transfer *in situ* and *in vitro* clearly demonstrate that *in vitro* studies of conjugation, transformation, and transduction, which usually use standardized and optimal growth conditions, are not adequate predictors of gene transfer in natural environments.

VI. Concluding Remarks

The possible adverse influence of genetically engineered bacteria, whether those that have escaped inadvertently or those that have been released purposefully, on the homeostasis of the biosphere is not known. However, the transfer of genes between bacteria, either by conjugation, transformation, or transduction, and the splicing of genes to create novel DNA sequences, either by transposons or insertion sequences, are probably natural events that occur in the biosphere, albeit rarely, independently of biotechnology research in the laboratory. Although the transfer of genes between bacteria can occur across interspecific and intergeneric boundaries, and gene splicings may result in novel nucleotide sequences, the genetic constitution of bacteria that have been isolated from natural habitats appears to have remained basically unchanged. As noted by Slater (1984), "it is quite apparent, especially to microbial ecologists, that genus and species boundaries are strenuously maintained and conserved: a *Flavobacterium* sp. isolated from one environment is much like that isolated somewhere else, even with respect to those properties that are not used as criteria for identification." However, anthropogenic activities have modified, to some extent, the bacterial gene pool. This is perhaps best illustrated by the increased incidence of bacteria containing multiple antibiotic resistance encoded on conjugative plasmids, as this increase appears to be correlated with the increased use of antibiotics, both as chemotherapeutic agents and as supplements to animal feeds. Furthermore, the relatively low population densities of appropriate hosts and donors in most natural environments, which restrict the probability of gene transfer, may be obviated when sizeable quantities of bacteria engineered for a specific function are introduced into these environments.

The lack of substantial data on the survival, establishment, and growth of, and on the transfer of genetic information by, genetically engineered bacteria in natural environments has hindered the assessment of the risk to the biosphere of the release, either accidental or planned, of these organisms. For example, plasmid pBR322 is extensively used as a vector for inserting engineered genes into bacteria, as this plasmid is both nonconjugative and poorly mobilizable and, hence, considered "safe" (i.e., has a low risk of being transferred to bacteria indigenous to natural habitats, including the human gastrointestinal tract). However, by means of triparental matings, transfer of this plasmid to bacteria indigenous to the human gastrointestinal tract has been demonstrated (Levine *et al.*, 1983). Furthermore, many of the data on gene transfer between bacteria have been obtained with laboratory strains and with the genetic recombination studies having been performed under optimal laboratory conditions for gene transfer. Such data, however, may not be directly applicable to genetic transfer *in situ*, e.g., in natural aquatic and terrestrial environments. For example, conjugation between environmental isolates of gram-negative bacteria appears to occur at temperatures that are suboptimal or even inhibitory for laboratory strains (Curtiss, 1976). Insufficient research has evaluated the mediating influence of environmental factors, both biotic and abiotic, on the survival, establishment, and growth of, and genetic transfer by, genetically engineered bacteria (see Stotzky and Krasovsky, 1981; Omenn and Hollaender, 1984). However, the limited data that are available indicate that abiotic factors—such as pH, salinity, aeration, water content, and particulates—and biotic factors—such as competition between the engineered microbe and the indigenous microbiota of the specific habitat being studied, generation time, and plasmid size—exert an influence on the survival of, and transfer of genetic information by, engineered microbes *in situ*.

Most studies on the survival of, and gene transfer by, genetically engineered bacteria in natural environments have been conducted with *E. coli*, usually debilitated strains, and a major environmental concern has been the accidental release of these organisms. Although debilitated strains of other species will, undoubtedly, also be used as hosts for engineered genes, there is insufficient information on the survival and genetic exchange, neither *in vivo* nor *in situ*, of bacteria other than *E. coli*. Moreover, it is highly probable that bacterial species engineered for a specific function (e.g., degradation of recalcitrant hazardous wastes, pesticides, and oil spills; prevention of ice nucleation on plants; pest control; nitrogen fixation) will not be debilitated and will be either derived from, or compatible with, the environment into which they will be introduced. Consequently, these deliberately released engineered microbes will have a vastly greater potential for survival *in situ* than enteric bacteria, and therefore, they will constitute a potentially greater risk to the biosphere. Currently, nondebilitated bacterial strains of MPCA are being released into the environment (Betz *et al.*, 1983), but there appears to be little information on the survival of, and genetic transfer by, these microbes.

Because of the uncertainty of the potential ecological impacts of the release into natural environments of genetically engineered microbes, the initial releases should perhaps be into isolated and insulated environments. Islands, considerably distant from the mainland, should be considered for such test releases, not only because of their isolation and insulation, but also because the relative simplicity of insular biotas should enable such potential impacts to be evaluated more easily than impacts on more complex mainland biotas. The successful use of islands in the initial testing of various biological control methods indicate their potential usefulness. For example, the initial efficacy of the release of sterile male screw-worms was first evaluated on Sanibel Island and then on Curacao (Baumhover, 1955), and the use of a molluscide to control the snail vector of the cattle liver-fluke was initially tested on the small island of Shapinsay in the Orkneys (Heppleston, 1972). Other potential ecological perturbants have also been evaluated initially on islands (*see* Simberloff, 1974).

However, because of the apparent rapid air-borne (as well as possible water-borne) spread of microbes between continents (see Babich and

Stotzky, 1974), testing on islands should not be considered as an absolute guarantee for the distributional restriction or containment of released genetically engineered microorganisms. Nevertheless, the potential biospheric dangers of such releases may be reduced if initial releases are to insular environments.

Another aspect about which there has been essentially no research is the potential effects of the release of genetically engineered microbes on the ecological structure and function of natural environments. Although it is obviously difficult (considering the current paucity of data on the detection, survival, and growth of, and genetic transfer by, such microbes in natural environments) to design experiments to test the potential perturbations that an engineered microbe might have on the multitude of ecological events (e.g., biogeochemical cycles, species diversity, epidemiology of plant and animal diseases) in any particular environment, this aspect is really the "bottom line" concern about the release of genetically engineered microbes. The "current wisdom" is that novel genetic information will have an ecological impact only if its information is expressed. However, the observations of Kozyrovskaya *et al.* (1984)—which indicated that the acquisition of a plasmid can result in a spectrum of unrelated, unanticipated, and nonpredicted biochemical and other physiological alterations in the recipient bacteria—suggest that studies designed to evaluate the survival of, and gene transfer by, genetically engineered bacteria in natural habitats be alert for such unanticipated and nonpredicted alterations and that the current wisdom may be erroneous.

The current data base does not appear to be sufficient to assess adequately the risk to the biosphere of the release, whether inadvertent or deliberate, of genetically engineered microbes. This apparent lack of basic, preliminary data emphasizes the acute need for additional scientific research to provide the data necessary for the formulation by regulatory agencies [such as the U.S. EPA, as mandated under the Toxic Substances Control Act (TSCA) and the Federal Insecticide, Fungicide, and Rodenticide Act (FIFRA) (Rissler, 1984)] of meaningful guidelines for the biotechnology industry. In addition to producing genetically engineered microbes, both academia and the biotechnology industry should assess the potential survival, establishment, and growth of, and the genetic transfer by, these engineered microbes *in situ*. Such a regulatory approach is not unique and has been applied to the chemical industry: as mandated under TSCA, before the marketing of a novel chemical or of an established chemical with a novel use, the manufacturer must provide the U.S. EPA with toxicological data, including the results of ecotoxicological studies, about the potential adverse effects of the chemical on the biosphere. A similar approach would seem to be warranted for the products of genetic engineering.

ACKNOWLEDGMENTS

The preparation of this article was supported, in part, by Grants R809067 and CR812484 from the U.S. Environmental Protection Agency. The views expressed in this article are not necessarily those of the Agency. The helpful suggestions of Dr. Monica A. Devanas are gratefully acknowledged.

REFERENCES

Aardema, B. W., Lorenz, M. G., and Krumbein, W. E. (1983). *Appl. Environ. Microbiol.* **46,** 417–420.
Altherr, M. R., and Kasweck, K. L. (1982). *Appl. Environ. Microbiol.* **44,** 838–843.
Anderson, J. D. (1974). *J. Med. Microbiol.* **7,** 85–90.
Anderson, E. S. (1975a). *Nature* **225,** 502–504.
Anderson, J. D. (1975b). *J. Med. Microbiol.* **8,** 83–88.
Anderson, J. D., Gillespie, W. A., and Richmond, M. H. (1973). *J. Med. Microbiol.* **6,** 461–473.
Aoki, T., Egusa, S., and Watanabe, T. (1973). *Jpn. J. Microbiol.* **17,** 7–12.
Armstrong, J. L., Shigeno, D. S., Calomiris, J. J., and Seidler, R. J. (1981). *Appl. Environ. Microbiol.* **42,** 277–283.
Avery, O. T., MacLeod, C. M., and McCarty, M. (1944). *J. Exp. Med.* **79,** 137–158.
Babich, H., and Stotzky, G. (1974). *Crit. Rev. Environ. Contr.* **4,** 353–421.
Babich, H., and Stotzky, G. (1980). *Water Res.* **14,** Res. 14, 185–187.
Baross, J. H., Liston, J., and Morita, R. Y. (1974). *In* "International Symposium on *Vibrio parahaemolyticus*" (T. Fujino, G. Sakaguchi, R. Sakazaki, and Y. Takeda, eds.), pp. 129–137. Saikon, Tokyo.
Baumhover, A. H. (1955). *J. Econ. Entomol.* **48,** 462–466.
Bell, J. B., Macrae, W. R., and Elliott, G. E. (1980). *Appl. Environ. Microbiol.* **40,** 486–491.
Bell, J. B., Macrae, W. R., and Elliott, G. E. (1981). *Appl. Environ. Microbiol.* **42,** 204–210.
Bell, J. B., Elliott, G. E., and Smith, D. W. (1983). *Appl. Environ. Microbiol.* **46,** 227–232.
Beringer, J. E., and Hirsch, P. R. (1984). *In* "Current Perspectives in Microbial Ecology" (M. J. Klug and C. A. Reddy, eds.), pp. 63–70. American Society for Microbiology, Washington, D.C.
Bestetti, G., Galli, E., Ruzzi, M., Baldacci, G., Zennaro, E., and Frontali, L. (1984). *Plasmid* **12,** 181–188.
Betz, F., Luvin, M., and Rogul, M. (1983). *Recomb. DNA Tech. Bull.* **6,** 135–141.
Bialey, H. (1984). *BioTechnology* **2,** 239.
Blanton, M. V., and Barnett, L. B. (1969). *Anal. Biochem.* **3,** 150–154.
Burman, L. G. (1977). *J. Bacteriol.* **131,** 69–75.
Burton, N. F., Day, M. J., and Bull, A. T. (1982). *Appl. Environ. Microbiol.* **44,** 1026–1029.
Cenci, G., Morozzi, G., Scazzocchio, F., and Morosi, A. (1982). *Zbl. Bakt. Hyg., I. Abt. Orig. C* **3,** 440–449.
Chatigny, M. A., Hatch, M. T., Wolochow, H., Adler, T., Hresko, J., Macher, J., and Besemer, D. (1979). *Recomb. DNA Tech. Bull.* **2,** 62–76.
Chatterjee, D. K., Kellogg, S. T., Furukawa, K., Kilbane, J. J., and Chakrabarty, A. M. (1981). *In* "Recombinant DNA" (A. G. Walton, ed.), pp. 199–212. Elsevier, Amsterdam.
Chopra, I., Bennett, P. M., and Lacey, R. W. (1973). *J. Gen. Microbiol.* **79,** 343–345.
Cohen, P. S., Pilsucki, R. W., Myahl, M. L., Rosen, C. A., Laux, D. C., and Cabelli, V. J. (1979). *Recomb. DNA Tech. Bull.* **2,** 106–113.

Cole, M. A., and Elkan, G. H. (1979). *Appl. Environ. Microbiol.* **37**, 867–870.

Conant, J. E., and Sawyer, W. D. (1967). *J. Bacteriol.* **93**, 1869–1875.

Corliss, T. L., Cohen, P. S., and Cabelli, V. J. (1981). *Appl. Environ. Microbiol.* **41**, 959–966.

Curtiss, R., III (1976). *Annu. Rev.Microbiol.* **30**, 507–533.

Curtiss, R., III, Clark, J. E., Goldschmidt, R., Hsu, J. C., Hull, S. C., Inoue, M., Maturnin, L. J., Moody, R., and Pereira, D. A. (1977). *In* "Plasmids. Medical and Theoretical Aspects" (S. Mitsuhashi, L. Rosival, and V. Krcmery, eds.), pp. 375–387. Springer-Verlag, New York.

Dancer, B. N. (1980). *J. Gen. Microbiol.* **121**, 263–266.

Davidson, M. S., and Summers, A. O. (1983). *Appl. Environ. Microbiol.* **46**, 565–572.

Devanas, M. A., Rafaelli-Eshkol, D., and Stotzky, G. (1986). *Appl. Environ. Microbiol.* (In press).

Devanas, M. A., and Stotzky, G. (1986). *Curr. Microbiol.* (In press).

Dobritsa, S. V. (1984). *FEMS Microbiol. Lett.* **23**, 35–39.

Don, R. H., and Pemberton, J. M. (1981). *J. Bacteriol.* **145**, 681–686.

Duboise, S. M., Moore, B. E., Sorber, C. A., and Sagik, B. P. (1979). *Crit. Rev. Microbiol.* **7**, 245–285.

Duggleby, C. J., Bayley, S. A., Worsey, M. J., Williams, P. A., and Broda, P. (1977). *J. Bacteriol.* **130**, 1274–1280.

Duval-Iflah, Y., and Chappuis, J. P. (1984). *In* "Current Perspectives in Microbial Ecology" (M. J. Klug and C. A. Reddy, eds.), pp. 264–272. American Society for Microbiology, Washington, D.C.

Dykhuizen, D., Campbell, J. H., and Rolfe, B. G. (1978). *Microbios* **23**, 99–113.

Edlin, G., Lin, L., and Kudrna, R. (1975). *Nature* **225**, 735–736.

Edlin, G., Tait, R. C., and Rodriguez, R. L. (1984). *BioTechnology* **2**, 251–254.

El Solh, N., and Ehrlich, S. D. (1982). *Plasmid* **7**, 77–84.

Farrar, W. E., Eidson, M., Guerry, P., Falkow, S., Drusin, L. M., and Roberts, R. B. (1972). *J. Infect. Dis.* **126**, 27–33.

Feary, T. W., Sturtevant, A. B., and Lankford, J. (1972). *Arch. Environ. Health* **25**, 215–220.

Fontaine, T. D., and Hoadley, A. W. (1976). *Health Lab. Sci.* **13**, 238–245.

Freter, R. (1984). *In* "Current Perspectives in Microbial Ecology" (M. J. Klug and C. A. Reddy, eds.), pp. 105–114. American Society for Microbiology, Washington, D.C.

Freter, R., Brickner, H., Fekete, J., O'Brien, P. C. M., and Vickerman, M. M. (1979). *Recomb. DNA Tech. Bull.* **2**, 68–76.

Freter, R., Freter, R. R., and Brickner, H. (1983). *Infect. Immun.* **39**, 60–84.

Gardner, P., and Smith, D. H. (1969). *Ann. Intern. Med.* **71**, 1–9.

Glassman, D. L., and McNicol, L. A. (1981). *Plasmid* **5**, 231.

Godwin, D., and Slater, J. H. (1979). *J. Gen. Microbiol.* **111**, 201–210.

Goldberg, R. B., Bender, R. A., and Streicher, S. L. (1974). *J. Bacteriol.* **118**, 810–814.

Gowland, P. C., and Slater, J. H. (1984). *Microb. Ecol.* **10**, 1–13.

Goyal, S. M., and Hoadley, A. W. (1979). *Rev. Microbiol.* **10**, 50–58.

Goyal, S. M., Gerba, C. P., and Melnick, J. L. (1979). *Water Res.* **13**, 349–356.

Grabow, W. O. K., and Prozesky, O. W. (1973). *Antimicrob. Agents Chemother.* **3**, 175–180.

Grabow, W. O. K., Middendorf, I. G., and Prozesky, O. W. (1973). *Water Res.* **7**, 1589–1597.

Grabow, W. O. K., Prozesky, O. W., and Smith, L. S. (1974). *Water Res.* **8**, 1–9.

Grabow, W. O. K., Prozesky, O. W., and Burger, J. S. (1975). *Water Res.* **9**, 777–782.

Grabow, W. O. K., van Zyl, M., and Prozesky, O. W. (1976). *Water Res.* **10**, 717–723.

Graham, J. B., and Istock, C. A. (1978). *Mol. Gen. Genet.* **166**, 287–290.

Graham, J. B., and Istock, C. A. (1979). *Science* **204**, 637–639.
Greaves, M. P., and Wilson, M. J. (1969). *Soil Biol. Biochem.* **1**, 317–232.
Greaves, M. P., and Wilson, M. J. (1970). *Soil Biol. Biochem.* **2**, 257–268.
Griffith, F. (1928). *J. Hyg.* **27**, 113–159 (reprinted: *J. Hyg.* **64**, 129–175, 1966).
Groves, D. J., Wilson, G. A., and Young, F. E. (1974). *J. Bacteriol.* **120**, 219–226.
Guinee, P. A. M. (1965). *Antonie van Leeuwenhoek J. Microbiol. Serol.* **31**, 314–321.
Guinee, P. A. M. (1970). *J. Bacteriol.* **102**, 291–292.
Gyles, C. L., Palchaudhuri, S., and Mass, W. K. (1977). *Science* **198**, 198–199.
Gyles, C., Falkow, S., and Robbins, L. (1978). *Am. J. Vet. Res.* **39**, 1438–1441.
Hada, H. S., and Sizemore, R. K. (1981). *Appl. Environ. Microbiol.* **41**, 199–202.
Haefeli, C., Franklin, C., and Hardy, K. (1984). *J. Bacteriol.* **158**, 389–392.
Hamer, D. H. (1977). *Science* **196**, 220–221.
Helling, R. B., Kinney, T., and Adams, J. (1981). *J. Gen. Microbiol.* **123**, 129–141.
Heppleston, P. B. (1972). *J. Anim. Ecol.* **9**, 235–248.
Hogrefe, C., and Friedrich, B. (1984). *Plasmid* **12**, 161–169.
Holloway, B. W. (1979). *Plasmid* **2**, 1–19.
Hooykaas, P. J. J., van Brussel, A. A. N., den Dulk-Ras, H., van Slogteren, G. M. S., and Schilperoort, R. A. (1981). *Nature* **291**, 351–353.
Hughes, V. M., and Datta, N. (1983). *Nature* **302**, 725–726.
Hummel, R. P., Miskell, P. W., and Altemeier, W. A. (1977). *Surgery (St. Louis)* **82**, 382–385.
Hyder, S. L., and Streitfeld, M. M. (1978). *J. Infect. Dis.* **138**, 281–286.
Inoue, M., and Mitsuhashi, S. (1975). *Virology* **68**, 544–546.
Ivarson, K. C., Schnitzer, M., and Cortez, J. (1982). *Plant Soil* **64**, 343–353.
Jarolmen, H., and Kemp, G. (1969). *J. Bacteriol.* **99**, 487–490.
Jarolmen, H., Bondi, A., and Cromwell, R. L. (1965). *J. Bacteriol.* **89**, 1286–1290.
Jukes, T. H. (1972). *BioScience* **22**, 526–534.
Kamp, P. F., and Chakrabarty, A. M. (1979). *In* "Plasmids of Medical, Environmental, and Commercial Importance" (K. N. Timmis, and A. Puhler, eds.), pp. 275–283. Elsevier, Amsterdam.
Kasuya, M. (1964). *J. Bacteriol.* **88**, 322–328.
Kelch, W. J., and Lee, J. S. (1978). *Appl. Environ. Microbiol.* **36**, 450–456.
Kelly, W. J., and Reanney, D. C. (1984). *Soil Biol. Biochem.* **16**, 1–8.
Kerr, A., Manigault, P., and Tempe, J. (1977). *Nature* **265**, 560–561.
Kleeberger, A., and Klingmuller, W. (1980). *Mol. Gen. Genet.* **180**, 621–627.
Kobori, H., Sullivan, C. W., and Shizuya, H. (1984). *Appl. Environ. Microbiol.* **48**, 515–518.
Kozyrovskaya, N. A., Gvozdyak, R. I., Muras, V. A., and Kordyum, V. A. (1984). *Arch. Microbiol.* **137**, 338–343.
Kralikova, K., Krcmery, V., and Krcmery, V. (1983). *Recomb. DNA Tech. Bull.* **69**, 98–100.
Krasovsky, V. N., and Stotzky, G. (1986). In preparation.
Lacey, R. W. (1975). *Bacteriol. Rev.* **39**, 1–32.
Lacey, R. W. (1984). *Br. Med. Bull.* **40**, 77–83.
Lacy, G. H., and Leary, J. V. (1975). *J. Gen. Microbiol.* **88**, 49–57.
Lacy, G. H., Stromberg, V. K., and Cannon, N. P. (1984). *Can. J. Plant Pathol.* **6**, 33–39.
LaFont, J.-P., Guillot, J. F., Chaslus-Dancla, E., Dho, M., and Eucher-Lahon, M. (1981). *Bull. Inst. Pasteur* **79**, 213–231.
Lafont, J.-P., Bree, A., and Plat, M. (1984). *Appl. Environ. Microbiol.* **47**, 639–642.
Laux, D. C., Cabelli, V. J., and Cohen, P. S. (1982). *Recomb. DNA Tech. Bull.* **5**, 1–5.
LeBlanc, D. J., Hawley, R. J., Lee, L. N., and St. Martin, E. J. (1978). *Proc. Natl. Acad. Sci. U.S.A.* **75**, 3484–3487.

Lereclus, D., Menou, G., and Lecadet, M.-M. (1983). *Mol. Gen. Genet.* **191**, 307–313.

Levine, M. M., Kaper, J. B., Lockman, H., Black, R. E., Clements, M. L., and Falkow, S. (1983). *Recomb. DNA Tech. Bull.* **6**, 89–97.

Levy, S. B., and Marshall, B. (1979). *Recomb. DNA Tech. Bull.* **2**, 77–80.

Levy, S. B., and Marshall, B. (1981). *Recomb. DNA Tech. Bull.* **4**, 91–97.

Levy, S. B., Marshall, B., and Rowse-Eagel, D. (1980). *Science* **209**, 391–394.

Liang, L. N., Sinclair, J. L., Mallory, L. M., and Alexander, M. (1982). *Appl. Environ. Microbiol.* **44**, 708–714.

Linton, K. B., Richmond, M. A., Bevan, R., and Gillespie, W. A. (1974). *J. Med. Microbiol.* **7**, 91–102.

Lipson, S. M., and Stotzky, G. (1984). *In* "Viral Ecology" (A. H. Misra and H. Polasa, eds.), pp. 165–178. South Asian Publ., New Delhi.

Lorenz, M. G., Aardema, B. W., and Krumbein, W. E. (1981). *Mar. Biol.* **64**, 225–230.

Lovett, P. S. (1973). *J. Bacteriol.* **120**, 488–494.

McDaniel, R. G. (1981). *In* "Recombinant DNA" (A. G. Walton, ed.), pp. 245–257. Elsevier, Amsterdam.

Mach, P. A., and Grimes, D. J. (1982). *Appl. Environ. Microbiol.* **44**, 1395–1403.

McNicol, L. A., Aziz, K. M. S., Huq, I., Kaper, J. B., Lockman, H. A., Remmers, E. F., Spira, W. M., Voll, M. J., and Colwell, R. R. (1980). *Antimicrob. Agents Chemother.* **17**, 477–483.

Marsh, E. B., and Smith, D. H. (1969). *J. Bacteriol.* **100**, 128–139.

Marshall, B., Schluederberg, S., Tachibana, C., and Levy, S. B. (1981). *Gene* **14**, 145–154.

May, J. W., Houghton, R. H., and Perret, C. J. (1964). *J. Gen. Microbiol.* **37**, 157–169.

Meckes, M. C. (1982). *Appl. Environ. Microbiol.* **43**, 371–377.

Moodie, H. S., and Woods, D. R. (1973). *J. Gen. Microbiol.* **76**, 437–440.

Morrison, W. D., Miller, R. V., and Sayler, G. S. (1978). *Appl. Environ. Microbiol.* **36**, 724–730.

Nakahara, H., Ishikawa, T., Sarai, Y., Kondo, I., and Mitsuhashi, S. (1977). *Nature* **266**, 165–167.

Nivas, S. C., York, M. D., and Pomeroy, B. S. (1976). *Am. J. Vet. Res.* **37**, 433–437.

Novick, R. P., and Morse, S. I. (1967). *J. Exp. Med.* **125**, 45–59.

Novotny, C. P., and Lavin, K. (1971). *J. Bacteriol.* **107**, 671–682.

Olson, B. H., Barkay, T., and Colwell, R. R. (1979). *Appl. Environ. Microbiol.* **38**, 478–485.

Omenn, G. S., and Hollaender, A., eds. (1984). "Genetic Control of Environmental Pollutants." Plenum, New York.

Orzech, K. A., and Burke, W. F. Jr. (1984). *FEMS Microbiol. Lett.* **25**, 91–95.

Ottolenghi-Nightingale, E. (1969). *J. Bacteriol.* **100**, 445–452.

Ottolenghi-Nightingale, E. (1972). *Infect. Immun.* **6**, 785–792.

Ottolenghi, E., and MacLeod, C. M. (1963). *Proc. Natl. Acad. Sci. U.S.A.* **50**, 417–419.

Pemberton, J. M., and Don, R. H. (1981). *Agric. Environ.* **6**, 23–32.

Pertsova, R. N., Kunc, F., and Golovleva, L. A. (1984). *Folia Microbiol.* **29**, 242–247.

Petrocheilou, V., Grinsted, J., and Richmond, M. H. (1976). *Antimicrob. Agents Chemother.* **10**, 753–761.

Pickup, R. W., Lewis, R. J., and Williams, P. A. (1983). *J. Gen. Microbiol.* **129**, 153–158.

Radford, A. J., Oliver, J., Kelly, W. J., and Reanney, D. C. (1981). *J. Bacteriol.* **147**, 1110–1112.

Reanney, D. C. (1976). *Bacteriol. Rev.* **40**, 552–590.

Reanney, D. C. (1977). *BioScience* **27**, 340–344.

Reanney, D. C., Gowland, P. C., and Slater, J. H. (1983). *In* "Microbes in their Natural Environments, Thirty-Fourth Symposium of the Society for General Microbiology" (J. H.

Slater, R. Whittenbury, and J. W. T. Wimpenny, eds.), pp. 379–421. Cambridge Univ. Press, London.

Reineke, W., and Knackmuss, H.-J. (1979). *Nature* **277**, 385–386.

Richter, M. W., Stotzky, G., and Amsterdam, D. (1973). *Abstr. Annu. Meet. Am. Soc. Microbiol.* p. 99.

Rissler, J. F. (1984). *Recomb. DNA Tech. Bull.* **7**, 20–30.

Roberts, M., and Falkow, S. (1979). *Infect. Immun.* **24**, 982–984.

Rosenberg, C., Casse-Delbart, F., Dusha, I., David, M., and Boucher, C. (1982). *J. Bacteriol.* **150**, 402–406.

Saftlas, H. B. (1984). Standard and Poor's Industry Surveys, July 19, pp. H1–5.

Sagik, B. P., and Sorber, C. A. (1979). *Recomb. DNA Tech. Bull.* **2**, 55–61.

Sagik, B. P., Sorber, C. A., and Morse, B. E. (1981). *In* "Molecular Biology, Pathogenicity, and Ecology of Bacterial Plasmids" (S. B. Levy, R. C. Clowes, and E. L. Koenig, eds.), pp. 449–460. Plenum, New York.

Salzman, T. C., and Klemm, L. (1968). *Proc. Soc. Exp. Biol. Med.* **128**, 392–394.

Saunders, J. R. (1979). *Nature* **278**, 601–602.

Saunders, J. R. (1984). *Br. Med. Bull.* **40**, 54–60.

Schilf, W., and Klingmüller, W. (1983). *Recomb. DNA Tech. Bull.* **6**, 101–102.

Schneider, H., Formal, S. B., and Baron, L. S. (1961). *J. Exp. Med.* **114**, 141–148.

Sharples, F. E. (1983). *Recomb. DNA Tech. Bull.* **6**, 43–56.

Shaw, D. R., and Cabelli, V. J. (1980). *Appl. Environ. Microbiol.* **40**, 756–764.

Shaw, P. D. (1986). *In* "Plant-Microbe Interactions" (E. Nester and T. Kosuge, ed.). McMillan, New York (in press).

Simberloff, D. S. (1974). *Annu. Rev. Ecol. Syst.* **5**, 161–182.

Simon, R. D., Shilo, M., and Hastings, J. W. (1982). *Curr. Microbiol.* **7**, 175–180.

Singleton, P. (1983). *Appl. Environ. Microbiol.* **46**, 756–757.

Singleton, P., and Anson, A. E. (1981). *Appl. Environ. Microbiol.* **42**, 789–791.

Singleton, P., and Anson, A. E. (1983). *Appl. Environ. Microbiol.* **46**, 291–292.

Slater, J. H. (1984). *In* "Current Perspectives in Microbial Ecology" (M. J. Klug and C. A. Reddy, eds.), pp. 87–93. American Society for Microbiology, Washington, D.C.

Smith, H. W. (1969). *Lancet* **1**, 1174–1176.

Smith, H. W. (1970a). *J. Med. Microbiol.* **3**, 165–180.

Smith, H. W. (1970b). *Nature* **228**, 1286–1288.

Smith, H. W. (1971). *Nature* **234**, 155–156.

Smith, H. W. (1972). *J. Med. Microbiol.* **5**, 451–458.

Smith, H. W. (1975a). *Nature* **255**, 500–502.

Smith, H. W., Parsell, Z., and Green, P. (1978). *J. Gen. Microbiol.* **190**, 37–47.

Smith, M. G. (1975b). *J. Hyg.* **75**, 363–370.

Smith, M. G. (1976). *Nature* **261**, 348.

Smith, P. R., Farrell, E., and Dunican, K. (1974). *Appl. Microbiol.* **27**, 983–984.

Stallions, D. R., and Curtiss, R., III. (1972). *J. Bacteriol.* **111**, 294–295.

Stewart, K. R., and Koditschek, L. (1980). *Mar. Pollut. Bull.* **11**, 130–133.

Stotzky, G. (1965). *Can. J. Microbiol.* **11**, 629–636.

Stotzky, G. (1973). *In* "Modern Methods in the Study of Microbial Ecology" (Th. Rosswall, ed.). *Bull. Ecol. Res. Commun.* **17**, 17–28, Stockholm.

Stotzky, G. (1974). *In* "Microbial Ecology" (A. I. Laskin and H. Lechevalier, eds.), pp. 57–135. Chemical Rubber Co., Boca Raton, Florida.

Stotzky, G. (1980). *In* "Microbial Adhesion to Surfaces" (R. C. W. Berkeley, J. M. Lynch, J. Melling, P. R. Rutter, and B. Vincent, eds.), pp. 231–247. Ellis Harwood, Chichester.

Stotzky, G. (1986). *In* "Interactions of Soil Minerals with Natural Organics and Microbes" (P. M. Huang, ed.). Soil Sci. Soc. Amer., Madison (In press).

Stotzky, G., and Krasovsky, V. N. (1981). *In* "Molecular Biology, Pathogenicity, and Ecology of Bacterial Plasmids" (S. B. Levy, R. C. Clowes, and E. L. Koenig, eds.), pp. 31–42. Plenum, New York.

Stotzky, G., Schiffenbauer, M., Lipson, S. M., and Yu, B. Y. (1981). *In* "Viruses and Wastewater Treatment" (M. Goddard and M. Butler, eds.), pp. 199–204. Pergamon, Oxford.

Sturtevant, A. B., and Feary, T. W. (1969). *Appl. Microbiol.* **18,** 918–924.

Talbot, H. W., Yamamoto, D. Y., Smith, M. W., and Seidler, R. J. (1980). *Appl. Environ. Microbiol.* **39,** 97–104.

Terawaki, Y., Takayasu, H., and Akiba, T. (1967). *J. Bacteriol.* **94,** 687–690.

Timoney, J. F., Port, J., Giles, J., and Spanier, J. (1978). *Appl. Environ. Microbiol.* **36,** 465–472.

van Larebeke, N., Genetello, C., Shell, R., Schilperoort, R. A., Hermans, A. K., Hernalsteens, J. P., and van Montagu, M. (1975). *Nature* **255,** 742–743.

Walmsley, R. H. (1976). *J. Bacteriol.* **126,** 222–224.

Walton, J. R. (1966). *Nature* **211,** 312–313.

Watanabe, T., Aoki, T., Ogata, Y., and Egusa, S. (1971). *Ann. N.Y. Acad. Sci.* **182,** 383–410.

Watson, B., Currier, T. C., Gordon, M. P., Chilton, M. D., and Nester, E. W. (1975). *J. Bacteriol.* **123,** 255–264.

Weinberg, S. R., and Stotzky, G. (1972). *Soil Biol. Biochem.* **4,** 171–180.

Williams, P. A. (1982). *Philos. Trans. R. Soc. London, Ser. B* **297,** 631–639.

Williams, P. A., and Worsey, M. J. (1976). *J. Bacteriol.* **125,** 818–828.

Yano, K., and Nishi, T. (1980). *J. Bacteriol.* **143,** 552–560.

Young, F. E., and Spizizen, J. (1963). *J. Bacteriol.* **86,** 392–400.

Zund, P., and Lebek, G. (1980). *Plasmid* **3,** 65–69.

Zurkowski, W., and Lorkiewicz, Z. (1979). *Arch. Microbiol.* **123,** 195–201.

Apparatus and Methodology for Microcarrier Cell Culture

S. REUVENY AND R. W. THOMA

The New Brunswick Scientific Company
Edison, New Jersey

I. Introduction

The need to improve methods for cultivation of anchorage-dependent cells (ADCs), especially to produce them in larger quantities and to use them for more efficient production of viral vaccines for therapeutic purposes, led a number of researchers in the 1960s and 1970s to look for efficient methods to replace the labor-intensive, multiple-process roller bottle system. Several techniques were devised and proposed: Microcarrier beads (van Wezel, 1967), a multiplate propagator (Schleicher and Weiss, 1968), spiral film bottles (House *et al.*, 1972), artificial capillary propagators (Knazek *et al.*, 1972), a packed glass bead system (Whiteside *et al.*, 1979), and membrane tubing

ADVANCES IN APPLIED MICROBIOLOGY, VOLUME 31

reels (Jensen, 1977). These systems and their relative advantages and disadvantages were reviewed recently by Spier (1980a). In general, all but the first of these suffer from the following shortcomings: Limited potential for scale-up, difficulties in taking cell samples, limited potential for monitoring and controlling the system, and difficulty in maintaining homogeneous environmental conditions throughout the culture. The microcarrier (MC) system has none of these drawbacks. In this system ADCs are propagated on the surface of small solid particles suspended in the growth medium by slow agitation. Cells are permitted to become attached to and spread on the surface of the MCs where they grow to confluence (Fig. 1). In fact, in this system the features of monolayer and suspension culture have been brought together. The surface that ADCs need is offered together with the advantages of homogeneous suspension culture known from traditional microbial and animal cell submerged culture systems.

The advantages of the MC cell culture system can be summarized as follows:

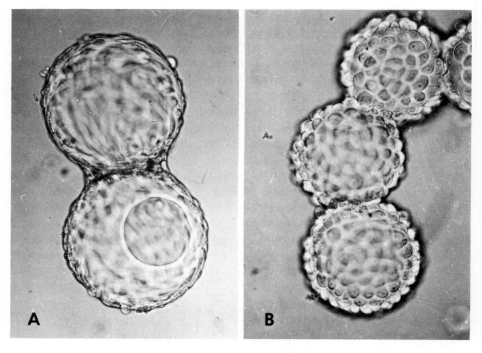

FIG. 1. Cell growth on an MC surface. (A) Cells with fibroblast morphology; (B) cells with epithelial morphology.

1. High surface-to-volume ratio which can be varied by changing the concentration of the MCs. This permits high cell yields per unit volume and the possibility of achieving highly concentrated cell products. Cell yields of up to 9×10^6/ml were obtained by Butler et al. (1983), who used MC-perfused cultures.

2. Cell propagation can be carried out in one high-productivity vessel instead of many small, low-productivity units with the consequent saving of space and culture medium. For example, a 2-liter MC cell culture using less than 5 liters of medium (with replenishment) gave cell and virus yields equivalent to those obtained with 250 Roux bottles using 25 liters of medium (Griffiths et al., 1980). Cell propagation in a single reactor usually saves laboratory space and always lowers the risk of contamination and the labor cost by reducing the number of manipulations required.

3. As the culture is well mixed and cells are distributed homogeneously, it is possible to monitor and control key environmental parameters and metabolic changes (e.g., pH, dissolved oxygen, and the concentration of medium components). As a consequence more reproducible cell propagation, product synthesis, and recovery are achieved.

4. It is possible to take a representative sample of cells for microscopic observation, chemical testing, or enumeration. This is an option not available with most other techniques.

5. As the MCs bearing the cells settle readily out of suspension, harvesting of cells and extracellular products can be done rather easily.

6. The availability of the MC system for propagation of ADCs opens possibilities for types of studies done otherwise only with difficulty, for example: Cell transfer without the use of proteolytic enzymes (Ryan et al., 1980), cocultivation of cells (Davis and Kerr, 1982), and perfusion of cell culture in columns (Bone and Swenne, 1982).

7. MC cell cultures can be scaled up readily using conventional equipment (fermentors) used for microbial processes.

Although the principle of MC cell culture had been disclosed by van Wezel in 1967, widespread application did not occur immediately, because van Wezel had observed and reported "toxic effects" that were manifested by a long lag period and complete destruction of the culture at MC concentrations over 2 mg(dry wt)/ml. The MCs van Wezel used were cross-linked dextran beads derivitized with diethylaminoethyl groups (DEAE Sephadex A-50, Pharmacia, Uppsala). Although efforts were made in several laboratories to find alternatives to these MCs or to modify their so called toxic properties, no one was completely successful until Levine et al. (1977) reported a dramatic improvement by reducing the ion-exchange capacity of the MC to 2 mEq/g dry wt of material. The Levine discovery and the increasing demand for biologicals produced by MCs have led to much research and have made available several commercial products. In parallel, several large-scale processes for production of biologicals in MC-based systems have been developed (Table I). A thorough review of the different MCs used for culturing animal cells with attention to the history of the technology, the physical and chemical properties of the MCs, their application in processes for production of biological products, as well as their use in basic studies in cell biology, has been published by Reuveny (1983b). Several other recent reviews (Clark and Hirtenstein, 1981a; Feder and Tolbert, 1983; Tolbert and Feder, 1984; Sinskey *et al.*, 1981; Spier, 1980a; van Wezel, 1982) have addressed to some extent the topic of selection or design of equipment and apparatus for suspension cell culture or more specifically for MC cell culture. We propose to present a state-of-the-art review and to project, on the basis of pragmatic considerations and our own observations, the direction of evolution of apparatus and methods for the special purpose of MC cell culture.

TABLE I

SOME LARGE-SCALE PROCESSES EMPLOYING MICROCARRIERS

Product	Cell type	Scale	Reference
Cells (for virus)	Primary chick embryo fibroblast	140 liter (batch)	Scattergood *et al.* (1982)
Foot-and-mouth disease virus	IBRS2 pig kidney cell line	235 liter (batch)	Meignier *et al.* (1980)
Herpes virus	MRC-5 human fibroblast diploid cells	10 liter (closed perfusion)	Griffiths and Thornton (1982)
Poliomyelitis virus	Primary monkey kidney cells	625 liter (batch)	van Wezel *et al.* (1980)
Polio virus	Vero cells	1000 liter (batch)	Montagnon *et al.* (1984)
Rabies virus	Primary dog kidney cells	40 liter (batch)	van Wezel and van Steems (1978)
Human fibroblast interferon (β)	FS-4 diploid foreskin cells	44 liter (batch)	Edy (1984)
IFN(β)	Foreskin	14 liter (batch)	Giard *et al.* (1981)
IFN(β)	Foreskin	50 liter (batch)	Morandi *et al.* (1984)
IFN(β)	Mouse cell line (genetically engineered)	20 liter (batch)	Delzer *et al.* (1984)
Angiogenesis factor and plasminogen activator	Human diploid fibroblast	44 liter (perfusion)	Tolbert and Feder (1983)
Plasminogen activator	Bowes melanoma cell line	40 liter (open perfusion)	Kluft *et al.* (1983)

II. Growth Systems: General Discussion

Several kinds of special systems have been devised for growth of animal cells on MCs. For stationary cultures at the laboratory scale the most common device is the Petri dish placed in an incubator in an atmosphere of 95% air and 5% CO_2. Roller bottles, used initially for cultivation of ADCs on the inner surface, have been used for cultivation of ADCs on MCs also (Pharmacia, 1982; Nielson and Johansson, 1980). Strand *et al.* (1981a,b) described a system in which human fibroblasts were propagated on MCs located in the extracapillary space of a hollow fiber matrix. The cells were nourished and induced to produce interferon by nutrients and inducers perfused through the capillaries. Clark and Hirtenstein (1981a) described a fluid lift system for propagation of cells on MCs in which the medium was conditioned in secondary vessels and the cells were retained in the primary vessel by screens with 100-μm openings.

The foregoing systems have been used for laboratory experimentation or production of small quantities of cells or cell products. With each of these, further studies are needed to develop their use for large-scale work. In practice the system employed by most investigators is the stirred tank reactor, or fermentor, which can be used at the smallest as well as the largest scale. The fermentor is simple in comparison with many of the other devices that have been used or proposed and provides a homogeneous, accessible culture that can be sampled, monitored by *in situ* sensors, and controlled readily. Moreover, much of the technology of microbial fermentation in fermentors can be transferred directly to processes for production of animal cells and their products. The advantages of a unit process over a system composed of multiple elements is discussed in depth by Spier (1980a).

Despite the similarity of cultivation of animal cells on MCs in fermentors to well-established microbial fermentation processes, and the even greater similarity to cultivation of freely suspended animal cells, there are peculiarities of MC systems that must be addressed in designing the vessel as well as in conducting the process. With MC systems two kinds of particles, the cells (10–35 μm in diameter) and the MCs (100–200 μm) must be managed separately and as a complex. During the first 6 or so hours after cell seeding conditions should favor attachment, with as even a distribution of cells among MCs as possible. During the attachment phase stirring should be gentle. After the attachment phase, conduction of the process should promote growth of the cells in a monolayer until they are confluent. Shearing forces resulting from rapid agitation must be avoided to prevent detachment or damage to the cells. MC particles, with or without cells attached, tend to settle rather quickly as a result of gravitational force. Special stirring devices and vessel configurations are used to get a homogeneous suspension of MCs without generating high shear force at any locus in the culture. The gentle mixing systems employed usually do not promote gas exchange sufficiently in fermentors of conventional configuration. Therefore changes in dimensions and other design features have been incorporated into new vessels built especially for MC cell culture, and existing vessels have been modified or used in unconventional ways.

A schematic representation of a laboratory glass vessel designed especially for MC cell culture is shown in Fig. 2. The bottom should be fully rounded and the walls should be smooth to avoid opportunities for local accumulation of MCs. As aeration (gas exchange) is effected often solely via the surface, i.e. the liquid–gas interface, the ratio of the height (h) to the diameter (d) should be 1:1, rather than the 2:1 or 3:1 ratio common with fermentors used for microbial fermentations. Note that the real consideration is the ratio of surface area (A) to liquid volume (V), which is inversely proportional to the depth of the liquid. Thus as the volume of the vessel and the culture is increased the h/d ratio must become smaller if A/V is to remain constant. The practical limit of h/d is not necessarily 1:1, but as this ratio becomes

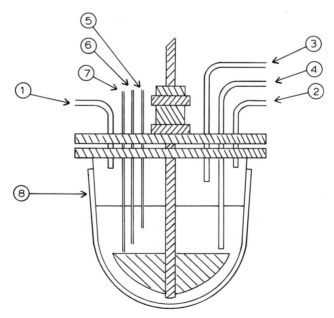

FIG. 2. Fermentor for cultivation of animal cells on MCs. (1) Gas inlet; (2) gas outlet; (3) inoculation port; (4) liquid withdrawal; (5) temperature probe; (6) DO probe; (7) pH probe; (8) heating jacket. Nutrient feed lines, acid–base addition ports, other sensors (e.g., ORP) not shown.

lower the vessel becomes increasingly difficult to construct and operate. Materials of construction are critical. Vessels are made of glass or 316L (low-carbon) stainless steel. Flexible parts, gaskets, etc., should be tested for toxicity. Silicone rubber is preferred. The inner parts of the vessel that are in contact with the growth medium should be siliconized in order to prevent cell attachment and sticking of the MCs.

The fermentor for MC culture is usually provided with ports for the following purposes: Seeding, harvesting, sampling, feeding, supernatant fluid outlet, base addition, gas inlet (to culture as well as headspace), gas outlet, and entry of probes (pH, dissolved oxygen, and oxidation–reduction). Special attention should be given to location and configuration of inlet and outlet ports. Although penetrations through the bottom of the vessel are undesirable, as they provide stagnant spots where MCs may accumulate, sometimes they are used. On the other hand too many pipes entering through the head plate and extending below the liquid surface create a baffling effect which disrupts the flow pattern of the MCs and may increase shearing forces; furthermore the pipes may present irregular surfaces on which the MCs may stick. If it is necessary to use a bottom valve for cleaning or emptying the vessel it should be a ball or ram valve so that as little dead space as possible is created. If penetrations are made through the side wall they should slope downward within the vessel at 30–45° to prevent accumulation of MCs on the intruding segment and permit working at low volumes. Changing medium or washing cells in batch or perfusion culture can be done easily by withdrawing liquid through an outlet pipe equipped with a 40- to 80-μm stainless steel screen. In order to prevent clogging of the filter, especially in perfused cultures, a cylindrical filter can be mounted on the stirrer shaft. Griffith *et al.* (1982) and Kluft *et al.* (1983) claim that the rotational movement of the shaft-mounted filter prevents its clogging.

III. Agitation Systems

A. General Requirements

Conventional systems for stirring microbial fermentations are not usually suitable for MC cultures. The following points serve to emphasize the special requirements of MC cultures and the design measures that have been taken to meet these requirements:

1. Stirring should be as gentle as possible, just sufficient to distribute the MCs homogeneously throughout the culture. de Bruyne and Morgan (1981) have discussed the problem in quantitative terms and shown the need for a vertical component in the circulating force.

2. Bearings or hard moving surfaces in contact with stationary surfaces should be located above the liquid surface to avoid mechanical attrition of the MCs. Thus, a hanging or top-entering agitator shaft is preferred to a bottom-entering type.

3. The stirrer should be driven by a motor that provides smooth and controlled rotation in the range of 10–150 revolutions/minute (rpm). During the cell attachment phase stirring should be just sufficient to suspend the particles and may be done intermittently in a timed cycle (Clark and Hirtenstein, 1981a). After the cell attachment phase, stirring speed may be increased to suspend the MCs homogeneously and increase oxygen supply. Later, when confluent growth is approached, stirring speed may be increased further to prevent overgrowth of cells from one MC to another and consequent aggregation of MCs.

4. The stirring system should be vibration free and without erratic motion, to avoid trauma to the cells. Changes in speed should be made gradually (Pharmacia, 1982; Hirtenstein et al., 1982).

5. The design of the stirrer should be such that a range of culture volumes can be accommodated, since there are procedures in which cells and MCs are seeded at a low volume and stepwise or gradual increases in volume are made.

6. Baffles, often provided in reactors used for microbial fermentations, should not be used, as turbulence and shearing forces must be avoided.

B. Types of Impellers

The type of impeller used for MC cell culture deserves detailed discussion in the light of the foregoing exposition of requirements and constraints for agitation systems.

1. The Turbine or Bar Impeller

The spinner flask, developed in the 1950s for growing animal cells in suspension, was used by early investigators for MC cell culture. In this device a Teflon-coated magnetic bar, suspended and free to rotate at the bottom of a fixed shaft suspended from the top of the vessel, is coupled magnetically to a motor beneath the vessel (Fig. 3A). At the 44- and 150-liter scales Edy (1984) and Scattergood et al. (1984) have reported the use of conventional turbine impellers. Although these systems support growth of cells on MCs, they have several disadvantages:

1. Relatively high speeds, with unwanted high shear forces, are required to keep the MCs in suspension.

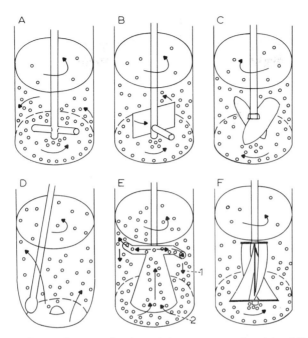

FIG. 3. Stirring devices for MC cell culture. (A) Bar type impeller; (B) paddle impeller; (C) tilted paddle impeller; (D) rod stirrer; (E) the Kedem pumping impeller: (1) MCs settling under effect of gravity; (2) pumping on medium containing MCs; (F) Feder and Tolbert impeller.

2. There is a stagnant point immediately below the shaft at the bottom of the vessel where MCs tend to accumulate.

3. There is no prevailing force moving the MCs upward.

Tyo and Wang (1981) have studied the shear forces generated in magnetic bar type spinner flasks (0.1- to 1.0-liter volumes) and their effects on cell growth and virus production in MC cell cultures. They defined an integrated shear factor as follows:

$$2\Pi \ ND_i/(D_t - D_i)$$

where N is rpm, D_i is the diameter of the impeller, and D_t is the diameter of the tank. They postulated that the factor should be kept below 40 in order to permit cell attachment and maximum cell yield. If attachment is carried out during a time when no agitation is imposed the factor may be increased to 80. As virus propagation is more sensitive to shear than is cell growth, the factor should be reduced below 20 in order to get maximal yield of virus. According to estimations made by Tyo and Wang (1981) the shear force generated by turbine impellers in large fermentors (1000 liters) is predicted to be significantly less than in small fermentors. The stirring speed required for homogeneous suspension of MCs changes according to the volume, the vessel dimensions, and the impeller size. Generally impellers with large diameters suspend the MCs at lower rpm rates, and thus with lower shear forces generated.

2. Special Agitation Systems

Although many practitioners of MC-based cell culture have used and continue to use agitation systems developed for cultivation of microorganisms or animal cells in free suspension, a growing appreciation of the special requirements for attachment to and growth of cells on MCs has led researchers around the world to develop new agitation systems for MC cell culture. Some of these are described below.

One of the first impellers designed especially for MC cell culture is the paddle and anchor Impeller. The rod or turbine is replaced by a relatively large flat vertical blade. With this type of device it is possible to generate a homogeneous suspension at a relatively low stirring speed without substantial shearing. In Fig. 3B a spinner flask with a Teflon paddle is shown. Hirtenstein et al. (1982) compared growth of human diploid fibroblast (MRC-5) and monkey kidney (Vero) cells in paddle and bar type spinners and found that the paddle type allowed 25–35% increase in cell yield. In their hands the stirring speed needed for obtaining an even suspension of MCs was 20–40 rpm as compared to 50–60 rpm with the bar-type spinner. Spier and Whiteside (1984) and Griffiths et al. (1982, 1984) have used Teflon paddle impellers in 20-liter carboys and 2- to 10-liter fermentors, respectively. Nielson and Johansson (1980) have used a 1-liter glass vessel equipped with a large silicone rubber plate for propagation of cells on Biosilon MCs. On a still larger scale a stainless steel anchor-shaped impeller has been used. Although the various paddle and anchor-type impellers have been shown to improve cell yield, they still suffer from the disadvantages that MCs accumulate in the dead spot immediately under the shaft, no upward movement of fluid is generated, and adverse shearing effects are seen when rotational speed is increased to attempt to get homogeneous suspension.

Improvement of the paddle-type impeller was achieved by twisting the paddles at a 30–45° angle from the vertical (Fig. 3C). This causes an upward movement of fluid and homogeneous dispersion of MCs at lower stirring speeds, a condition that favors higher cell yields. Griffiths and Thornton (1982) have used a fermentor with a silicone rubber blade angled at 30° from the vertical and obtained an even suspension of Cytodex MCs at 30 rpm. Similar impellers that produce an upward movement of fluid and thereby a more efficient dispersion of MCs are the screw type described by Clark and

Hirtenstein (1981a) and the plough impeller of Hirtenstein *et al.* (1982). The latter were able to generate an even suspension of cell-bearing Cytodex MCs at 15–30 rpm and to accomplish at least a 50% increase in cell yield as compared to that achieved with the traditional bar-type spinner. At the present time several commercial companies are producing MC fermentors with screw-type impellers.

A rod with a bulb at the lower end, suspended from the top of the vessel and rotated so that the bulb moves in a circle in the low part of a contoured Pearson flask with the motion of the rod describing a cone, was described by de Bruyne and Morgan (1981) (Fig. 3D). Upward movement of the MCs is generated and there is no accumulation of MCs at the bottom. An even suspension of MCs can be generated at 15–30 rpm. Hirtenstein *et al.* (1982) have found that 50–70% greater cell yields can be achieved with this type of spinner as compared with the conventional bar-type spinner. In spite of its efficiency in small glass laboratory fermentors with contoured bottoms, the use of this agitator in large-scale fermentors presents an engineering problem not as yet solved.

Feder and Tolbert (1983) and Tolbert and Feder (1984) have described an agitation system that consists of four flexible sheets held vertically, spanning the depth of the culture fluid (Fig. 3F). These sheets need only to be turned slowly to disperse the MCs adequately.

Reuveny (1983a) tested a new impeller developed by A. Kedem for use in MC cell culture (Israeli and U.S. Patents pending). The basic principle of this impeller is that the culture fluid containing MCs is pumped upward from the bottom of the vessel through the hollow impeller shaft. The fluid leaves the shaft near its top but below the surface of the culture through two short outlet pipes, on opposite sides of the shaft, pointing backward tangentially to the motion of the shaft (Figs. 3E and 4). Circulation of the culture and an even suspension of the MCs can be obtained with this device at rpm values of 10–30. There is no stagnant spot where MCs might accumulate, there is minimal collision of MCs with the impeller, and the movement of MCs relative to the immediate environment is slow. Reuveny studied propagation of several types of cells on Cytodex 1 MCs using this stirring system and found in each case a significant increase in cell yield as opposed to that with a paddle type spinner flask (Fig. 5). The main disadvantage of this stirring system is that it fixes the volume that must be used, and thus precludes the useful procedure of "building" the culture in the growth vessel.

The New Brunswick Scientific Company is planning to market a draft tube stirring system especially designed for MC culture. In operation it is similar to the Kedem impeller (see Fig. 6). Growth of baby hamster kidney (BHK) cells attached to Cytodex 1 MCs in an aerated or nonaerated 1.5-liter vessel

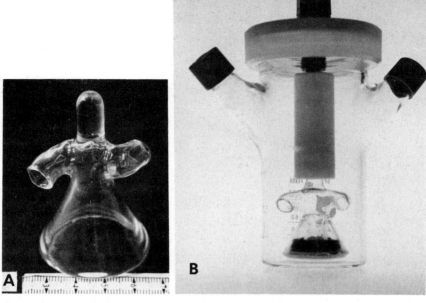

FIG. 4. (A) The Kedem pumping impeller. (B) Installed in a 100-ml spinner.

provided with this stirring system is described in Fig. 7. Similar results were obtained at the 5-liter scale with Cytodex 3 or Biosilon MCs. This new stirring system was tested at a larger scale and was found to be capable of lifting heavy Biosilon beads to a height of about 1.5 m at a relatively slow stirring speed (about 35 rpm).

Bartling *et al.* (1984) described a scull-type fermentor which they found generated significant increases in yields of BHK cells grown in free suspension as compared with yields realized in regular fermentors equipped with turbine impellers. In the scull fermentor intense agitation can be achieved by the movement of four branch anchor-type stirrers. In our opinion this stirring system should be tested with MC cell cultures. Himmler *et al.* (1984) have described a bulk flow draft tube fermentor in which oxygen is sparged beneath the tube and circulation is assisted further by a marine propeller inside the tube. This fermentor was found effective in propagating hybridoma cells encapsulated in agarose beads. In our opinion this type of agitation system should be tested for its utility in cultivating cells on MCs, especially as the delivery of oxygen to the cells would no longer be dependent on diffusion through the surface interface.

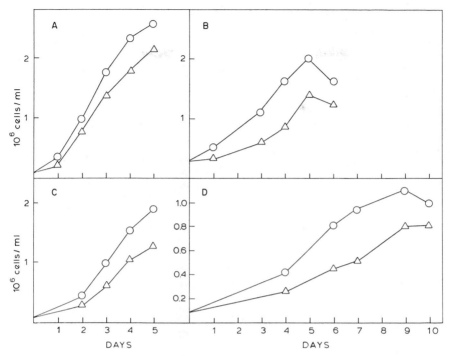

Fig. 5. Cell growth on Cytodex 1 MCs in spinners equipped with a paddle impeller (△) and with the Kedem pumping impeller (○). Culture volume: 100 ml; MC concentration: 3 mg dry wt/ml; stirring speed: 40 rpm with the paddle, 25 rpm with the Kedem impeller. (A) baby hamster kidney cell line; (B) primary chick embryo fibroblasts; (C) canine kidney epithelial cell line; (D) human foreskin diploid cells.

C. Reactors without Impellers

There have been several publications dealing with growth of cells on MCs in vessels without mechanical stirring devices. Clark and Hirtenstein (1981a) described a fluid lift system in which the MCs with cells are retained by filter screens with 100-μm openings. The filtered fluid is circulated through a medium reservoir in which it is aerated. Better cell yields than those with bar spinners were achieved. Although the potential advantages of better control of environmental parameters and gentle mixing are evident, the system is not truly homogeneous and is not readily scaled up. Strand *et al.* (1984a,b) have reported using a hollow fiber cartridge in which cells were grown on MCs entrapped in the intercapillary space. The cells were fed by diffusion of nutrients through the capillary walls, as medium was circulated

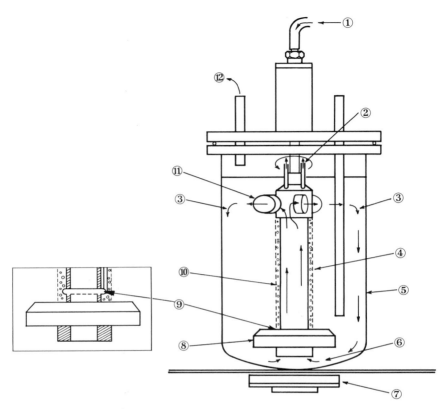

Fig. 6. New Brunswick Scientific draft tube impeller. (1) Air inlet; (2) air outlets from the draft tube; (3) MCs settling by gravity; (4) 400-mesh stainless steel screen; (5) glass vessel; (6) medium containing MCs pumped into the draft tube; (7) magnet in the stirrer; (8) cylindrical magnet; (9) sparger, inside the draft tube; (10) air bubbles generated by the sparger; (11) side-arm outlets for medium carrying MCs; (12) air outlet from the vessel.

through the capillaries. Tolbert *et al.* (1984) have mentioned a proprietary device, a "static maintenance reactor," in which a fully grown MC cell culture was kept at high density (10^8 cells/ml) while medium was perfused through the system.

IV. Monitoring and Controlling Parameters

Those parameters that are customarily controlled in microbial fermentation and animal cell suspension culture are usually controlled in MC cell culture in stirred vessels. Special requirements of a number of more critical parameters will be discussed in turn.

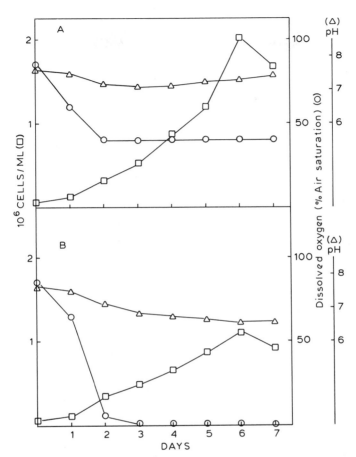

FIG. 7. BHK cell growth on Cytodex 1 MCs in a vessel with the draft tube (NBS) impeller. Vessel volume, 1.5 liters; culture volume, 1.2 liters; MC concentration, 3 mg dry wt/ml. (A) Aerated culture (40% saturation); (B) nonaerated culture.

A. STIRRING SPEED

Stirring speeds for MC cell culture are relatively low, in the range of 10–150 rpm. Several researchers have suggested initiating the culture in a small volume without agitation (Clark *et al.*, 1980) or with intermittent mixing (1 minute in every 35–45 minutes) for several hours followed by dilution of the culture (Griffith and Thornton, 1982). These measures allow more uniform and efficient cell attachment and increase cell yields in small fermentors. However, this procedure seems impractical in larger fermentors as it can create problems in controlling temperature, supplying oxygen to the cells, and distributing the cells evenly among the MCs which settle at a different rate. Edy (1984) reported that setting up the culture at the final volume and stirring continuously from the beginning had no deleterious effect on cell growth and that the actual cell distribution on the MCs correlated well with the theoretical Poisson distribution. van Wezel (1982) also reported that cells became attached well to MCs even though the culture was stirred continuously during the attachment phase. However, it is generally agreed that during attachment stirring should be minimal, only sufficient to keep the cells in suspension. Fleischaker *et al.* (1981a) estimated a mere 4% increase in viscosity occurred during growth in MC cell culture. Thus, an increase in power consumption by the motor driving the agitator is not a good index of cell growth, as it is in some bacterial or fungal cultures.

B. Temperature Control

Harakas *et al.* (1984) reported the sensitivity of animal cells to small fluctuations in temperature. At temperatures below 37°C cells grew very slowly; at temperatures above 37°C, cells lost viability. His report reinforced the opinion of most workers in the field that animal cells *in vitro* have a more stringent requirement for temperature constancy than do most microbial cultures. Moreover, as the transfer functions of other sensors, e.g., dissolved oxygen (DO), oxidation–reduction potential (ORP), and pH, are affected by temperature, close control of temperature is needed for proper monitoring and control of a number of parameters. For the level of sensitivity, response time, and stability required, platinum resistance sensors are preferred.

The difficulty of monitoring and controlling temperature in MC cell culture is aggravated because stirring speed is slow and therefore heat transfer is poor. An unacceptable temperature gradient may be set up from the outside wall to the inner part of the culture. One way of mitigating this effect is to use fully jacketed fermentors and to measure the temperature in the jacket as well as near the middle of the fermentor, and control accordingly. In contrast with many microbial fermentations, cooling to remove metabolic heat is not a problem in MC cell culture. It was estimated by Fleischaker *et al.* (1981a) to be only 0.017 kcal/liter-hour with a cell concentration of 10^6/ml. The need for jacketed fermentors is illustrated further by the work of Harakas *et al.* (1984) who located a 100-liter stainless steel reactor in a room at 37°C and observed a 24-hour lag in temperature equilibration and cell growth, thus dramatizing the poor heat conductivity of air.

C. pH Control

pH is a key parameter in MC cell culture as it affects cell viability, cell attachment to the MCs, cell growth, and metabolism. The optimal pH for cell growth varies with the type of cell, but is usually in the range of 7.0–7.5. The optimal pH for cell attachment may be different from that for cell growth as shown by Clark and Hirtenstein (1981a) and Manousas et al. (1980). Ceccarini and Eagle (1971) and Gailiani et al. (1976) showed that pH fluctuation during the cell growth period had an adverse affect on cell yield. The importance of monitoring and controlling pH accurately in MC cell culture has been stressed by Foehring et al. (1980).

In the practice of animal cell culture the buffering system is usually CO_2-bicarbonate. In this system the pH is determined by the ratio in solution of CO_2 and bicarbonate according to the following expression:

$$pH = -\log k \, [CO_2]/[HCO_3^-]$$

where k is the dissociation constant of carbonic acid. It should be mentioned that many times CO_2 and bicarbonate play an essential role for cell growth in addition to their pH buffering action (McLimans, 1972). During the first few hours of growth pH tends to remain alkaline. CO_2 production is low because cell concentration is low, and mixing and exposure to air cause dilution of the dissolved CO_2. During the later phases of growth when cell concentration and metabolic activity increase, pH tends to drop as CO_2 and lactic acid accumulate. In MC cell culture when cell density is high, relatively large amounts of CO_2 and lactic acid are formed and the tendency for the culture to become acidic is strong; thus, pH control becomes especially important.

pH is usually measured by a steam-sterilizable combination glass and reference electrode (West et al., 1961). The signal from this sensor is conditioned, fed to a controller, compared to a set point, and an appropriate signal is sent to a pump or valve as required. pH is corrected by addition of acid (usually HCl) or base (usually bicarbonate or sodium hydroxide), or by changing the concentration of CO_2 in the gas phase. The acid/base addition method entails the risk of producing high local concentrations and extremes of pH, especially at low stirring speeds. Moreover, positive and negative ions from acid and base may accumulate to the degree that osmotic pressure is changed unfavorably. Control of CO_2 concentration in the medium is effected by changing the proportion of CO_2 in a mixture that is delivered constantly to the reactor (Sinskey et al., 1981; Edy, 1984; Scattergood et al., 1983). Sometimes control of pH is imposed by both CO_2 control and base addition.

CO_2 generated in excess by the cells can be stripped out of the system by gassing the culture. However, lactic acid once formed accumulates in the medium. Researchers have tried to solve the lactate problem in several ways: Addition of buffer, e.g., N-2-hydroxyethylpiperazine-N'-2-ethanesulfonate (HEPES) (Gebb et al., 1980); increasing oxygen tension, thereby encouraging aerobic metabolism and lower production of lactate (Balin et al., 1977); continuous feeding of glucose to the culture at physiological concentration (5.5 mM) to reduce accumulation of lactate (Gebb *et al.*, 1980); using a carbohydrate other than glucose in the medium formula. Imamura *et al.* (1982) and Edy (1984) have used fructose instead of glucose to support the growth of Madin–Darby canine kidney (MDCK) and foreskin cells in MC culture and found that only about 25% of the lactic acid formed with glucose was generated with fructose. Fleischaker *et al.* (1981a) have monitored the consumption of base from a reservoir mounted on a load cell to estimate the amount of metabolic lactate formed.

D. OXYGEN MEASUREMENT AND CONTROL

Oxygen is a key nutrient in cell metabolism. The concentration of dissolved oxygen (DO) in the culture can be measured by steam-sterilizable probes of the galvanic or polarographic type (Johnson *et al.*, 1964). DO concentration can affect cell yield and thus directly or indirectly affect product expression (Spier and Griffiths, 1984). At low oxygen tension, cells tend to grow at a low rate and produce a high concentration of lactic acid, while too high an oxygen tension is toxic to the cells. The optimal concentration of DO for cell growth has been found to vary from 3–20%, 15–100% saturation with air (Spier and Griffiths, 1984; Pharmacia, 1982). Oxygen utilization rates (OUR) vary among cell types in the range of 0.05–0.5 mmol $O_2/10^9$ cells-hour (Fleischaker and Sinskey, 1981; Spier and Griffiths, 1984). OURs are measured by on-stream analysis of the composition of the inflowing and outflowing gas. The difference in the oxygen concentration multiplied by the gas flow rate gives the OUR. Oxygen content of the gas can be measured by polarographic or paramagnetic analyzers (Ingham *et al.*, 1984) or more elegantly by mass spectrometry. Mass spectrometry has a faster response time than the other methods and is more reliable. Fleischaker *et al.* (1981a) have estimated that in a 5-liter fermentor with 10^6 cells/ml, a flow rate of 840 ml of gas/minute is needed to sense a 2% difference in oxygen concentration which is the limit of sensitivity with the paramagnetic analyzer. At this flow rate the paramagnetic analyzer reacts unduly to small changes in flow rate. However, in larger scale cultures or at higher cell densities (and constant flow rate) the paramagnetic analyzer can be used (Ingham *et al.*, 1984). Fleischaker *et al.* (1981a) have used the dynamic method of estimating OUR. With a DO electrode in place they stopped the oxygen supply to the culture briefly and measured the decrease in DO vs time. Measurement or OUR provides a means of estimating cell mass on-line (Lydersen *et al.*, 1985; Fleischaker, 1981a) and in combination with CO_2 evolution data permits calculation of the respiratory quotient (RQ), i.e., millimoles of CO_2 produced/millimoles O_2 consumed by the cells. Together with cell density the RQ gives an appreciation of the metabolic activity of the cells. The optimal

DO level usually decreases during a period of virus propagation, reflecting a lower OUR (Tyo and Wang, 1979; van Hemert *et al.*, 1969; van Wezel, 1984).

In microbial fermentations the level of DO in the culture medium can be controlled over wide limits by changing the agitation speed, aeration rate, or both. The limit to the speed at which animal cells or MCs with cells can be agitated and the reasons for this have been discussed in an earlier section of this article. Aeration in microbial cultures is usually effected by sparging air through an open pipe or perforated ring near the bottom of the vessel beneath the impeller. If this type of aeration must be used with cell cultures to maintain minimal DO concentrations it must be used with caution, as sparging air invariably generates foam, especially with the high protein concentration (5–10% serum) common to most cell culture media. MCs tend to float in the foam, and bubbles can actually damage cells according to Kilburn and Webb (1968). The damage to cells by air bubbles was postulated by Spier and Griffiths (1984) to be due to the interfacial tension of the bubble in contact with the cell. This latter damaging effect can be overcome to some extent by increasing the concentration of serum, adding surface active agents such as pluronic polyol F-68, or by ensuring that the bubbles are large (Spier and Griffiths, 1984). Suppressing foam with chemical antifoam agents, a practice common in microbial fermentations, cannot usually be done with cell cultures as these agents are toxic to the cells. All of the problems related to controlling DO with suspension culture of animal cells are aggravated with MC cell culture, for the following reasons: Stirring speeds are significantly lower, MCs have a tendency to float at the liquid-foam interface, and cell densities, and therefore oxygen demands, are relatively high. Tyo and Wang (1981) found a 30% decrease in final cell yield resulted from increasing the culture volume from 100 to 1000 ml. This decrease was attributed by them to a deficiency in oxygen supplied to the cells. Even at the 1-liter scale, it would seem, oxygen depletion can be a growth-limiting factor with MC cell culture.

Researchers in many laboratories have taken conventional as well as unconventional approaches to solution of the problem of supplying oxygen to MC cell cultures. All have found that oxygen transfer is affected drastically by the rate of stirring and by the ratio of the area of the gas–liquid interface to the culture volume. Fleischaker and Sinskey (1981) found that in a 14-liter fermentor in which oxygen was supplied to the cells by surface aeration only the oxygen mass transfer coefficient (k_La) was directly proportional to the stirring speed in the range of 15–100 rpm and inversely proportional to the culture volume, according to the expression:

$$k_La = 0.414N/V^{2.05}$$

where N is the stirring speed and V is the culture volume. Spier and Griffiths (1984) found that in an unsparged 4-liter fermentor fitted with an eight-bladed turbine impeller the $k_L a$ was affected even more by stirring speed. The following relationship was proposed:

$$k_L a = 1.5 \times 10^{-5} \times \text{rpm}^{1.85}$$

Tyo and Wang (1981) have calculated $k_L a$ values for cultures in 100-ml (S/V, 0.24 cm^{-1}), 1-liter (S/V, 0.11 cm^{-1}), and 10-liter vessels (S/V, 0.031 cm^{-1}) and found them to be 1.3, 0.83, and 0.27, respectively. A 30% decrease in cell yield was observed in the 1-liter fermentor compared with the 100-ml spinner while an 80% decrease in cell yield was predicted for the 10-liter fermentor. Realization of the importance of having a high ratio of surface area to volume has led most workers to choose vessels with a height-to-diameter ratio of 1:1. As the surface/volume ratio is inversely proportional to the height, maintaining a constant ratio as volume is increased requires maintaining a constant height, and this leads to a highly unconventional vessel and soon becomes completely impractical.

Another means of increasing the effectiveness of surface aeration is to enrich the atmosphere in the head space with oxygen. Glacken *et al.* (1983) calculated that if pure oxygen were used instead of air in the head space, oxygen limitation would be seen at the 3.5-liter scale. However, several other workers have reported that dissolved oxygen can be controlled by surface oxygenation of cultures in 100- to 200-liter fermentors. Hirtenstein and Clark (1981) anticipated that oxygen limitation would be seen above the 200- to 300-liter scale. van Wezel (1982) saw the effects of oxygen limitation at the 50-liter scale, while Scattergood *et al.* (1983) encountered the problem of 140 liters. These differences in perception of the scale at which surface aeration is no longer effective in maintaining DO at or above the critical level derive from differences in MC concentration, cell type, fermentor configuration, and mixing system.

Van Wezel (1982) used a special perforated ring, like a sparger, located 2 cm above the liquid to direct a gas mixture (oxygen, nitrogen, and carbon dioxide) downward on the surface and accelerate its rate of solution. By use of this device DO could be controlled in cultures up to 350 liters in volume. In general it is safe to say that DO can be controlled in MC cell cultures up to 300 liters by controlling the composition of the head space gas. This is usually done by mixing oxygen and nitrogen in the inlet stream, with a provision for adding air or replacing the artificial mixture by air entirely in the early stage of cell growth, for economy. van Wezel (1984) has described a similar system which included also regulating stirring speed within a limited range. Another means of controlling DO is by increasing the head space gas pressure to increase the solubility of oxygen. However, as mentioned by

Spier and Griffiths (1984) this measure should be taken only in the latter stages of growth, as the toxic limit of DO may be reached in early stages when OUR is low. A further drawback to elevating head space pressure is that CO_2 solubility will be increased also and pH will be depressed as a consequence.

Sparging small amounts of air or oxygen into large fermentors has been used by van Wezel (1982) and by Delzar et al. (1984) with MC cell cultures. Successful sparging with little or no foaming is accomplished usually with a single-orifice sparger generating large bubbles. Spier and Griffiths (1984) have found with a sparged suspension culture of BHK cells, k_La values of 1–25 hour^{-1} as opposed to 0.1–4 with unsparged cultures. They described the relationship between k_La, stirring speed, and air flow rate in a 4-liter vessel with eight turbine blades on the impeller as follows:

$$k_La = 3 \times 10^{-5} \times \text{rpm}^{1.58} \times \text{AFR}^{0.58}$$

when AFR is air flow rate in ml/minute. Thus k_La can be seen to depend strongly on stirring speed and air flow rate. Katinger et al. (1979) have used an airlift fermentor for propagation of animal cells in suspension. They achieved gentle mixing by sparging large bubbles upward through a draft tube, which stabilized liquid flow and minimized foam formation. This system would appear to be applicable to oxygenation in large reactors. Katinger (1980) was not able to grow cells on MCs in airlift fermentors, but Himmler et al. (1980) have been able to propagate hybridoma cells entrapped in agarose beads in a modified bulk flow draft tube fermentor in which pure oxygen was sparged intermittently on demand. The similarities in physical sensitivities and physiological requirements of entrapped hybridoma cells and MCs with attached cells suggests that the Himmler system should be tried with MC cell cultures.

Circulation systems, in which cell-free medium in an external loop or reservoir is aerated in order to supply oxygen to the cells, have been used by several groups. Griffiths and Thornton (1982) used a closed perfusion system for propagation of MRC-5 cells on Cytodex 1 MCs in reactors up to 10 liters. Dissolved oxygen was monitored and controlled by surface aeration in a reservoir. Clark and Hirtenstein (1981a) have reported the use of a similar system. However, Katinger (1980) and separately Glacken et al. (1983) have concluded that this method of DO control is not practical for the large scale as high circulation rates, with resultant foaming and medium denaturation, are required to maintain the critical DO in all parts of the vessel containing the cells.

Several research groups have found that it is possible to oxygenate cell cultures using thin-walled, oxygen-permeable silicone tubing immersed in the culture medium. Oxygen diffuses through the wall of the tube without producing bubbles. Sinskey *et al.* (1981) found that 5 m of silicone tubing (i.d. 0.147, o.d. 0.196, wall 0.025 cm) submerged in a 10-liter vessel, with pure oxygen in the tube and air in the head space, could supply 0.25–0.33 mmol O_2/liter-hour at stirring speeds of 15–120 rpm. The oxygen transfer rate was less sensitive to changes in agitation than it is with surface aeration. The possibility of supplying oxygen at low stirring rates is evident. Moreover, as silicone tubing is permeable to CO_2 also, it would seem possible to control pH by delivering CO_2 through the same tube. However, Spier and Griffiths (1984) indicated that difficulty in controlling pH might be expected this way because of the difference in diffusion rate through the tube wall of the two gases. The feasibility of using silicone tubing for oxygenation of large-scale MC cell cultures is not clear at this time because of the diversity of data and opinions that appear in the literature. Fleischaker *et al.* (1981a), Fleischaker and Sinskey (1981), Hirtenstein and Clark (1981), Griffiths *et al.* (1980), Sinskey *et al.* (1981), Tyo and Wang (1981), and Glacken *et al.* (1983) have reported widely different results and conclusions. The estimates for length of tubing to provide oxygen requirements range from 1 meter of tubing for 10 liters, i.e., 10 cm/liter (Fleischaker and Sinskey, 1981) to 0.5 m/liter (Sinskey *et al.*, 1981) and 1 m/liter of culture (Griffiths *et al.*, 1980). In all three cases pure oxygen was passed through the tube. Glacken *et al.* (1983) have calculated a need for 30 m of 2.5 cm o.d. silicone tubing to provide oxygen for 1000 liters of cell culture, while Hirtenstein and Clark (1981) have calculated that 500 m of 0.196 o.d., 0.025-cm wall tubing, or 1000 m of 0.15 o.d., 0.05-cm wall tubing would be required for effective oxygenation of 1000 liters of MC cell culture. Evidently more empirical data from large-scale experimental systems are needed to allow firm conclusions to be made about the practicality of using silicone tubing in large fermentors. It seems to us that a more fruitful approach to providing a high surface area in a compact volume would be to employ a cartridge-mounted pleated silicone membrane. An interesting approach was suggested by Glacken *et al.* (1983) in which the inside wall of the culture vessel would be lined with a silicone membrane.

Whiteside *et al.* (1984) and Spier and Whiteside (1984) have described a novel caged aeration system. In this system air is sparged into a closed wire mesh cage attached to and rotating with the stirrer shaft. The penetrations in the screen are small enough to exclude the MCs. Bubbles and foam are generated inside the cage only, while oxygen is delivered to the cells by the medium which flows freely in and out of the cage. Whiteside *et al.* (1984) reported that with this system they were able to control DO in a 10-liter baby hamster kidney (BHK) MC culture. Recently at the New Brunswick Scientific Company the cage and rotating draft tube principles have been combined. A ring sparger was introduced into the rotating double-walled draft tube, the outer wall of which is made of 400-mesh wire cloth. The air (or gas mixture) is pressurized through the stirring shaft and out through the ring sparger, generating bubbles inside the double-walled draft tube and exiting through the two outlets at the top of the draft tube (Fig. 6); 1-liter cultures of BHK cells were aerated without significant foaming (Fig. 7).

In summary, it should be clear that DO is the most critical parameter for scaling up MC cell cultures. In designing an aeration system for MC cell culture any one or any combination of the above approaches, viz. control of surface aeration, delivery of oxygen by sparging, and controlled stirring speed may be used to maintain critical DO concentration. Spier and Griffiths (1984) have written a computer program in BASIC which permits calculation of the effects of changes in rate of stirring, sparging, and other variables on oxygen transfer rate in a specified fermentor, if the specific OUR of the cells is known. In spite of the interest and research activity in this area there is still a lack of good systems for supplying oxygen to large-scale MC cell cultures. This is the main reason that with few exceptions the scale of MC culture is limited to 300–400 liters. Montagnon *et al.* (1984) have reported producing killed polio virus vaccine using Vero cells at the 1000-liter scale. However, they employ a relatively low (1–1.5 g/liter) MC concentration and only moderate cell yields (10^6/ml) were realized. They provided no information about the type of oxygenation system they used.

E. Oxidation–Reduction Potential

Oxidation reduction potential (ORP) is the electrical charge of the medium sensed by a platinum electrode vs a reference half-cell, e.g., a calomel electrode. The biological significance of this is not clear but it has been suggested that electron transport in the respiratory chain is impaired by an excess or deficiency of electron-donating molecules in the environment. The optimal potential of the medium for cell growth was found to be +75 mV (Klein *et al.*, 1971; Taylor *et al.*, 1971; Toth, 1977). The desired potential can be achieved by gassing the medium with a mixture of 95% air and 5% CO_2 prior to seeding, a treatment known as poising.

Toth (1977) obtained improved cell yields by monitoring and controlling ORP during growth of mouse cells in suspension culture. He obtained the desired effect by varying the composition of the gas introduced, changing the stirring speed, or adding chemical reducing agents. He observed that ORP fell during the growth phase and rose during the stationary phase and suggested that monitoring ORP can serve to forecast transitions in the growth cycle. Griffiths (1984) monitored ORP during growth of BHK and MRC-5 cell cultures and found that the ORP leveled off 24 hours before the end of the exponential growth phase and thus can be used as an indicator of the optimal time to infect cells with Herpes virus. The changes in ORP were less definite with MRC-5 diploid cells than with a BHK cell line. Limited experimentation by Griffiths (1984) showed no improvement in cell yield by controlling ORP in MC cell cultures, but found more consistent growth and a shortened lag by poising the medium.

ORP values measured are markedly affected by pH and DO. Lengyel and Nyiri (1965) maintained that in an aerated system the ORP measured was largely a function of the DO. It can be argued that in a culture where pH and DO are monitored and controlled ORP measurement is of little significance.

The intracellular concentration of reduced nicotinamide adenine dinucleotide (NADH) can be estimated with an autoclavable electrode (Arminger *et al.*, 1984). The fluorescence of NADH (emmission of 460 nm when irradiated at 360 nm) gives an on-line measure of NADH concentration and thus viable cell concentration. There is no information about the effectiveness of this electrode in MC cell culture. However, it was reported that fluorescence was a linear function of NADH concentration in RPM11640 growth medium with and without serum. Moreover, fluorescence was not affected by pH over the physiological range (Arminger *et al.*, 1984).

F. CARBON DIOXIDE CONCENTRATION

The importance of CO_2 in the CO_2–bicarbonate buffering system is described in an earlier section. McLimans (1972) confirmed that it is required for metabolism by animal cells and estimated that the optimal CO_2 concentration for cell growth was 0.2–2%. Nevertheless, in most small-scale uncontrolled cell cultures good cell yields are achieved although CO_2 concentrations may reach 5–10%. It should be mentioned that in most large-scale cell cultures CO_2 is added to control pH; dissolved CO_2 is not measured and controlled as such, although autoclavable CO_2 electrodes are available. The effect of controlling the partial pressure of CO_2 on large scale cell culture has yet to be determined. However, Delzer *et al.* (1984) have reported that by changing CO_2 concentration in the head space in the range of 6–36%, a slow decrease in β-interferon production by "genetically engineered" mouse cells was observed.

The rate of CO_2 production can be used as an indication of cell concentration, if the cell-specific CO_2 production rate is known. Moreover, the respiratory quotient (RQ), moles CO_2 produced/mole O_2 utilized, can serve as a valuable indicator of the metabolic state of the cells. Fleischaker *et al.* (1981a) estimated that 16.8 ml of CO_2/hour were generated in a 5-liter culture when the cell concentration was 10^6/ml. Although the amount of CO_2 in the exit gas stream can be measured continuously by mass spectrometry or infrared absorption spectrophotometry (Ingham *et al.*, 1984). Fleischaker *et al.* (1981a) envision a response time problem with on-line analysis because of slow desorption of CO_2 from the medium at low flow rates, and an accuracy problem as the proportion of CO_2 in the inlet stream is changed to correct pH. Thus, they have recommended the dynamic method of measuring CO_2 production rate.

G. Advanced Systems for Process Control and Optimization

Until recently the most common control variables in animal cell culture, i.e., temperature, stirring speed, DO, and pH have been monitored and controlled manually or with simple, noninteractive analog controllers. Attempts to optimize conditions have been empirical for the most part, although many have recognized the potential value of more highly sophisticated computer-aided control systems. A few (Spier, 1980b; Nyiri, 1977) have described in some detail the use of advanced computerized instrumentation to achieve more accurate on-line monitoring and control, to vary set points of key parameters according to a predetermined program, and ultimately to acquire and process data in time to adjust set points to achieve optimal growth or productivity.

If more sophisticated process control is to be achieved progress will have to be made on the theoretical side in the development of mathematical models for growth of animal cells, viruses, and metabolic pathways; and on the analytical side in development of more and better on-line *in situ* sensors for cell mass, substrates, intermediate metabolites, and specific products. Fleischaker *et al.* (1981a) have reviewed the application of on-line sensors to the study of animal cell culture and have suggested some new ones. They have used a microcomputer and suitable sensors to calculate OUR from DO (by the dynamic method), lactic acid production from base utilization rate, RQ, from OUR, gas flow rate, and CO_2 concentration in the influent and effluent gas streams, and glucose consumption by on-line analysis on a continuous sample stream. From these data ATP flux in mmol/g dry cell wt-hour was calculated and from this, cell density in the culture was estimated. The rate of lactic acid production was used to calculate glucose utilization rate, and glucose was fed accordingly to maintain a concentration hypothesized to keep lactic acid production minimal. The overall objective of this sophisticated control regime was to ensure that induction of the culture was timed to elicit maximal production of interferon.

V. Inoculum

On a volume/volume basis the inoculum size in animal cell culture is usually greater than that used in microbial fermentations. The range is 5–30% and depends markedly on the plating efficiency of the cells used (Pharmacia, 1982). As a rule continuous cell lines do not require as much inoculum as do diploid cell strains or primary cells. Hu and Wang (1983) have found that by determining the optimal size of the MC, the amount of inoculum (cells/MC) can be decreased. By optimizing bead size they were able to decrease the critical number of cells/bead to one-fifth of that used previously.

The ratio of number of MCs to number of cells is in the broad range of 1:5 to 1:30, although a ratio of 1:5 to 1:10 is preferred (Butler and Thilly, 1982; Clark *et al.*, 1980; Clark and Hirtenstein, 1981b; Pharmacia, 1982). The concentration of MCs used in batch culture is in the range of 1–5 mg dry wt/ml (with Cytodex MCs) which provides about 5–25 cm^2 surface area/ml culture. The MC concentration most commonly used in batch cultures is 3 mg/ml (15 cm^2 of surface/ml). Higher MC concentrations, i.e., higher cell concentrations, can be used but then simple batch culture is no longer possible; frequent medium replenishment is necessary. With MC cell cultures in packed or concentrated perfusion mode still higher concentrations, up to 12 mg/ml, providing about 60 cm^2/ml, can be used (Tolbert and Feder, 1984). Apparently a minimal initial cell concentration of about 5×10^4 to 5×10^5/ml is needed to initiate rapid cell growth and insure a relatively short lag period (van Wezel, 1984).

To ensure as even a distribution of cells as possible on the MCs and to permit complete utilization of cells in the inoculum, the concentration of MCs should be greater during the attachment phase than later in the fermentation. For this reason several researchers have suggested seeding the cells and MCs into a reduced volume, about one-third that of the intended operating volume. After the attachment phase medium can be added to bring the volume to that required (Griffiths and Thornton, 1982; Pharmacia, 1982; Clark *et al.*, 1980). These workers agreed that during the attachment phase the culture should be essentially static, with only intermittent stirring. Others have found this procedure impractical for large-scale operation (van Wezel, 1982; Edy, 1984). Edy has found that an even distribution of cells can be achieved although the culture is stirred constantly during the attachment phase.

The physiological state of the culture is important if the best utilization of cells in the inoculum is to be achieved. Cells should be harvested for use as inoculum during the late exponential phase and definitely before onset of the stationary phase. It is preferable for the MCs to be placed in the medium prior to its seeding with cells to allow the MCs time to adsorb proteins. Some workers recommend adding amino acids and vitamins to the medium at seeding to increase plating efficiency of the cells (Clark and Hirtenstein, 1982; Griffiths and Thornton, 1982). All of these precautions are more important with cells of low plating efficiency, e.g., diploid cell strains, or when cells must be seeded at low density for pragmatic reasons.

VI. Harvesting of Cells

A. Release of Cells from Microcarriers

In a serial transfer procedure to build up volume for the final production stage, the ratio of the size of one fermentor (or its contents) to the next in line may be 1:20 to 1:4. The number of cells actually transferred depends not only on the volume of the culture but also on the efficiency with which the cells are harvested. The harvesting procedure should produce a well-dispersed, single-cell suspension in order to achieve an even distribution of cells among the MCs. Cells are usually released from the MCs by treatment with proteolytic enzymes, e.g., trypsin, collagenase, pronase, or hyaluronidase with or without the addition of a chelating agent such as EDTA. These treatments cause injury to the cell and affect the integrity of the plasma membrane to some degree (Anghileri and Dermietzel, 1976). Thus, enzymatic treatment should be avoided if possible and controlled carefully if it must be used. RDB, a protein of plant origin with no untoward effect on cells and no susceptibility to inhibition by serum, has been used by Ben Nathan *et al.* (1984) and by Fiorentini and Mizrahi (1984) instead of the former proteinases with good results. Hu (1984) reported that he was able to detach cells from DEAE–dextran MCs by employing trypsin and adjusting the pH above the physiological extreme (about 8.2), and to reattach the cells to fresh as well as the old MCs by readjusting conditions. Tolbert and Feder (1984) have reported that cell harvesting with proteolytic enzymes is more efficient with cell-MC aggregates, which are formed when cells have overgrown the MCs and have bridged between them, than it is with cells grown in a monolayer on the MCs.

The efficiency of recovery of cells from MCs is often very low. van Wezel *et al.* (1980) reported 53–63% efficiency from Cytodex MCs. Scattergood *et al.* (1982) obtained a 78% yield from Cytodex 3 MCs. Gebb *et al.* (1982, 1984) have shown that at the laboratory scale up to 95% efficiency can be achieved in harvesting cells from collagen coated MCs without cell separation. The efficiency of recovery of cells from MCs is dependent on the type of MC, the type of cell, its physiological state, and the harvest procedure. The efficiency of harvest is usually high when collagen coated or denatured gelatin beads are used (Gebb *et al.*, 1982). They reported that cells recovered with collagenase from collagen-coated MCs (Cytodex 3) showed minimal membrane damage and higher plating efficiency in comparison with cells harvested from ionically charged MCs.

Harvesting cells from MCs is a risky operation at best in scaling up MC cell cultures. Different researchers have used different means to circumvent this problem. One approach is to use primary cells to reduce the number of stages. Primary cells, obtained directly from animal tissue, organs, or embryos by applying a mixture of enzymes and a chelating agent, can be obtained in sufficient quantity to permit seeding a rather large fermentor directly. Scattergood *et al.* (1983) used a cell suspension obtained from 10 chick embryos to seed a 9-liter fermentor and on this basis have calculated a need for about 70 embryos to seed a 60-liter fermentor, an operation that seems feasible. By the same reasoning 1100 embryos, an impractical number, would be needed to seed a 1000-liter fermentor. At this scale it seems that at least one step of cell harvesting from MCs and transfer to fresh beads in a larger fermentor would be necessary. The need for an intermediate step is shown even more dramatically in another process described by van Wezel *et al.* (1981). They propagated monkey kidney primary cells in MC culture for production of polio vaccine. Two kidneys taken from one monkey supplied only enough cells to seed a 10-liter fermentor. Because of the high cost of breeding monkeys and the necessity to adhere to strict and extensive quality control measures, it is more economical to subculture the cells, even up to 12 generations in culture, than to use primary cells for production of this virus. Except when inexpensive primary cells can be used, all other systems in which ADCs are grown on MCs almost invariably require at least one intermediate harvesting step in the scale-up train.

Another way to avoid the operation of stripping the cells from the MCs between stages is to use devices other than MCs in the early stages. Suspensions prepared by trypsinization of cell layers grown in roller bottles are used to provide the seed for relatively small fermentations (Edy, 1984; Morandi *et al.*, 1984). However, cell seeding from multiple unit propagators involves a high labor cost and more importantly increases significantly the risk of contamination. Consider that sometimes as many as 100–200 roller bottles are needed to seed a single 10-liter fermentor. Single-unit propagators designed to avoid a pooling operation include three proprietary devices: Multitory trays (Nunc), the Gyrogen (Chemap), and the Opticell (KC Biologicals).

Several researchers have shown that under some circumstances it is possible to scale up MC cell cultures without the conventional harvesting operations. Crespi and Thilly (1981) transferred Chinese hamster ovary (CHO) and monkey kidney (LLC-MK2) cells directly from MCs to MCs while stirring in a medium low in calcium content. Manousas *et al.* (1980) added fresh MCs to heavily sheathed MC cell cultures to generate a new wave of cell proliferation and virus production. Delzer *et al.* (1984) added fresh MCs to mouse L-cell cultures grown to confluence and observed movement of cells to the fresh MCs. Kluft *et al.* (1983) expanded the volume of an MC culture of human melanoma cells from 3 to 10 to 40 liters by adding fresh medium and MCs incrementally. The small fraction of cells free of MCs became attached to the fresh MCs and grew on them. MC to MC transfer of cells, when it can be done, not only saves material and labor costs, but also avoids the risk of contamination entailed in the sequence of steps in conventional harvesting. Generally cell transfer from MC to MC is achieved more easily with epithelial cell lines than with elongated fibroblast primary cells or diploid cell strains.

B. SEPARATION OF CELLS FROM MICROCARRIERS

Cells are harvested from MCs either in the vessel in which they have been grown or in a separate vessel. If harvesting is carried out in the growth vessel the MCs with adhering cells are washed by permitting them to settle, decanting the supernatant medium, adding buffer, and repeating the sequence of steps if desired. Dissociation agents are added in minimal concentration and volume, stirring is implemented for a brief period, medium containing serum is added to stop the action of the proteolytic enzyme, and the suspension of cells and MCs (or cells only if stripped MCs are retained by a filter) is transferred to the next larger vessel. Several disadvantages attend this procedure: The minimal volume that can be used is the smallest working volume of the fermentor; all of the dissociating agent used is transferred to the next fermentor; washing by sedimentation and decantation is slow; washing and separation are inefficient unless a filter (60–100 μm to retain MCs but not cells) is used on the input of the transfer line.

The procedure for harvesting cells from the MCs in a separate vessel is similar in principle but operationally different. A principle advantage is that the volume of the harvest vessel may be chosen to permit the smallest possible operating volume, that which is limited by the total mass of MCs. One gram of Cytodex MCs may be expected to give a settled bed volume of 14–18 ml. Scattergood et al. (1983) reported that MCs from a 9-liter fermentor gave a bed volume of 350 ml. At minimal volume washing, treating, and separating can all be done more efficiently. van Wezel et al. (1980) have described a special apparatus for harvesting monkey kidney cells from MCs. Agitation is provided by a Vibromixer and the MCs are retained in the vessel by a stainless steel screen with 60-μm openings at the bottom outlet port. This apparatus has been used by Scattergood et al. (1983) for harvesting chick embryo fibroblasts from MCs. This apparatus is supplied commercially by Contact, Holland. Spier et al. (1977) have described the use of a small bore tube (1.2 mm) for continuous stripping of cells from MCs.

In order to obtain an even distribution of cells on MCs in the next (larger) stage, the freed cells should be separated from the used MCs, which still retain a variable number of viable cells even after the harvesting treatment (van Wezel et al., 1980). Used MCs, still retaining viable cells, if transferred to the next stage in effect are seeded more heavily than the new MCs and thus will achieve confluence earlier.

Cells harvested (released) from the MCs can be separated from them by gravity sedimentation, filtration, or by density gradient centrifugation. Billig *et al.* (1984) described testing (at the laboratory scale) these three techniques for cell separation after trypsinization. Gravity sedimentation is based on the difference in settling rates between cells and MCs with no agitation. Filtration was done through an 88-μm pore size nylon screen. Density gradient centrifugation in Ficoll Paque at low speed sent the MCs to the bottom of the tube while the cells were located in an intermediate band. From the standpoint of efficiency of cell recovery the latter two methods were superior (65–75% vs 35–45%). However, the filtration method of van Wezel *et al.* (1980) is the only one of the three that has been tested at the large scale. The other two need further development before large scale application can be made.

MCs separated from cells can be reused only when they are the ionically charged type. Collagen or gelatin coated beads cannot of course be reused after exposure to proteolytic enzymes.

VII. Mode of Cell Propagation and Product Production

A. Batch and Modified Batch Mode

Until recent years the most important products obtained from ADC cultures were the viral antigens. The viruses propagated on the cells are usually cytopathogenic (e.g., poliomyelitis or foot-and-mouth disease virus), i.e., the virus kills the host cell during its growth. Thus the most efficient process consists of two batch phases, first a phase when maximal cell density is achieved and then a phase when virus propagation is effected. Another product produced in batch mode from ADCs is human fibroblast β-interferon. After a first phase in which cells are grown to optimal cell density induction is carried out for interferon production.

In the simple batch mode of operation temperature, pH, and DO are usually controlled, but nutrients, e.g., glucose and glutamine, are depleted continuously and inhibitory waste products, e.g., lactic acid and ammonia, accumulate. Thus cell propagation slows and stops and cell yield is limited (Butler, 1984). Improvement in cell yield is obtained by periodically replacing a portion of the culture supernatant with fresh medium, thereby restoring nutrients and removing waste products. Clark and Hirtenstein (1981a) took another approach, feeding essential depleted nutrients (cystine, glutamine, inositol, glucose, choline, and pyridoxine) to chick embryo fibroblasts after 3 days growth on MCs. By feeding a concentrate of essential nutrients the cost of adding whole medium was avoided. A disadvantage of shot or slug feeding, as pointed out by van Wezel (1984) is that a sudden change in environment can cause separation of cells from MCs. A more sophisticated approach to controlling cell propagation is to feed essential nutrients on demand, to maintain predetermined concentrations and concomitantly to limit accumulation of waste products. Fleischaker *et al.* (1981a) fed glucose to an MC cell culture on demand to maintain a concentration of 0.5 mM (vs 20 mM normally batched) and achieved higher cell density and lower accumulation of lactic acid. Glacken *et al.* (1983) kept the glutamine concentration in an MC cell culture as low as 0.2 mM (vs 4 mM) by continuous feeding and thus reduced the level of ammonium ion in the culture by over 60%. Although the fed-batch mode of process conduction generally leads to a higher cell density and viability, inevitably cells do begin to die, probably because of waste product accumulation (Reuveny, unpublished data).

An interesting different approach to maintaining MC cell cultures in a viable state for extended periods was taken by Morandi *et al.* (1982). They propagated MRC-5 cells on Cytodex 1 MCs in 1.2- and 5-liter fermentors while dialyzing the medium against 5 and 20 liters, respectively, of fresh serum-free medium. On the smaller scale the dialysis was done with 150 cm^2 of dialysis tubing; on the larger scale dialysis was done with hollow fibers providing 8000 cm^2 of surface. This system provided low-molecular-weight nutrients continuously and waste products were diluted continuously. High cell yields were realized.

B. Continuous or Extended Operation

Truly continuous or fed-batch methods for propagating animal cells are not practical in MC cell culture as they would entail continuous or repeated harvesting of cells and replenishment of MCs. However, in recent years a need has arisen for products that are secreted continuously from genetically engineered ADCs without affecting their viability. Examples of such products are tissue plasminogen activator, viral antigens, hormones, and human plasma proteins which are secreted from genetically engineered cell lines. An attractive mode of producing such products is to propagate the cells to achieve high density, and afterward to establish conditions such that viability is maintained, significant growth does not occur, and product is secreted continuously. Product can be produced by recovering supernate free of MCs periodically, or by perfusing the system continuously.

Manousas *et al.* (1980) have propagated oncoviruses in MC culture by intermittant withdrawal of supernatant. Reuveny (1983) produced carcinoembryonic antigen in an MC culture of adenocarcinoma cells by periodic harvesting of culture supernate. The main advantage of the pseudocontinuous, repeated harvest system is its simplicity. Product is produced in the same vessel in which cells are propagated, and harvest is accomplished by periodically letting the MCs settle and withdrawing supernatant through a closed system. However, in contrast with perfusion systems, relatively low concentrations of cells and products are obtained.

The main advantage of perfusion culture is that there is a constant supply of medium to the cells and a constant removal of waste products. The environment of the cells remains constant, although not necessarily uniform throughout the culture. Butler *et al.* (1984) achieved a density of 9×10^6 MDCK cells/ml using DEAE–dextran MCs at a concentration of 7.5 mg/ml in perfused culture. Tolbert and Feder (1984) realized cell concentrations of about 10^7/ml with polyacrylamide MCs in a concentration of 12 mg/ml. MC cell cultures, in comparison with microbial or animal cell suspension culture systems, are especially well suited to management in perfusion systems, as the MCs are large enough to be retained by screens that offer little resistance to liquid flow, and dense enough to settle under the force of gravity alone in a reasonable period of time.

A number of special devices for accomplishing perfusion have been described. van Wezel *et al.* (1984), Kluft *et al.* (1983), and Griffiths and Thornton (1982) have all used a steel 100-μm pore size cage mounted on the stirring shaft and rotating with it. MCs are excluded from the cage and thus clear culture medium can be pumped out of the vessel from inside the cage while fresh medium is supplied continuously to the culture outside the cage. Tolbert and Feder (1984) added a settling bottle, clarifying vessel, effluent reservoir, and medium reservoir all external to the growth vessel. After multilayer growth was achieved, perfusion was effected by pumping MCs and cell aggregates to a settling vessel, where settling rate exceeded upward flow rate and supernatant virtually free of MCs was sent to a filtration vessel. From there, the stream was split, some being returned to the medium reservoir and thence to the main vessel, and some being harvested continuously through a spin filter. An additional advantage to the settling bottle is that cells grew not only on the surface of the MCs but also between them, giving a higher cell concentration than would be expected if the cells formed only a single confluent monolayer (see also Reuveny, 1983, for multilayering on DEAE-cellulose MCs). Butler *et al.* (1983) have used a device conceptually similar in which culture supernatant was pumped through a column separator located above the liquid level. More recently, Tolbert *et al.* (1984) have described another system in which cells were grown in a conventional reactor until maximal density was achieved, and then transferred at still higher density (10^8/ml) into a static maintenance reactor in which the cells were perfused while kept in a nonproliferating state. Cells were maintained in this state up to 2 months under low-shear conditions. Strand *et al.* (1984a,b) have used a hollow fiber cartridge in which MC-bearing cells were entrapped in the interfiber space with no agitation. The cells were fed by medium passed through the capillaries. High cell density and interferon production were demonstrated in this system. Clark and Hirtenstein (1981a) used a fluid lift system for perfusing MC cell cultures continuously.

In the foregoing paragraphs perfusion systems have been discussed mainly for the advantages they offer for production of products secreted during a nonproliferating stage. However, several groups (Griffiths and Thornton, 1982; Strand et al., 1984a,b; Feder and Tolbert, 1984; and others) have used closed perfusion systems as a means of obtaining good oxygen control and high cell yields in low volume reactors. After high cell density was achieved the cells were used for production of a biological product (e.g., a virus or interferon). Among the characteristics of perfusion systems that have been cited as advantageous for economical, efficient cell production are the following: Tolbert and Feder (1984) have shown that in perfusion systems medium is utilized four times as efficiently as in roller bottles. Kluft et al. (1983) reduced the serum content of the medium to 0.5% during the maintenance period. The concentration of nutrients in contact with the cells can be changed by varying either the flow rate or the concentrations in the feed stream. Griffiths and Thornton (1982) used a closed (recirculation) perfusion system for MRC-5 cell growth on MCs at a concentration of 5 mg/ml and found that after 110 hours of operation, even at the maximal flow rate of which the system was capable the nutrient concentration in the growth vessel was 20% lower than that in the reservoir. Under these conditions increased delivery of nutrient would have to be done by increasing the concentration in the feed. Butler (1984) reported that at a feed rate of 2 ml/liter-minute in an open (nonrecirculated) system ammonia accumulated to a concentration of 2.3 mM, comparable to that seen in a batch system, when growth stopped. It seems that at the high concentration of metabolically active cells achieved a higher flow rate would have been necessary to reduce accumulation of inhibitory end products. In an open system, the efficiency of utilization of medium would have been reduced. In closed perfusion systems the efficiency of medium utilization can be increased and the product can be accumulated in relatively small volume, but accumulation of toxic metabolic products will occur as well. In general, control of the environment is more difficult in high density perfusion systems, closed or open, than in closed batch systems.

The ease with which the nutrient composition of the environment of the cells can be changed in a perfusion system makes it possible to study the effects of step or pulse changes on growth and metabolism, for basic information as well as to move toward optimal operating conditions (Spiess et al., 1982; Salstrom et al., 1984). Unlike closed batch systems the perfusion system permits calculation of oxygen consumption, CO_2 production, glucose consumption, lactic acid production, etc., by on-line measurements made on the input and output streams. For many measurements *in situ* sensors may be used (Lydersen *et al.*, 1985).

The foregoing discussion of the characteristics of perfusion systems stresses for the most part advantages. It is worthwhile to review some of the disadvantages of perfusion systems. Perfusion systems are not suitable for products produced during active growth or during the stage of declining metabolic activity and decreasing viability. There is a risk of genetic instability during the maintenance period, which may last for months. The apparatus is relatively complicated. Pumps (which may fail), filters (which may clog), level sensors, auxiliary vessels for medium and effluent, and a requirement for better control instrumentation are all features of the perfusion system which add to capital costs and may contribute contamination susceptibility or operating problems.

VIII. Medium

The media that are used for growth of cells in stationary monolayer cultures are usually suitable for use in MC cell culture. Several modified formulas or nutrient feeding regimens have been devised and are discussed in the following paragraphs.

Griffiths and Thornton (1982) and Clark *et al.* (1980) have suggested enriching the medium during the initial cell attachment phase. Several changes related to the higher cell density and consequent more vigorous metabolic activity have been made. Organic buffers, e.g., HEPES, have been added to provide pH control to supplement the CO_2–bicarbonate system. Glucose has been replaced by galactose or fructose to reduce lactic acid production (Imamura *et al.*, 1982). High-molecular-weight nonprotein polymers have been added to low-serum media to reduce turbulence, to raise the viscosity and protect cells from mechanical damage, and to promote cell attachment (Nunc, 1981). van Wezel and van der Welden de Groot (1980) and Clark *et al.* (1982) have pointed out that MC cell cultures are more sensitive to the quality of the growth medium, especially the serum, than are stationary monolaryer cultures. Clark *et al.* (1982) proposed that stringent control measures be used in selection of each batch of serum.

Clark *et al.* (1980) have warned that antibiotics, used commonly to protect cell cultures from microbial infection, can affect cell growth rate and saturation density adversely in MC cell culture, especially with cells of low plating efficiency.

Clark and Hirtenstein (1981a) have indicated that reducing the serum concentration in the medium after confluent growth has been attained seems to prevent sloughing off of cells from the MCs. Clark *et al.* (1982) and Crespi *et al.* (1981) have demonstrated that it is possible to grow cells in MC cell culture in serum-free medium.

The quality of water poses a problem in cell culture systems in general (Girard, 1977) and especially in MC cell culture (Morandi *et al.*, 1984). Usually in large-scale operations reverse osmosis (RO) systems are used to provide water of specific resistivity of at least 1 MΩ/cm. Purified water is stored at 80°C before use to prevent bacterial growth (Morandi *et al.*, 1984). In all cases attention should be paid to the quality of the tap water available and possible seasonal or erratic fluctuations in its composition (Girard, 1977; Morandi *et al.*, 1984).

One of the most critical problems in cell culture generally and MC cell culture especially, because it is aggravated by high cell density, is the inhibitory effects of accumulated products. Gaseous products, e.g., CO_2 and to some extent NH_3, can be purged by the gas stream through the system. Highly soluble, nonvolatile products such as lactic acid and NH_4^+ will accumulate in the culture medium (Butler, 1984). Accumulation of lactic acid can be diminished by changing the carbon source or by feeding glucose slowly. The generation of NH_4^+ ion, which seems to have a drastic adverse effect on cell growth, is related to metabolism of glutamine (Butler, 1984). Glacken *et al.* (1983) have suggested slow feeding glutamine to cell cultures to reduce the amount of NH_4^+ ion that accumulates. The possibility that other waste products may reach growth-limiting concentrations was suggested by Birch and Cartwright (1982). On the other hand, the possibility that growth slows or ceases because the medium becomes depleted of essential nutrients has been explored also. Polastri *et al.* (1984) and Butler (1984) have analyzed MC cell cultures for amino acid utilization and found some differences related to cell type. However, in all cases glutamine was the amino acid consumed most rapidly.

On the large scale medium is prepared in a mixing vessel and sterilized by filtration into a sterile holding tank. The holding tank is refrigerated. During storage the medium is tested intensively for sterility and its ability to support cell growth (Morandi *et al.*, 1984). The capital cost of the large storage vessels can be eliminated if a minimum-risk sterilization procedure is used. Morandi *et al.* (1984) claim to have reduced the risk and frequency of contamination drastically by filter-sterilizing twice between the mixing vessel and the fermentor. Another way to prepare medium on the large scale is to batch an incomplete formula of the heat-stable components, carry out sterilization by heat, and add the heat-labile components (e.g., serum, glutamine) in concentrated solution through a sterilizing filter (Keay, 1974; van Wezel, 1984).

IX. Conclusions

Interest in development of improved apparatus and methodology for MC cell culture is strong at this time as a result of recognition of a growing number of useful products for human and animal therapy that can be produced by ADCs. Just as the discovery of the MC technique was motivated by the need to provide more efficient ways of cultivating ADCs for production of viral vaccines, better ways of managing cell growth and product synthesis in MC cell culture have continued to be sought by researchers interested for the most part in production of products for therapy: viruses, viral antigens, interferons, tissue plasminogen activator, hormones, and human plasma proteins. Some of these products are secreted from genetically engineered cell lines.

MC cell culture is the method of choice for producing useful products from ADCs, although cell propagators providing large continuous surface areas may have value in certain applications. It is generally accepted that MC cell culture offers the following principal advantages: a high ratio of growth surface to culture volume is provided; a single large production unit can replace a number of smaller units; mixed, homogeneous MC cell cultures can be sampled, monitored, and controlled more easily than inhomogeneous systems; harvesting of cells and extracellular products can be done easily; and MC cell cultures can be scaled up readily using conventional equipment (fermentors) used for microbial processes.

Although MC cell culture technology has borrowed heavily from fermentation technology, several properties of MCs and ADCs have made variations in conventional fermentation apparatus necessary. MCs must be agitated gently, just enough to maintain a homogeneous suspension, to avoid damaging the cells or dissociating the cells from the MCs. Many agitation systems that are effective on the small scale have been devised; a few offer scale-up potential. Maintaining DO at a critical level is important, but conventional sparging is not feasible with MC cell cultures, as the cells are damaged by foam and the integrity of the MC-cell complex is destroyed easily by shear forces. Although surface aeration is effective at the small scale, and a variety of other experimental systems for oxygenation have been tested, the lack of an effective aeration system is still the most important factor limiting MC cell cultures to 300- to 400-liter volumes, with rare exceptions.

Problems unique to MC cell culture systems, not shared with suspension cell culture or microbial systems, are related to the need to manage the initial attachment of cells to MCs as well as to manage the detachment, or harvest of cells from MCs at the end of one stage for inoculation of the next stage. Special apparatus for cell harvesting has been devised, but researchers continue to seek means of circumventing or simplifying the harvesting procedure.

A number of environmental parameters, e.g., pH, DO, ORP, CO_2, and temperature are monitored and controlled in advanced MC cell culture processes just as they are in microbial fermentations. MC cell culture is subject to the same kind of limitations as is fermentation in respect to availability and quality of specific sterilizable *in situ* sensors and other instrumentation. Serious studies of on-line analysis and computerized control of MC cell culture systems have been initiated and rapid advances may be expected.

The mode of operation of an MC cell process may be simple batch, fed batch, or extended operation with continuous or intermittent separation and withdrawal of secreted product. If the product is secreted by nonproliferating cells, various types of mixed or packed, closed or open perfusion systems may be employed. The more complex systems entail greater operational risks; simpler systems are often more reliable.

Growth media for MC cell culture are essentially the same as for suspension cell culture. In contrast to media for microbial fermentations they are costly, require filter sterilization, and must be subjected to rigorous quality control. Formulation of serum-free, low-cost, heat-sterilizable media for cell culture is the subject of research in many laboratories.

Five to 10 years ago many believed that the development of genetically engineered microbial cells would eliminate the need for large-scale mammalian cell culture. It was assumed that any product produced by animal cells could be produced by microbial cells into which mammalian genes had been cloned. However, in recent years it has become evident that there are mammalian cell products that cannot be produced efficiently by microbial cells. Genetically engineered animal cell lines are more efficient for these more complex products; consequently, we are witnessing a vigorous development of methods and apparatus for MC cell culture.

ACKNOWLEDGMENTS

The authors wish to express their gratitude to L. Miller and D. Freedman for their review and criticism of the manuscript, and to L. de Nome for her technical help in testing the NBS draft tube fermentor. The testing of the Kedem impeller was done by S. Reuveny, A. Mizrahi, L. Silberstein, and M. Klein.

REFERENCES

Anghilieri, L. J., and Dermietzel, P. (1976). *Oncology* 33, 11–23.
Arminger, W. B., Zabriskie, D. W., Maenner, G. F., and Farro, J. F. (1984). *Proc. Biotechnol.* 84, 621–628.

Balin, A. K., Goodman, D. B. P., and Rasmussen, H. (1976). *J. Cell Physiol.* **89**, 235–250.

Bartling, S. J. (1984). *Dev. Biol. Standard.* **55**, 143–147.

Ben Nathan, D., Barzilai, R., Lazar, A., and Shahar, A. (1984). *Dev. Biol. Standard.* (In press).

Billig, D., Clark, J. M., Ewell, A. J., Carter, C. M., and Gebb, C. (1984). *Dev. Biol. Standard.* **55**, 67–75.

Birch, J. R., and Cartwright, T. (1982). *J. Chem. Technol. Biotechnol.* **32**, 313–317.

Bone, A. J., and Swenne, I. (1982). *In Vitro* **18**, 141–148.

Butler, M. (1984). *Dev. Biol. Standard.* (In press).

Butler, M., and Thilly, W. G. (1982). *In Vitro* **18**, 213–219.

Butler, M., Imamura, T., Thomas, J., and Thilly, W. G. (1983). *J. Cell Sci.* **61**, 351–363.

Carter, C. M., and Ewell, A. J. (1982). *In Vitro* **18**, 312.

Ceccarini, C., and Eagle, H. (1971). *Proc. Natl. Acad. Sci. U.S.A.* **68**, 229–233.

Clark, J. M., and Hirtenstein, M. D. (1981a). *Ann. N.Y. Acad. Sci.* **369**, 33–46.

Clark, J. M., and Hirtenstein, M. D. (1981b). *J. Interferon Res.* **4**, 391–400.

Clark, J., Hirtenstein, H., and Gebb, C. (1980). *Dev. Biol. Standard.* **46**, 117–124.

Clark, J., Gebb, C., and Hirtenstein, M. D. (1982). *Dev. Biol. Standard.* **50**, 81–92.

Crespi, C. L., and Thilly, W. G. (1981). *Biotechnol. Bioeng.* **23**, 983–994.

Crespi, C. L., Imamura, T., Leong, P., Fleischaker, R. J., Brunengraber, H., and Thilly, W. G. (1981). *Biotechnol. Bioeng.* **23**, 2673–2689.

Davis, P. F., and Kerr, C. (1982). *Exp. Cell Res.* **141**, 455–459.

de Bruyne, N. A., and Morgan, B. J. (1981). *Am. Lab.* **13**, 52–61.

Delzer, J., Hauser, H., and Lehmann, J. (1984). *Dev. Biol. Standard.* (In press).

Edy, V. G. (1984). *Adv. Exp. Med. Biol.* **172**, 169–178.

Edy, V. G., Augenstein, D. C., Edwards, C. R., Gruttenden, V. F., and Lubiniecki, A. S. (1982). *Tex. Rep. Biol. Med.* **41**, 169–174.

Ewell, A. J., and Carter, C. M. (1982). *In Vitro* **18**, 312–313.

Feder, J., and Tolbert, W. R. (1983). *Sci. Am.* **248**, 24–31.

Fiorentini, D., and Mizrahi, A. (1984). *Dev. Biol. Standard.* (In press).

Fleischaker, R. J., and Sinskey, A. J. (1981). *Eur. J. Appl. Microbiol. Biotechnol.* **12**, 193–197.

Fleischaker, R. J., Weaver, J. C., and Sinskey, A. J. (1981a). *Adv. Appl. Microbiol.* **27**, 137–167.

Fleischaker, R. J., Giard, D. J., Weaver, J., and Sinskey, A. J. (1981b). In "Advances in Biotechnology" (E. M. Young, C. W. Robinson, and C. Vezina, eds.), Vol. I, pp. 425–430. Pergamon Press, New York.

Foehring, B., Tija, S. T., Zenke, W. M., Sauer, G., and Doerfler, W. (1980). *Proc. Soc. Exp. Biol. Med.* **164**, 222–228.

Frappa, J., Beaudry, Y., Quillan, J. P., and Fontanges, R. (1979). *Dev. Biol. Standard.* **42**, 153–158.

Gailiani, S., McLimans, W. F., Nussbaum, A., Robinson, F., and Rohot, O. (1976). *In Vitro* **12**, 363–372.

Gebb, C., Clark, J. M., and Hessle, H. (1980). *Eur. J. Cell Biol.* **22**, 601.

Gebb, C., Clark, J. M., Hirtenstein, M. D., Lindgren, G., Lindgren, U., Lindskog, U., Lundgren, B., and Vertblad, P. (1982). *Dev. Biol. Stand.* **50**, 93–102.

Gebb, C., Lundgren, B., Clark, J., and Lindskog, U. (1984). *Dev. Biol. Standard.* **55**, 57–65.

Giard, D. J., and Fleischaker, R. J. (1980). *Antimicrob. Agents Chemother.* **18**, 130–136.

Giard, D. J., Fleischaker, R. J., Sinskey, A. J., and Wang, D. I. C. (1981). *Dev. Ind. Microbiol.* **22**, 299–309.

Girard, H. C. (1977). *In* "Cell Culture and Its Applications" (R. T. Acton and J. D. Lynn, eds.), pp. 111–127. Academic Press. New York.

Glacken, M. W., Fleischaker, R. J., and Sinskey, A. J. (1983). *Ann. N.Y. Acad. Sci.* **413**, 355–372.

Griffiths, B. (1984). *Dev. Biol. Standard.* **55**, 113–116.

Griffiths, B., Thornton, B., and McEntee, I. (1980). *Eur. J. Cell Biol.* **22**, 606.

Griffiths, B., Atkinson, T., Electricwala, A., Latter, A., Ling, R., McEntee, I., Riley, P. M., and Sutton, P. M. (1984). *Dev. Biol. Standard.* **55**, 31–36.

Griffiths, J. B., and Thornton, B. (1982). *J. Chem. Technol. Biotechol.* **32**, 324–329.

Griffiths, J. B., Thornton, B., and McEntee, I. (1982). *Dev. Biol. Standard.* **50**, 103–110.

Harakas, N. K., Lewis, C., Bartram, R. D., Wildi, B. S., and Feder, J. (1984). *Adv. Exp. Med. Biol.* **172**, 1119–138.

Himmler, G., Palfi, G., Rüker, F., Katinger, H., and Scheirer, W. (1984). *Dev. Biol. Standard.* (In press).

Hirtenstein, M. D., and Clark, J. M. (1981). *ESCAT Newslett.* **6**, 9–17.

Hirtenstein, M. D., Clark, J., Lindgren, G., and Vretblad, P. (1980). *Dev. Biol. Standard.* **46**, 109–116.

Hirtenstein, M. D., Clark, J. M., and Gebb, C. (1982). *Dev. Biol. Standard.* **50**, 73–80.

House, W., Shearer, M., and Maroudas, G. (1972). *Exp. Cell Res.* **71**, 293–296.

Hu, W. S. (1984). *Abstr. 188th Annu. Meet. Am. Chem. Soc., Philadelphia.*

Hu, W. S., and Wang, D. I. C. (1983). *Abstr. 187th Annu. Meet. Am. Chem. Soc., Washington, D.C.*

Imamura, T., Crespi, C. L., Thilly, W. G., and Brunengraber, H. (1982). *Anal. Biochem.* **124**, 353–358.

Ingham, J., Piehl, H., Dittmar, K. E. J., and Lehman, (1984). *Proc. 3rd Eur. Congr. Biotechnol.* **I**, 173–178.

Jensen, M. D. (1977). *In* "Cell Culture and Its Applications" (R. T. Acton and J. D. Lynn, eds.), pp. 589–602. Academic Press, New York.

Johnson, M. J., Borkowski, J., and Engblom, C. (1964). *Biotechnol. Bioeng.* **6**, 457–468.

Katinger, H. W. D. (1980). *ESCAT Newslett.* **5**, 5–8.

Katinger, H. W. D., Scheirer, W., and Kramer, E. (1979). *Ger. Chem. Eng.* **2**, 31–38.

Keay, L. (1977). *In* "Cell Culture and Its Applications" (R. T. Acton and J. D. Lynn, eds.), pp. 513–532. Academic Press, New York.

Kilburn, D. G., and Webb, F. C. (1968). *Biotechnol. Bioeng.* **10**, 801–814.

Klein, F., Jones, W. I., Mohlandt, B. G., and Lincoln, R. E. (1971). *Appl. Microbiol.* **21**, 265–271.

Kluft, C., van Wezel, A. L., van der Welden, C. A. M., Emeis, J. J., Verheijen, J. H., and Wijngoards, G. (1983). *In* "Advances in Biotechnological Processes" (A. Mizrahi and A. L. van Wezel, eds.), Vol. 2, pp. 97–100. Liss, New York.

Knazek, R. A., Kohler, P., and Dedrick, R. (1972). *Science* **178**, 65–67.

Lengyel, Z. L., and Nyiri, L. (1965). *Biotechnol. Bioeng.* **7**, 91–100.

Levine, D. L., Wong, J. S., Wang, D. I. C., and Thilly, W. G. (1977). *Somat. Cell Genet.* **3**, 149–155.

Lydersen, B. K., Pugh, G. G., Paris, M. S., Sharma, B. P., and Noll, L. A. (1985). *Biotechnology* **3**, 63–67.

McLimans, W. F. (1972). *In* "Growth, Nutrition and Metabolism of Cells in Culture" (G. H. Rothblat and V. J. Gristofolo, eds.), Vol. I, pp. 137–162. Academic Press, New York.

Manousos, M., Ahmed, M., Torchio, C., Wolfi, J., Shibley, G., Stephens, R., and Mayyasi, S. (1980). *In Vitro* **16**, 507–515.

Meignier, B. (1979). *Dev. Biol. Standard.* **42**, 141–145.
Meignier, B., Maugeot, H., and Favre, H. (1980). *Dev. Biol. Standard.* **46**, 249–256.
Montagnon, B. J., Vincent-Falquez, J. C., and Fanget, B. (1984). *Dev. Biol. Standard.* **55**, 37–42.
Morandi, M., Valeri, A., and Scalvo, I. S. V. T. (1982). *Biotechnol. Lett.* **7**, 465–468.
Morandi, M., Stanghellini, L., and Valeri, A. (1984). *Dev. Biol. Standard.* (In press).
Nielsen, V., and Johansson, A. (1980). *Dev. Biol. Standard.* **46**, 131–136.
Nunc (1981). *Biosilon Bulletin No. 1: Cultivation Principles and Working Procedure.* Trade Publication.
Nyiri, L. K. (1977). *In* "Cell Culture and Its Applications" (R. T. Acton and J. D. Lynn, eds.), pp. 161–189. Academic Press, New York.
Pharmacia Fine Chemicals (1982). *Microcarrier Cell Culture: Principles and Methods.* Trade publication.
Polastri, G. D., Friesen, H. J., and Mauler, R. (1984). *Dev. Biol. Standard.* **55**, 53–56.
Puhar, E., Einsele, A., Buehler, H., and Ingold, W. (1980). *Biotechnol. Bioeng.* **22**, 2411–2416.
Reuveny, S. (1983a). Ph. D. Thesis, Hebrew University, Jerusalem.
Reuveny, S. (1983b). *In* "Advances in Biotechnological Processes" (A. Mizrahi and A. L. van Wezel, eds.), Vol. 2, pp. 1–32. Liss, New York.
Ryan, U. S., Mortara, M., and Whitaker, C. (1980). *Tissue Cell* **12**, 619–636.
Salstrom, J. S., Chadfield, R. C., and Murphy, T. J. (1984). *Biotechniques* **March/April**, 103–105.
Scattergood, E. M., Schlabach, A. J., McAleer, W. J., and Hilleman, M. R. (1983). *Ann. N.Y. Acad. Sci.* **413**, 332–339.
Schleicher, J. B., and Weiss, R. E. (1968). *Biotechnol. Bioeng.* **10**, 617–624.
Sinskey, A. J., Fleischaker, R. J., Tyo, M. O., Giard, D. J., and Wang, D. I. C. (1981). *Ann. N.Y. Acad. Sci.* **369**, 47–59.
Spier, R. E. (1980a). *Adv. Biochem. Eng.* **14**, 119–162.
Spier, R. E. (1980b). *Dev. Biol. Standard.* **46**, 159–162.
Spier, R. E. (1982). *J. Chem. Technol. Biotechnol.* **23**, 304–312.
Spier, R. E., and Griffiths, B. (1984). *Dev. Biol. Standard.* **55**, 81–92.
Spier, R. E., and Whiteside, J. P. (1969). *Biotechnol. Bioeng.* **18**, 659–667.
Spier, R. E., and Whiteside, J. P. (1984). *Dev. Biol. Standard.* **51**, 151–152.
Spier, R. E., Whiteside, J. P., and Bolt, K. (1977). *Biotechnol. Bioeng.* **19**, 1735–1738.
Spiess, Y., Smith, M. A., and Vale, W. (1982). *Diabetes* **31**, 189–193.
Strand, M. S., Quarles, J. M., and McConnel, S. (1984a). *Biotechnol. Bioeng.* **26**, 503–507.
Strand, M. S., Quarles, J. M., and McConnel, S. (1984b). *Biotechnol. Bioeng.* **26**, 508–512.
Taylor, G. W., Kondig, J. P., Nagle, S. C., and Higuchi, K. (1971). *Appl. Microbiol.* **21**, 923–933.
Tolbert, W. R., and Feder, J. (1984). *Annu. Rep. Ferm. Process.* **6**, 35–74.
Tolbert, W. R., Lewis, C., Jr., White, P. J., and Feder, J. (1984). *Abstr. 188th Annu. Meet. Am. Chem. Soc., Philadelphia.*
Toth, G. M. (1977). *In* "Cell Culture and Its Applications" (R. T. Acton and J. D. Lynn, eds.), pp. 617–636. Academic Press, New York.
Tyo, M. A., and Wang, D. I. C. (1979). *Abstr. 183rd Annu. Meet. Am. Chem. Soc. Washington, D.C.*
Tyo, M. A., and Wang, D. I. C. (1981). *In* "Advances in Biotechnology" (E. M. Young, C. W. Robinson, and C. Vezina, eds.), Vol. I, pp. 141–146. Pergamon Press, New York.
van Hemert, P., Kilburn, D. G., and van Wezel, A. L. (1969). *Biotechnol. Bioeng.* **11**, 875–885.

van Wezel, A. L. (1967). *Nature (London)* **216,** 64–65.

van Wezel, A. L. (1982). *J. Chem. Technol. Biotechnol.* **32,** 318–323.

van Wezel, A. L. (1984). *Dev. Biol. Standard.* **55,** 3–9.

van Wezel, A. L., and van der Welden-de Groot, C. A. M. (1980a). *Process Biochem.* **13,** 6–8.

van Wezel, A. L., van der Welden-de Groot, C. A. M., and van Herwaarden, J. A. M. (1980b). *Dev. Biol. Standard.* **46,** 151–158.

van Wezel, A. L., van der Welden-de Groot, C. A. M., de Haan, H. H., van den Heuvel, N., and Schasfoort, R. (1984). *Dev. Biol. Standard.* (In press).

West, J. M., Stickle, G. P., Walter, K. D., and Brown, W. E. (1961). *J. Biochem. Microbiol. Technol. Eng.* **3,** 125–137.

Whiteside, J. P., Whiting, B. R., and Spier, R. E. (1979). *Dev. Biol. Standard.* **42,** 113–120.

Whiteside, J. P., Farmer, S., and Spier, R. E. (1984). *Dev. Biol. Standard.* (In press).

Naturally Occurring Monobactams

WILLIAM L. PARKER, JOSEPH O'SULLIVAN, AND RICHARD B. SYKES

The Squibb Institute for Medical Research
Princeton, New Jersey

I. Introduction

For over 40 years, the eukaryotic fungi were the only known producers of naturally occurring β-lactam antibiotics. By a chance observation in 1928, Fleming discovered penicillin as a product of the common mold, *Penicillium notatum*. From a number of related structures produced by the molds, benzyl penicillin emerged as the antibiotic compound of choice and became generally available for the treatment of infectious diseases by the late 1940s. With the identification and isolation of the penicillin nucleus (6-aminopenicillanic acid) in the late 1950s it was possible for a family of semisynthetic penicillins to emerge. The general structure of penicillin along with the other currently known classes of β-lactam antibiotics is shown in Fig. 1.

The discovery of cephalosporin C from a species of *Cephalosporium acremonium* in 1955 (Newton and Abraham) was the culmination of work started by Brotzu in 1947. From cephalosporin C, the chemically derived nucleus (7-aminocephalosporanic acid) opened the way to the development of the semisynthetic cephalosporins.

By the early 1970s, when these two major classes of clinically useful antibiotics had been well established, novel β-lactam molecules began to make their appearance. However, unlike the penicillins and cephalosporins, these new compounds were not the products of fungi but were secondary metabolites of the prokaryotic streptomycetes. The first compounds to be identified were the cephamycins (Nagarajan *et al.*, 1971), similar in structure

ADVANCES IN APPLIED MICROBIOLOGY, VOLUME 31

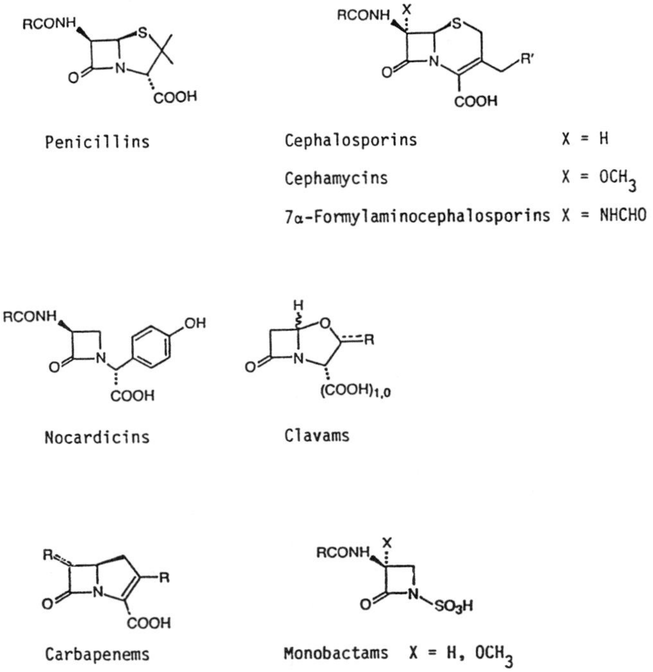

Cephalosporins	X = H
Cephamycins	X = OCH_3
7α-Formylaminocephalosporins	X = NHCHO

Monobactams	X = H, OCH_3

FIG. 1. β-Lactam structural types.

to the cephalosporins but with a methoxyl substituent in the 7α-position on the β-lactam ring. These streptomycete-produced molecules made a dramatic impact on β-lactam development in the early 1970s. From the cephamycins came the knowledge that cephalosporin-like molecules could be stabilized to the action of gram-negative β-lactamases while retaining activity against gram-negative bacteria. From these natural molecules came the semisynthetic cefoxitin, cefmetazole, and cefotetan, broad spectrum antibiotics showing a high degree of stability to hydrolysis by β-lactamases.

A novel series of monocyclic β-lactams, the nocardicins, was identified by workers at Fujisawa from a strain of *Nocardia uniformis* (Aoki *et al.*, 1976). Although of great academic interest, these molecules exhibited no antibacterial activity of practical significance (Kurita *et al.*, 1976). A large series of semisynthetic derivatives failed to improve on the poor biological activities of the nocardicin molecules.

One of the most significant developments of natural product research during the 1970s was the discovery of β-lactam-containing molecules possessing the ability to irreversibly inhibit β-lactamases. Among the first to emerge was clavulanic acid, a product of *Streptomyces clavuligerus* (Howarth *et al.*, 1976). The successful development of this molecule has led to the marketed product Augmentin, a combination of amoxicillin and clavulanic acid. The molecular structure of clavulanic acid is unique among the naturally occurring bicyclic β-lactams in having an oxygen in place of sulfur in the five-membered ring. The chemical manipulation of clavulanic acid has been pursued in depth by both Beecham and Glaxo, but to date there is no indication that compounds superior to clavulanic acid have emerged.

Within the class of naturally occurring β-lactamase inhibitors, by far the most common are the carbapenems. The carbapenems comprise a burgeoning series of molecules produced by streptomycetes and include olivanic acids (Brown *et al.*, 1976), thienamycins (Albers-Schonberg *et al.*, 1978), epithienamycins (Stapley *et al.*, 1977), PS compounds (Okamura *et al.*, 1979), asperenomycins (Tanabe *et al.*, 1982), and the carpetimycins (Nakayama *et al.*, 1981). Unlike the clavulanic acids, carbapenems exhibit potent broad spectrum antibacterial activity in addition to their inhibitory activity on the action of β-lactamases. Research and development of carbapenems has taken high priority in many pharmaceutical companies looking for potent broad spectrum agents. One compound, imipenem (*N*-formimidoyl thienamycin), has been studied extensively in the clinic, and there is every reason to believe that Merck will bring this compound to the market place in the near future.

Following the discovery of β-lactam antibiotics as the products of fungi and actinomycetes, the first reports of β-lactam production by bacteria came in 1981 (Imada *et al.*, 1981; Sykes *et al.*, 1981). These antibiotics are *N*-acyl derivatives of 3-amino-2-oxo-1-azetidinesulfonic acid (3-aminomonobactamic acid, Fig. 2) and thus constitute a structurally novel β-lactam class. These antibiotics, called monobactams (monocyclic β-lactam antibiotics produced by bacteria), have been found as products of six bacterial genera (Table I) and differ in the nature of the acyl substitutent and also in the presence or absence of a methoxyl group at the 3α-position (Fig. 1).

This structural type is unusual in that it contains a sulfamic acid function rarely found in nature. Most importantly, it supports the concept of an "activated" β-lactam (Sweet, 1972) demonstrating an elegant and simple way of achieving this activation. The electronegative sulfonic acid group attached

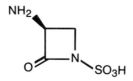

FIG. 2. 3-Aminomonobactamic acid.

TABLE I

SOURCES OF MONOBACTAMS

Monobactam	Structure	Producing organism
SQ 26,180	Figure 3	*Chromobacterium violaceum* ATCC 31,532
Sulfazecin	Figure 5	*Pseudomonas acidophila* G-6302
		Gluconobacter sp. ATCC 31,581
		G. oxidans subsp. *oxidans* ATCC 15,178 and ATCC 19,357
		G. oxidans subsp. *suboxidans* ATCC 19,441 and ATCC 23,773
		G. oxidans subsp. *industrius* ATCC 11,894
		Acetobacter sp. ATCC 21,760 and ATCC 21,780
		A. aceti subsp. *aceti* ATCC 15,973
		A. aceti subsp. *liquifaciens* ATCC 23,751
		A. peroxidans ATCC 12,874
		A. pasturianus subsp. *pasturianus* ATCC 6033
Isosulfazecin	Figure 5	*Pseudomonas mesoacidophila* SB-72310
EM5400 group	Figure 6	*Agrobacterium radiobacter* SC 11,742[a]
SQ 28,332	Figure 7	*Flexibacter* sp. ATCC 35,208
SQ 28,502	Unknown	*Flexibacter* sp. ATCC 35,103
SQ 28,503	Unknown	*Flexibacter* sp. ATCC 35,103

[a] Squibb Culture number.

at the β-lactam nitrogen simultaneously facilitates nucleophilic attack at the β-lactam carbonyl (Proctor *et al.*, 1982) while providing an anionic site which, from its ubiquity among β-lactam antibiotics, appears to be a prerequisite for binding with the target enzymes.

With a new focus on bacteria as producers of novel β-lactams, it was soon realized that these organisms are as rich a source of novel compounds as are actinomycetes and fungi. Species of *Flavobacterium* and *Xanthomonas* have been shown to produce desacetoxycephalosporin C (Singh *et al.*, 1982), and a family of 7α-formylaminocephalosporins has been reported from *Flavobacterium* fermentations (Singh *et al.*, 1984; Shoji *et al.*, 1984). Both cephalosporins and 7α-formylaminocephalosporins are produced by *Lysobacter* and *Xanthomonas* species (Ono *et al.*, 1984). Strains of *Erwinia* and *Serratia* have been shown to produce a carbapenem (Parker *et al.*, 1982b). Tabtoxin, a phytotoxin containing a β-lactam structure (Stewart, 1971), had previously been reported as a metabolite of *Pseudomonas tabaci*. This agent is toxic to a variety of organisms, including bacteria. However, it acts by a mechanism different from that of the β-lactam antibiotics.

In this article, we describe the detection, isolation, structure determination, and biological properties of the monobactams.

II. Detection of Monobactams

The rapid growth in the discovery of novel β-lactams can largely be accounted for by the employment of novel screening techniques. The first β-lactams were discovered because of the zones of inhibition they caused against susceptible wild-type bacteria. With the intensive efforts applied to this form of screening between 1940 and 1960, many of the agents produced in sufficient quantity to give a zone of killing were described by investigators in industry and elsewhere.

The growth in our understanding of the mode of action of β-lactam antibiotics and of their interaction with β-lactamases and the penicillin-binding proteins in the cell membrane led to different screening approaches. Cephamycins (7α-methoxycephalosporins), which are produced by streptomycetes (Nagarajan et al., 1971; Stapley et al., 1972), can be monitored by their morphological effects on sensitive test organisms. A screen was described by Brown et al. (1976) in which Klebsiella aerogenes became sensitive to penicillin when the β-lactamase produced by this organism was inhibited; use of this screen led to the detection of clavulanic acid and the olivanic acids.

The use of very sensitive mutants of Pseudomonas aeruginosa derived by successive rounds of mutation was described by Kitano et al. (1974) to screen for β-lactams produced by fungi and streptomycetes. Aoki et al. (1976) used a sensitive strain of Escherichia coli to detect a family of nocardicins produced by Nocardia uniformis subsp. tsuyamanensis. The use of an E. coli mutant lacking chromosomal β-lactamase and penicillin-binding protein 1B led to the discovery by Imada et al. (1981) of the monobactams sulfazecin and isosulfazecin in bacterial fermentations.

The screen used by Sykes et al. (1981), which led independently to the discovery of the monobactams, was a novel departure from earlier methods that depended on a visible zone of killing. The ability of Bacillus licheniformis to produce β-lactamase in the presence of β-lactams is the basis for this screen (Sykes and Wells, 1985). When B. licheniformis is grown in the presence of trace amounts of various β-lactams, β-lactamase is induced and secreted into the surrounding medium. If a chromogenic cephalosporin such as nitrocefin is now added, cleavage of the β-lactam ring occurs leading to a strongly colored product.

This method works only for intact β-lactam rings and is not dependent on intrinsic antibacterial activity. All classes of naturally occurring β-lactam antibiotics induce the β-lactamase activity, and the only non-β-lactams discovered to date that do so are the 3-acylamino-β-lactones (Sykes et al., 1982; Parker et al., 1982c; Wells et al., 1982c, 1984). The method, therefore, has marked specificity and sensitivity and has been used by Squibb workers to screen large numbers of bacteria. Ten of the 11 monobactams discussed in Section III were discovered using this method as well as a carbapenem produced by *Serratia* and *Erwinia* (Parker *et al.*, 1982b), bacterially produced cephalosporins (deacetoxycephalosporin C and 7α-formylaminocephalosporins) (Singh *et al.*, 1982, 1984), and the β-lactones referred to above.

III. Individual Monobactams

A. SQ 26,180

SQ 26,180, Fig. 3, the simplest of the naturally occurring monobactams, was isolated by workers at Squibb (Sykes *et al.*, 1981; Wells *et al.*, 1982a; Parker *et al.*, 1981, 1982a). The producing organism, a strain of *Chromobacterium violaceum*, was found in 1978 in a soil sample from the New Jersey Pine Barrens using the β-lactamase induction screen described in Section II. Although *C. violaceum* is widely dispersed in nature, SQ 26,180-producing strains have been found in only a limited number of habitats (Wells *et al.*, 1982c). SQ 26,180 is produced in sufficient quantity for isolation by culturing *C. violaceum* in standard complex media at pH 6.5 to 7 at 25°C with agitation.

Preliminary characterization of SQ 26,180 showed that it is a strongly acidic substance, having an electrophoretic mobility that is independent of pH from pH 7 to 2. The antibiotic is fairly stable at pH 5 but decomposes rapidly at pH 9 and 1. Isolation of the antibiotic was carried out as shown in Scheme 1.

The small size and simplicity of the antibiotic allowed straightforward deduction of the structure from the elemental analysis, spectroscopic characterization, and a few simple degradative reactions. The proposed structure, Fig. 3, was confirmed by a synthesis from 7-aminodesacetoxycephalosporanic acid (Fig. 4) that also established the R configuration at the 3-position of the azetidinone ring.

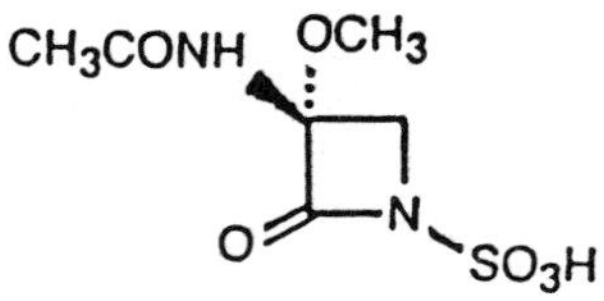

FIG. 3. SQ 26,180.

Chromobacterium violaceum ATCC 31,532 broth filtrate

(1) Ion-pair extraction at pH 5 into CH_2Cl_2 with cetyldimethylbenzylammonium chloride
(2) Back extraction into water with NaI
(3) Chromatography on Sephadex G-10 in aqueous methanol
(4) Chromatography on Whatman DE52 cellulose with a gradient of pH 5 sodium phosphate buffer or (for large-scale isolation) chromatography on Bio-Rad AG MP-1 resin with a NaI gradient in pH 5 phosphate buffer
(5) Chromatography on Sephadex LH-20 in water or Sephadex G-10 in water–methanol mixtures
(6) Chromatography on MCI GEL CHP20P resin eluting with water
(7) Conversion to the potassium salt on Dowex 50 (K^+) resin and crystallization from aqueous methanol

SQ 26,180

SCHEME 1. Isolation of SQ 26,180.

SQ 26,180 shows weak antibacterial activity (Table II). It is relatively stable to β-lactamase action, displaying no significant affinity for the penicillinase type β-lactamases, K1, TEM-2, and that produced by *Staphylococcus aureus*, nor to the β-lactamase found in the producing organism, *C. violaceum*. SQ 26,180 is a reversible competitive inhibitor of the Class I P-99 cephalosporinase produced by *Enterobacter cloacae* with a K_i of 80 μM. The compound shows good inhibitory activity against *Streptomyces* R61 DD-carboxypeptidase, having an I_{50} of 3 μM with no preincubation and 1 μM at 30 minutes of preincubation. As implied in the MIC data (Table II), binding to the essential penicillin-binding proteins (PBPs) of *E. coli,* and *S. aureus* is poor. However, it does bind PBP1a, PBP4 (DD-carboxypeptidase 1B) and PBP5/6 (DD-carboxypeptidase 1A) of *E. coli* and PBP1 (DD-carboxypeptidase) of *S. aureus*. SQ 26,180 induces irregular spheroplasts in *Proteus mirabilis*.

The antimicrobial activity of SQ 26,180 is clearly not strong enough to be of practical utility, but this antibiotic along with sulfazecin opened a new

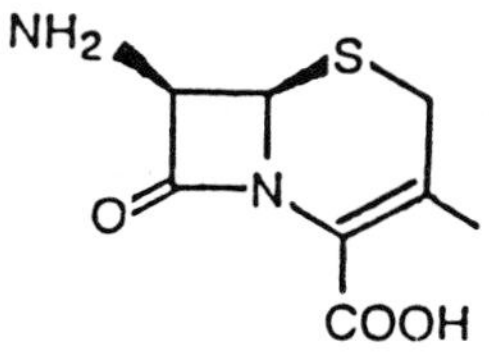

FIG. 4. 7-Aminodesacetoxycephalosporanic acid.

TABLE II

ANTIBACTERIAL ACTIVITY OF SQ 26,180[a]

Organism	SC number[b]	MIC (μg/ml)[c]
Staphylococcus aureus	1,276	50
Streptococcus faecalis	9,011	>100
Streptococcus agalactiae	9,287	12.5
Micrococcus luteus	2,495	25
Escherichia coli	8,294	>100
Escherichia coli	10,896	25
Klebsiella aerogenes	10,440	>100
Proteus mirabilis	3,855	>100
Enterobacter cloacae	8,236	>100
Pseudomonas aeruginosa	9,545	3.1
Pseudomonas aeruginosa	8,329	50

[a] Wells *et al.* (1982a).
[b] Squibb Culture number.
[c] Minimum inhibitory concentrations were determined by twofold agar dilution assay.

area of β-lactam research that has yielded clinically useful synthetic analogs as discussed in Section VI.

B. SULFAZECIN AND ISOSULFAZECIN

Sulfazecin and an epimer, isosulfazecin (Fig. 5), were reported in 1981 (Imada *et al.*, 1981; Kintaka *et al.*, 1981a,b) as bacterial fermentation products. These antibiotics were discovered by workers at Takeda Chemical Industries, Japan, as metabolites of acidophilic bacteria using a screen based on *Pseudomonas aeruginosa* PsC^ss^ (Kitano *et al.*, 1976, 1977) and *Escherichia coli* PG8, strains that are hypersensitive to β-lactam antibiotics. The β-lactam character of these antibiotics was also supported by morphological effects in *E. coli* LD-2 (elongation and bulge formation) and *Proteus mirabilis* ATCC 21,100 and by a slight lability to cephalosporinase.

Sulfazecin and isosulfazecin are produced by novel *Pseudomonas* species, *P. acidophila* strain G-6302 and *P. mesoacidophila* strain SB-72310, respectively. The isolation of sulfazecin (Asai *et al.*, 1981) is outlined in Scheme 2. Chemical and spectroscopic characterization established that sulfazecin is an acidic peptide having the empirical formula $C_{12}H_{20}N_4O_9S$ and containing sulfo, methoxyl, and β-lactam moeities. Acid hydrolysis yielded D-alanine, D-glutamic acid, and 2-oxo-3-sulfoaminopropionic acid. The structure (Fig. 5)

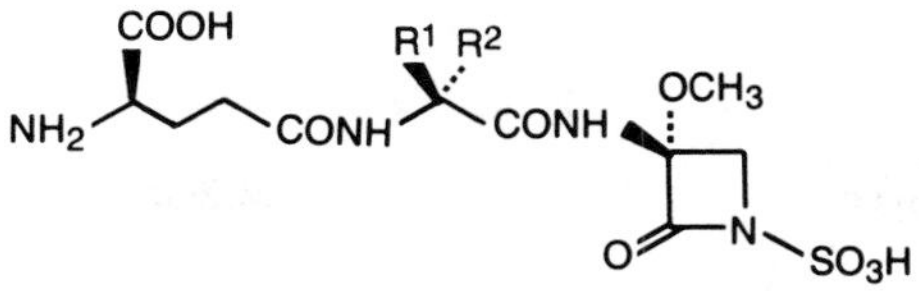

FIG. 5. Sulfazecin (R^1 = H, R^2 = CH_3) and isosulfazecin (R^1 = CH_3, R^2 = H).

was established by X-ray crystallography of the methanol solvate (Kamiya *et al.*, 1981).

Sulfazecin was discovered independently by workers at Squibb, United States (Sykes *et al.*, 1981; Liu *et al.*, 1982; Parker *et al.*, 1981, 1982a) using the *Bacillus licheniformis* β-lactamase induction screen described in Section II. This antibiotic (designated SQ 26,445 and EM5210) was produced by strains of *Gluconobacter* and *Acetobacter* and was the most commonly encountered monobactam (Wells *et al.*, 1982c).

Sulfazecin has moderate activity against some gram-negative organisms and weak activity against gram-positive organisms but is inactive against yeasts and fungi (Imada *et al.*, 1981; Sykes *et al.*, 1981; Kintaka *et al.*, 1981a). As noted above, sulfazecin shows detectable lability to cephalosporinase, but it is much less labile than benzylpenicillins, cephalosporin C, or cephamycin C (Kintaka *et al.*, 1981a). This relative stability to β-lactamase was ascribed by Imada *et al.* (1981) to the presence of the 3-methoxy group.

Although the activity of sulfazecins against gram-negative bacteria is quite modest, the antibiotic was efficacious *in vivo* against an *E. coli* infection in mice upon both subcutaneous and oral administration and was very nontoxic, having an LD_{50} > 10 g/kg by intravenous administration (Imada *et al.*, 1981; Kintaka *et al.*, 1981a).

Isosulfazecin was isolated from *Pseudomonas mesoacidophila* SB-72310 fermentations as shown in Scheme 3 (Kintaka *et al.*, 1981b). This antibiotic has a lower optical rotation than sulfazecin ($[a]_D$ +4.5° vs +94° in water) and

Pseudomonas acidophila G-6302 broth filtrate

(1) Sorption at pH 4 on charcoal and elution with 7% isobutanol
(2) Chromatography on Dowex 1 resin eluting with 1% NaCl
(3) Chromatography on charcoal as in step (1)
(4) Chromatography on DEAE Sephadex A-25 eluting with 0.1% NaCl in 0.05 *M*, pH 6, phosphate buffer
(5) Sorption on charcoal at pH 3.5 and elution with 50% aqueous acetone
(6) Crystallization from aqueous methanol

crystalline sulfazecin

SCHEME 2. Isolation of sulfazecin.

Pseudomonas mesoacidophila SB-72310 broth supernatant

> (1) Sorption on charcoal at pH 4.5 and elution with 50% aqueous acetone
> (2) Sorption on Dowex 1 resin and elution with 5% NaCl
> (3) Desalting on charcoal eluting with 20% aqueous methanol
> (4) Chromatography on DEAE Sephadex eluting with pH 6.6 phosphate buffer containing 0.5% NaCl
> (5) Sorption on charcoal at pH 3.2 and elution with aqueous methanol and aqueous acetone

isosulfazecin

SCHEME 3.   Isolation of isosulfazecin.

yields D-glutamic acid and L-alanine upon acid hydrolysis. The spectroscopic properties of isosulfazecin and sulfazecin are otherwise nearly identical and thus the structure shown in Fig. 5 was assigned. The antimicrobial activity of isosulfazecin is similar to, but weaker than the antimicrobial activity of sulfazecin.

C. EM5400

A mixture of β-lactamase-inducing substances, EM5400 (Fig. 6), is produced by *Agrobacterium radiobacter* (Wells *et al.*, 1982b). Producing strains are apparently quite rare, being found in only three locations (Wells *et al.*, 1982c). Like SQ 26,180, the EM5400 components were recognized as being strong acids by electrophoresis at pH 1.9 which showed three anionic β-lactamase-inducing zones.

Fermentation of the organism was carried out by standard procedures. From the culture broth, five components were ultimately isolated as outlined in Scheme 4 and their structures, Fig. 6, deduced from spectroscopic characterization (Parker and Rathnum, 1982). Other minor components of the same type were also observed but not fully characterized.

The structure of SQ 26,700, the first naturally occurring monobactam to lack a methoxyl group at the 3-position, was verified by synthesis from *N*-acetyl-D-tyrosine and (*S*)-3-aminomonobactamic acid (Fig. 2). SQ 26,812 and SQ 26,970 are derivatives of β-hydroxytyrosine, an unusual amino acid which had previously been reported only in the vancomycin group of antibiotics and in cutinase (Lin and Kolattukudy, 1980).

From antimicrobial data for the EM5400 compounds listed in Table III, it is apparent that they are very weak antibiotics, displaying some activity against gram-positive bacteria.

The 3α-methoxy compounds are quite resistant to hydrolysis by β-lac-

Agrobacterium radiobacter SC 11,742 broth supernatant

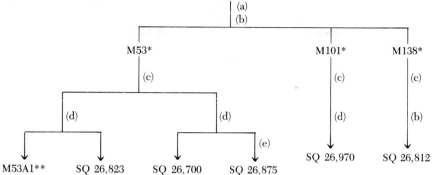

(a) Ion-pair extraction into methylene chloride with cetyldimethylbenzylammonium chloride and back-extraction into water with sodium thiocyanate
(b) Chromatography on Sephadex G-10 in aqueous methanol and in water
(c) Chromatography on QAE Sephadex with a $NaNO_3$ gradient
(d) Chromatography on MCI GEL CHP20P with water
(e) Chromatography of the tetrabutylammonium salt on silica gel with a gradient of methanol in dichloromethane

*Fractions with electrophoretic mobilities at pH 1.9 of 53, 101, and 138 relative to vitamin B_{12} (0) and *p*-nitrobenzenesulfonate anion (100).
**Not fully characterized.

SCHEME 4.   Isolation of EM5400 components.

tamases with the exception of SQ 26,823 which is hydrolyzed by the penicillinase of *Staphylococcus aureus* (Table IV). SQ 26,700, the nonmethoxylated compound, is hydrolyzed by all of the β-lactamases tested. Table IV shows that significant binding is only observed with the Class I cephalosporinase P-99 which is strongly inhibited by the methoxylated monobactams. The inhibition is readily reversed by dialysis of the enzyme–inhibitor complex.

The EM5400 monobactams were also tested for their ability to inhibit

	X	Y	Z
SQ 26,700	H	H	OH
SQ 26,812	OCH_3	OSO_3H	OSO_3H
SQ 26,823	OCH_3	H	H
SQ 26,875	OCH_3	H	OH
SQ 26,970	OCH_3	OH	OSO_3H

FIG. 6. EM5400 components.

TABLE III

ANTIBACTERIAL ACTIVITY OF THE EM5400 MONOBACTAMS[a]

Organism	SC number[c]	MIC (μg/ml)[b]				
		SQ 26,700	SQ 26,812	SQ 26,823	SQ 26,875	SQ 26,970
Staphylococcus aureus	1,276	25	>100	12.5	25	>100
Streptococcus faecalis	9,011	100	>100	>50	>100	>100
Streptococcus agalactiae	9,287	12.5	>100	25	50	>100
Micrococcus luteus	2,495	12.5	>100	12.5	100	>100
Escherichia coli	8,294	>100	>100	>50	>100	>100
Escherichia coli	10,896	>100	>100	25	6.3	25
Klebsiella aerogenes	10,440	>100	>100	>50	>100	>100
Proteus mirabilis	3,855	>100	>100	>50	>100	>100
Enterobacter cloacae	8,236	>100	>100	>50	>100	>100
Pseudomonas aeruginosa	9,545	>100	>100	50	50	>100
Pseudomonas aeruginosa	8,329	>100	>100	>50	>100	>100

[a] Wells et al. (1982b).

[b] Minimum inhibitory concentrations were determined by twofold agar dilution assay.

[c] Squibb Culture number.

DD-carboxypeptidase from *Streptomyces* R61 (Table V). All the compounds display good inhibitory activity; methoxylation (e.g., SQ 26,875) significantly increased inhibition and anionic charges in the side chain (e.g., SQ 26,812 and SQ 26,970) decreased activity.

D. SQ 28,332

SQ 28,332, Fig. 7, is a metabolite of *Flexibacter* sp. ATCC 35,208 (Singh *et al.*, 1983) that was discovered using the β-lactamase induction screen described in Section II. Of about 10^6 bacterial strains examined, this is the only producing strain found and thus appears to be quite rare. Like SQ 26,180 and the EM5400 antibiotics, SQ 28,332 is strongly acidic and remains anionic with no change in electrophoretic mobility between pH 7.0 and 2.3. The isolation of SQ 28,332 is outlined in Scheme 5.

SQ 28,332 is a water-soluble antibiotic with the empirical formula $C_{16}H_{26}N_6O_{12}S$ established by mass spectrometry and with infrared absorption (KBr) characteristic of β-lactam carbonyl (1760), carbamate (1715), secondary amide (1650), and sulfamate groups (1250 and 1030 cm^{-1}). The presence of a desmethoxy monobactam structure was evident from the ^{1}H and ^{13}C NMR spectra and from the production of L-2,3-diaminopropionic acid upon acid hydrolysis. Acid hydrolysis also yielded D-glyceric acid, D-alanine,

TABLE IV

INTERACTIONS OF β-LACTAMASES WITH THE EM5400 MONOBACTAMS[a]

β-Lactamase	Compound	Relative V_{max}	K_m (μM)	I_{50} (μM)[b]
S. aureus	Penicillin G	100	30	—
	SQ 26,700	12	980	—
	SQ 26,812	ND	ND	ND
	SQ 26,823	2	470	>500
	SQ 26,875	<0.02	—	>400
	SQ 26,970	<0.05	—	>400
TEM-2	Penicillin G	100	80	—
	SQ 26,700	7.1	280	—
	SQ 26,812	<0.02	—	>800
	SQ 26,823	<0.01	—	>50
	SQ 26,875	<0.01	—	>400
	SQ 26,970	<0.01	—	>400
K1	Penicillin G	100	170	—
	SQ 26,700	57	1280	—
	SQ 26,812	<0.03	—	>80
	SQ 26,823	<0.02	—	>500
	SQ 26,875	<0.01	—	>40
	SQ 26,970	<0.01	—	>40
P-99	Cephaloridine	100	580	—
	SQ 26,700	0.15	190	—
	SQ 26,812	<0.02	—	2
	SQ 26,823	<0.01	—	0.8
	SQ 26,875	<0.01	—	2
	SQ 26,970	<0.01	—	0.7

[a] Wells *et al.* (1982b).

[b] All studies were performed by spectrophotometric assay at 25°C and pH 7. I_{50} values were determined using 0.5 mM penicillin G as substrate for *S. aureus*, TEM-2, and K1 β-lactamases and 1.0 mM cephaloridine as the substrate for P-99 β-lactamase.

glycine, and N-methyl-L-serine. The arrangement of these residues in the C-3 side chain as shown in Fig. 7 was established by ^{1}H NOE experiments and by mass spectrometry of the product resulting from opening of the β-lactam ring with mild acid treatment. SQ 28,332 is the second desmethoxy monobactam to be isolated, and is the most complicated monobactam structure to be elucidated to date.

SQ 28,332 has weak activity against gram-positive bacteria and a trace of activity against a wall-permeable mutant of *E. coli*.

TABLE V

THE INHIBITION OF DD-CARBOXYPEPTIDASE
BY THE EM5400 MONOBACTAMS[a]

Monobactam	I_{50} (μM)[b]	
	$t = 0$ minutes	$t = 30$ minutes
SQ 26,700	37	2.2
SQ 26,812	200	33
SQ 26,823	3.5	1.0
SQ 26,875	2.4	0.48
SQ 26,970	37	22

[a] Wells *et al.* (1982b).

[b] Concentration causing 50% inhibition. Partially purified
Streptomyces R61 DD-carboxypeptidase was incubated at
30°C with the appropriate monobactam for 0 and 30 minutes. [^{14}C]Diacetyl-L-Lys-D-Ala-D-Ala was then added and
the extent of hydrolysis determined after 30 minutes.

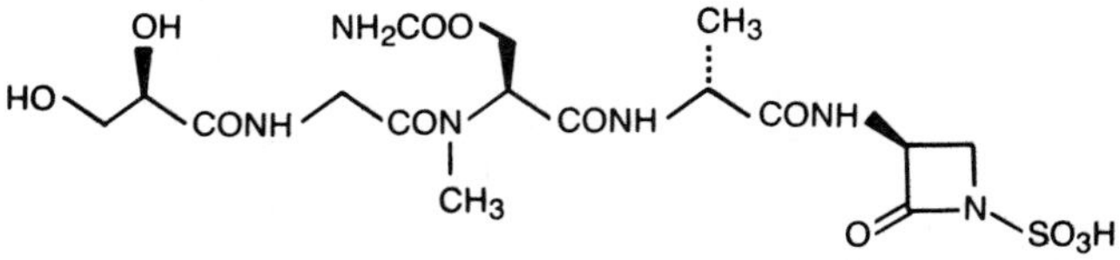

FIG. 7.   SQ 28,332.

Flexibacter sp. ATCC 35,208 broth supernatant

   (1) Ion-pair extraction into dichloromethane with cetyldimethylbenzylammonium
        chloride

   (2) Back extraction into water with sodium thiocyanate

   (3) Sorption on charcoal and elution with aqueous pyridine

   (4) Chromatography on MCI GEL CHP20P resin eluting with a water–methanol
        gradient

   (5) Chromatography on Bio-Rad AG1 resin with a pyridinium acetate gradient

   (6) Chromatography at pH 5 on cellulose with an acetonitrile–water gradient

SQ 28,332 Na salt

SCHEME 5.   Isolation of SQ 28,332.

E. SQ 28,502 AND SQ 28,503

Two β-lactamase-inducing substances are produced by *Flexibacter* sp. ATCC 35,103 (Cooper *et al.*, 1983). These metabolites are amphoteric peptides with isoelectric points near pH 7 and infrared absorption at 1760 cm^{-1} (KBr) characteristic of a β-lactam carbonyl group. They were isolated as shown in Scheme 6. Acid hydrolysis of each yields glycine, serine, isoleucine, methionine, arginine, glutamic acid, 2,3-diaminopropionic acid, sulfate, and an unidentified amino acid containing a *p*-hydroxyphenyl residue. Molecular weights indicated by mass spectrometry are 1462 for SQ 28,502 and 1446 for SQ 28,503. Because the monobactam residue constitutes a relatively small portion of these metabolites, the characteristics of this residue were somewhat less dominant in their spectra than in the spectra of their low-molecular-weight congeners. To aid in the recognition of the monobactam nucleus, a study of the infrared spectra, mass spectra, and nitrous acid cleavage of monobactams was undertaken (Cooper, 1983). This class of antibiotic shows fine structure in the infrared absorption in water between 1200 and 1350 cm^{-1}, neutral loss of SO_3 from the M + H quasimolecular ion in the positive FAB mass spectrum (Cohen *et al.*, 1982), and release of sulfate upon treatment of the opened β-lactam with HNO_2 (Feigl, 1966). The hydrolytic conditions necessary for opening the β-lactam can be used to distinguish desmethoxy from methoxymonobactams since the latter are distinctly more acid labile. SQ 28,502 and SQ 28,503 are both desmethoxy monobactams and clearly have much more complicated structures than any other antibiotics of this class reported to date. Their structures have not yet been determined.

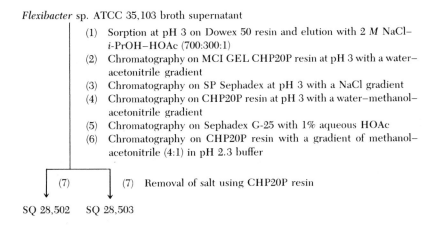

Flexibacter sp. ATCC 35,103 broth supernatant

(1) Sorption at pH 3 on Dowex 50 resin and elution with 2 *M* NaCl–*i*-PrOH–HOAc (700:300:1)
(2) Chromatography on MCI GEL CHP20P resin at pH 3 with a water–acetonitrile gradient
(3) Chromatography on SP Sephadex at pH 3 with a NaCl gradient
(4) Chromatography on CHP20P resin at pH 3 with a water–methanol–acetonitrile gradient
(5) Chromatography on Sephadex G-25 with 1% aqueous HOAc
(6) Chromatography on CHP20P resin with a gradient of methanol–acetonitrile (4:1) in pH 2.3 buffer

(7)   (7) Removal of salt using CHP20P resin

SQ 28,502   SQ 28,503

SCHEME 6.   Isolation of SQ 28,502 and SQ 28,503.

SQ 28,502 and SQ 28,503 have weak activity against both gram-positive and gram-negative bacteria. Both compounds are only weakly hydrolyzed by TEM-2 and K1 β-lactamases, the relative rates of hydrolysis being less than 0.02% of that observed with cephaloridine. Whereas neither enzyme is inhibited by the compounds at 150 μM concentration, P-99 β-lactamase is strongly inhibited by both. Inhibition of P-99 is time and concentration dependent. Second-order rate constants for inactivation are 3.1×10^4 liters/mol/minute for SQ 28,502 and 3.2×10^2 liters/mol/minute for SQ 28,503. When samples of totally inhibited P-99 β-lactamase (e.g., 2.5 mol of SQ 28,502 per mole of enzyme) are dialyzed for 30 hours at 25°C, no recovery of enzyme activity is achieved. These compounds are therefore potent irreversible inhibitors of the P-99 β-lactamase.

IV. Biosynthesis of Monobactams

The biosynthesis of monobactams in *C. violaceum* ATCC 31,532, *Acetobacter* sp. ATCC 21,780, and *A. radiobacter* ATCC 31,700 was studied by O'Sullivan *et al.* (1982, 1983). In defined minimal media, *C. violaceum* grew on glucose, trehalose, fructose, lactate, pyruvate, succinate, or malate as sole carbon source but SQ 26,180 was only detectable with glucose, trehalose, lactate, or pyruvate. The level of monobactam produced was about 2 to 10 μg/ml as compared with complex media where levels of about 20 μg/ml were produced.

Acetobacter spp. grew well with glucose, sucrose, glycerol, and ethanol as single carbon sources and produced up to 50 μg/ml on all of these with the exception of glycerol. Production of sulfazecin was not significantly improved by growth of *Acetobacter* on complex media.

A. radiobacter could grow on glucose, sucrose and glycerol but SQ 26,812 was only produced at detectible levels on glycerol.

It was clear from these studies that different pathways of carbon metabolism can lead to monobactam production in these bacteria. The pattern of monobactam production was also of interest in terms of biosynthetic regulation: monobactam was produced early in the logarithmic phase of growth for both *C. violaceum* and *Acetobacter;* however, in *A. radiobacter* monobactam production could not be detected until the stationary phase.

The form of regulation of a biosynthetic pathway is often revealed by the stimulation or inhibition of end product formation by key intermediates. It was of interest, therefore, to see if monobactam production could be significantly stimulated or inhibited by amino acids from which such compounds might be expected to arise. O'Sullivan *et al.* (1982) found that no evidence of stimulation or inhibition unrelated to growth could be demonstrated by the addition of single L- or D-amino acids to the growing cultures. It was postulated that this may be due to rapid interconversion and turnover of the small intracellular amino acid pools in rapidly growing gram-negative bacteria. This conclusion was supported by studies with radioactively labeled sugars and amino acids; scrambling of label was extensive and incorporation of radioactivity into monobactam from numerous sources was found.

In order to study the biosynthetic origin of the carbon atoms in the β-lactam ring of monobactams, short-term incorporation studies were designed. ^{14}C-Labeled amino acids were added to small samples of logarithmically growing cells and aliquots were removed at intervals up to about 1 hour. These samples were rapidly centrifuged and the supernatants analyzed by high-voltage electrophoresis followed by fluorography (Laskey and Mills, 1975) or by direct counting. In this way it was possible to determine when the radioactive amino acid disappeared from the medium and when radioactively labelled monobactam appeared.

In general, uptake of radiolabel was rapid and usually complete within 5 to 10 minutes after addition. Incorporation of radiolabel into monobactam, however, was not extensive, rarely accounting for more than 1% of the label added. It was possible, nonetheless, to establish a slightly preferential incorporation of radiolabel into SQ 26,180 from the amino acids, L-serine, L-cysteine, and glycine (Table VI). When a fivefold molar excess of unlabeled L-cysteine was added together with labeled L-serine, a stimulation effect on L-[U-^{14}C]serine incorporation was observed. Conversely, when unlabeled L-

TABLE VI

INCORPORATION OF AMINO ACIDS INTO SQ 26,180 BY *C. violaceum*[a]

	Uptake (%)	Incorporation into SQ 26,180 (%)	Relative incorporation
L-[U-^{14}C]Serine	>98	0.28	(1.0)
L-[U-^{14}C]Cysteine	90	0.21	0.8
L-[U-^{14}C]Alanine	>98	0.10	0.4
L-[U-^{14}C]Aspartate	>98	0.08	0.3
[U-^{14}C]Glycine	>98	0.37	1.3
[1-^{14}C]Glycine	>98	0.21	0.8
[2-^{14}C]Glycine	>98	0.51	1.8
L-[U-^{14}C]Serine + L-cysteine	>98	0.67	2.4
L-[U-^{14}C]Cysteine + L-serine	90	0.14	0.5
[U-^{14}C]Glycine + L-serine	>98	0.15	0.5

[a] O'Sullivan *et al.* (1982). One microcurie of labeled amino acid (50–60 μM) and 300 μM of unlabeled amino acid were added and uptake was determined at 20 minutes. The relative incorporation value of 1.0 was assigned to L-[U-^{14}C]serine.

serine was added together with labeled L-cysteine, incorporation of label was reduced over that with L-[U-^{14}C]cysteine alone (Table VI) (O'Sullivan *et al.*, 1982). These results were taken to indicate that L-serine was being incorporated into the β-lactam ring of SQ 26,180 and L-cysteine stimulated this incorporation by sparing L-serine from metabolism into other compounds.

Proof of the serine origin of the carbon atoms of the β-lactam ring was provided in another experiment in which mixtures of L-[U-^{14}C]serine and L-[3-^{3}H]serine or L-[3-^{14}C]serine and L-[3-^{3}H]serine were added to *C. violaceum* cells and the SQ 26,180 produced was analyzed for incorporation of both labels. It was found that the ratio of ^{3}H to ^{14}C in the electrophoretically separated SQ 26,180 was the same as that in the starting amino acid mixture (Table VII).

In contrast to the serine experiment, a similar approach using L-[U-^{14}C] cystine and L-[3,3'-^{3}H]cystine showed extensive loss of the ^{3}H in the monobactam produced. These experiments showed that ^{3}H at the C-3 position of serine was retained upon incorporation into SQ 26,180 by *C. violaceum* and that cysteine was metabolized prior to incorporation into the β-lactam ring.

The high incorporation of glycine (Table VI) into SQ 26,180 was explained by conversion into serine. [2-^{14}C]Glycine was better incorporated into SQ 26,180 than [1-^{14}C]glycine, and this was probably because C-2 of glycine is

TABLE VII

RETENTION OF ^{3}H-C3 ATOMS FROM SERINE AND CYSTINE
UPON INCORPORATION INTO MONOBACTAMS[a]

Precursor mixture	Starting ratio of ^{3}H to ^{14}C	Ratio of ^{3}H to ^{14}C in product	Monobactam produced
L-[U-^{14}C]Serine L-[3-^{3}H]Serine	7.4	7.6	SQ 26,180
L-[3-^{14}C]Serine L-[3-^{3}H]Serine	7.8	6.5	SQ 26,180
L-[U-^{14}C]Cystine L-[3,3'-^{3}H]Cystine	6.7	0.9	SQ 26,180
L-[3-^{14}C]Serine L-[3-^{3}H]Serine	10.0	10.8	Sulfazecin
L-[3-^{14}C]Serine L-[3-^{3}H]Serine	7.5	7.3	SQ 26,812

[a] O'Sullivan *et al.* (1982); conditions were the same as for Table VI.

more efficiently metabolized in the single carbon pool as methylenetetrahy-drofolate than the C-1 of glycine (Sagers and Gunsalus, 1961). Thus, the methylenetetrahydrofolate can condense with glycine to give serine (Huen-nekens and Osborn, 1959), and therefore incorporation from [2-^{14}C]glycine should be at least twice that of [1-^{14}C]glycine.

The labeled glycine incorporation data made the possibility of a condensa-tion of two glycine units to form the β-lactam ring unlikely and this leaves the origin of the nitrogen atom of the β-lactam ring still in question.

Using a mixture of ^{14}C- and ^{3}H-labeled serine, the origin of the carbon atoms in the β-lactam rings of the monobactams produced by *Acetobacter* and *A. radiobacter* was studied. Since both these monobactams (sulfazecin and SQ 26,812) have larger N-acyl groups, time course experiments were undertaken. It was found that serine was incorporated into sulfazecin by *Acetobacter* with excellent retention of the ^{3}H at the C-3 position. In *A. radiobacter* there was also sufficiently good retention of [3-^{3}H]serine in the SQ 26,812 produced to suggest that here too, the carbon atoms of the β-lactam ring were derived from serine (Table VII).

A similar conclusion regarding the origin of the carbon atoms of the β-lactam ring of sulfazecin was reached by workers at Takeda in Japan, using ^{13}C NMR studies (A. Imada, presented at the Twenty-first Interscience Conference on Antimicrobial Agents and Chemotherapy, Chicago, 1981).

The origin of the methyl moiety of the methoxyl group of sulfazecin was studied using L-[*methyl*-^{14}C]methionine (O'Sullivan *et al.*, 1982). Labeled methionine was added to a growing culture of *Acetobacter* sp. and the monobactam was recovered after 4 hours incubation. After purification by chromatography and electrophoresis, the label in the 3α-methoxyl group was determined. About 87% of the total label in sulfazecin was recovered from the methoxyl group, indicating that the ^{14}CH$_3$ was derived from meth-ionine; this was in agreement with earlier studies (O'Sullivan and Abraham, 1980) on the origin of the methoxy methyl group of cephamycin C.

In a study to establish the origin of the sulfamate sulfur of monobactams it was found that each of the three organisms studied, *C. violaceum*, *Acetobacter* and *A. radiobacter*, behaved differently in terms of sulfur uti-lization (O'Sullivan *et al.*, 1983). With *Acetobacter* sp. a wide range of organic and inorganic sulfur sources could act as precursor of the monobac-tam sulfur. With *C. violaceum* the list was somewhat more restricted, and in *A. radiobacter* only inorganic sulfur was incorporated into monobactam. If a common mechanism of sulfur donation exists, then only inorganic sulfur satisfies this requirement, and a high-energy form of sulfur such as phos-phoadenosyl phosphosulfate might well be involved as a sulfur donor prior to ring closure. The question of whether the sulfur was added before or after monobactam ring formation was not unambiguously resolved. However, the existence of a sulfated peptide in the sulfazecin producer, *Acetobacter*, would suggest sulfation prior to ring closure. The peptide contained alanine and N-terminal glutamate and incorporated serine with retention of the hydrogen at the C-3 position. It also contained sulfur and the mass spectral and electrophoretic data suggested a molecule with the properties one would expect of a substrate for a monobactam synthase. However, it was not established that this compound could be converted to sulfazecin by cell free extracts of *Acetobacter*, and its status in the biosynthetic pathway is therefore in question.

The variation in the N-acyl side chains of monobactams prompts a number of questions about the relatedness of the biosynthetic pathway in different organisms. Studies *in vitro* failed to indicate the existence of an acylase capable of modifying the side chain, such as exists in *Penicillium chrysogenum*. Monobactam producers always produced the same compound, and none of the known producers gave rise to more than one class of side chain. This suggests that the substrate for the ring closing enzyme may be different in each organism, and only the serine origin of the β-lactam carbon atoms and possibly the sulfur source are common factors. Further work is necessary before these issues can be resolved.

In a later study, O'Sullivan and Aklonis (1984) showed that penicillin G acylase could deacylate synthetic monobactams with nonpolar N-acyl groups such as aminothiazole acetic acid, providing that the α-carbon of the side chain was unsubstituted. They also reported that none of the naturally produced monobactams acted as substrates for this enzyme.

V. Bacterially Produced Agents That Synergize with β-Lactams

It is known that binding of a β-lactam to a given penicillin-binding protein (PBP) in the cell membrane correlated well with certain morphological features. Thus agents binding to PBP3 result in filamentous growth, those binding to PBP2 result in ovoid cells, and when both PBP3 and PBP2 are bound, bulge formation takes place (Spratt, 1980).

Recently agents have been described from the sulfazecin producer *Pseudomonas acidophila*, the isosulfazecin producer *P. mesoacidophila* (Imada *et al.* 1982; Shinagawa *et al.*, 1984), and from the monobactam producer *Chromobacterium violaceum* (Cooper *et al.*, 1985) that induce bulge formation in the presence of certain β-lactam antibiotics. Bulgecins were first reported from cultures of *P. acidophila* (Imada *et al.*, 1982) where it was observed that the culture filtrate caused bulge formation in *E. coli* and *Proteus mirabilis*. However, when sulfazecin was isolated from the filtrate in pure form it was

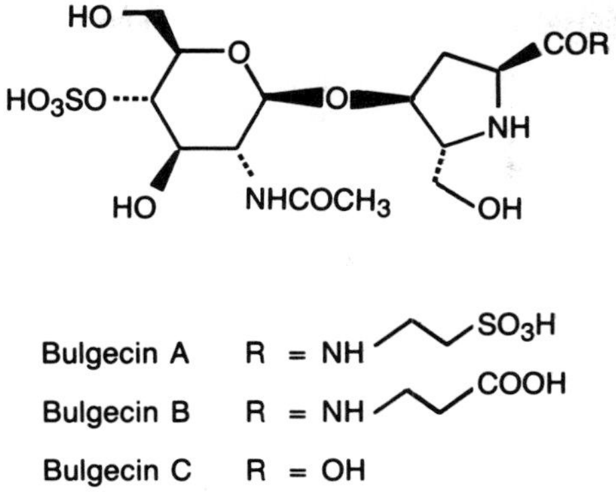

FIG. 8. Bulgecins.

found not to cause bulge formation. Further isolation studies resulted in the identification of a glycopeptide that when added together with sulfazecin or other β-lactams resulted in the formation of bulges. Cefmenoxime shows affinity for PBP3 (Tsuchiya *et al.*, 1981) and mecillinam for PBP2 (Spratt, 1980) and together they give rise to bulge formation (Otsuki, 1981). When bulgecin is added instead of mecillinam, the combination results in bulge formation, whereas cefmenoxime alone results in filamentation and bulgecin alone has no effect.

The shape, size, and rate of bulge formation was not the same for bulgecin as for mecillinam, indicating a different mechanism of action, and this was borne out by the lack of affinity of bulgecin for PBP2 (Imada *et al.*, 1982). More recently the structures of three bulgecins have been determined, Fig. 8 (Shinagawa *et al.*, 1984). All three contain D-glucosamine and a novel proline derivative; bulgecin A, additionally, has a taurine residue and bulgecin B has a β-alanine residue.

C. violaceum has been shown to produce two glycopeptides, SQ 28,504 and SQ 28,546, that synergize with β-lactam antibiotics. Molecular weights of 963 and 1179 are indicated by their mass spectra and thus they are substantially larger than the bulgecins. Complete structures have not yet been determined, but a relationship to the bulgecins is suggested by their spectra and by analysis of their hydrolysates (Cooper *et al.*, 1985).

VI. Conclusions

The naturally occurring monobactams discovered to date are uniformly poor antibiotics that afford in themselves no opportunity for clinical utility.

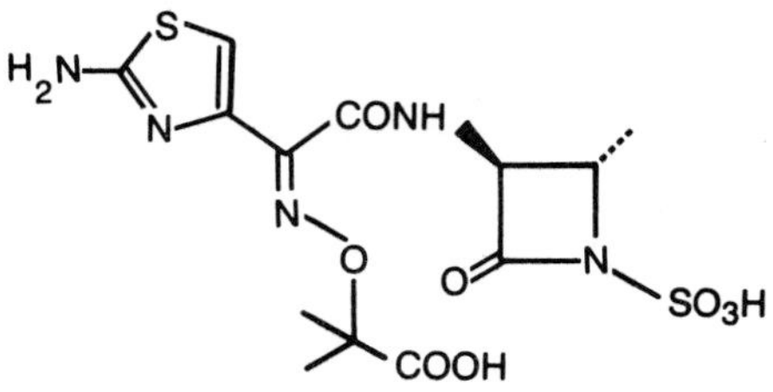

FIG. 9. Aztreonam.

However, the monobactam nucleus and analogs with a variety of substituents in the 1-, 3-, and 4-positions of the azetidinone ring are readily accessible by synthesis. This highly active area of research has been the subject of recent reviews (Cimarusti and Sykes, 1984; Koster *et al.*, 1982). From this work a number of synthetic analogs that have excellent antimicrobial activity *in vivo* have been developed. As of this writing, aztreonam (Fig. 9), the first synthetic monobactam to undergo clinical trials, has entered the market, and a second synthetic monobactam, AMA 1080 (Fig. 10), is presently undergoing clinical trials by Takeda/Roche. Both of these antimicrobial agents are useful for treating gram-negative bacterial infections by parenteral administration. An orally active prodrug, SQ 82,531 (Fig. 11), has recently been described (Breuer *et al.*, 1984; Koster *et al.*, 1984; Tanaka *et al.*, 1984; Jules *et al.*, 1984; Clark *et al.*, 1984) and is presently undergoing clinical evaluation. This agent illustrates an alternative mode of providing activation of the β-lactam and (after enzymatic hydrolysis to SQ 82,291) an anionic binding site. Because of the great synthetic flexibility inherent in the monobactam system, it is probable that further clinically useful antimicrobial agents will follow.

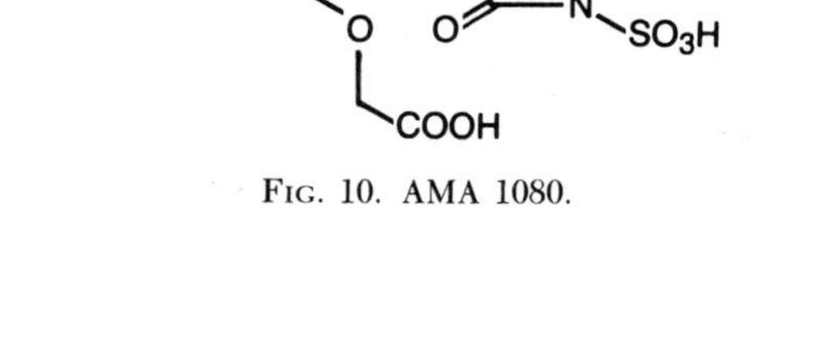

FIG. 10. AMA 1080.

FIG. 11. SQ 82,291 (R = H) and SQ 82,531 (R = CH$_2$COO-t-Bu).

REFERENCES

Albers-Schönberg, G., Arison, B. H., Hensens, O. D., Hirshfield, J., Hoogsteen, K., Kaczka, E. A., Rhodes, R. E., Kahan, J. S., Kahan, F. M., Ratcliffe, R. W., Walton, E., Ruswinkle, L. J., Morin, R. B., and Christensen, B. G. (1978). *J. Am. Chem. Soc.* **100**, 6491–6499.

Aoki, H., Sakai, H., Kohsaka, M., Konomi, T., Hosoda, J., Kubochi, Y., Iguchi, E., and Imanaka, H. (1976). *J. Antibiot.* **29**, 492–500.

Asai, M., Haibara, K., Muroi, M., Kintaka, K., and Kishi, T. (1981). *J. Antibiot.* **34**, 621–627.

Breuer, H., Straub, H., Treuner, U. D., Drossard, J.-M. Höhn, H., and Lindner, K. R. (1984). *Intersci. Conf. Antimicrob. Agents Chemother.* Abstract 135.

Brown, A. G., Butterworth, D., Cole, M., Hanscomb, G., Hood, J. D., Reading, C., and Rolinson, G. N. (1976). *J. Antibiot.* **29**, 668–669.

Cimarusti, C. M., and Sykes, R. B. (1984). *Med. Res. Rev.* **4**, 1–24.

Clark, J. M., Weinberg, D. L., Olsen, S. J., Bonner, D. P., and Sykes, R. B. (1984). *Intersci. Conf. Antimicrob. Agents Chemother.* Abstract 139.

Cohen, A. I., Funke, P. T., and Green, B. N. (1982). *J. Pharm. Sci.* **71**, 1065–1066.

Cooper, R. (1983). *J. Antibiot.* **36**, 1258–1262.

Cooper, R., Bush, K., Principe, P. A., Trejo, W. H., Wells, J. S., and Sykes, R. B. (1983). *J. Antibiot.* **36**, 1252–1257.

Cooper, R., Wells, J. S., and Sykes, R. B. (1985). *J. Antibiot.* **38**, 449–454.

Feigl, F. (1966). "Spot Tests in Organic Analysis," 7th Ed., p. 306. Elsevier, Amsterdam.

Howarth, T. T., Brown, A. G., and King, T. J. (1976). *J. Chem. Soc. Chem. Commun.* 266–267.

Huennekens, F. M., and Osborn, M. J. (1959). *In* "Advances in Enzymology" (F. F. Nord, ed.), Vol. XXI, pp. 369–446. Wiley (Interscience), New York.

Imada, A., Kitano, K., Kintaka, K., Muroi, M., and Asai, M. (1981). *Nature (London)* **289**, 590–591.

Imada, A., Kintaka, K., Nakao, M., and Shinagawa, S. (1982). *J. Antibiot.* **35**, 1400–1403.

Jules, K., Chin, N. X., Labthavikul, P., and Neu, H. C. (1984). *Intersci. Conf. Antimicrob. Agents Chemother.* Abstract 138.

Kamiya, K., Takamoto, M., Wada, Y., and Asai, M. (1981). *Acta Crystallogr.* **B37**, 1626–1628.

Kintaka, K., Kitano, K., Nozaki, Y., Kawashima, F., Imada, A., Nakao, Y., and Yoneda, M. (1981a). *J. Ferment. Technol.* **59**, 263–268.

Kintaka, K., Haibara, K., Asai, M., and Imada, A. (1981b). *J. Antibiot.* **34**, 1081–1089.

Kitano, K., Kintaka, K., Suzuki, S., Katamoto, K., Nara, K., and Nakao, Y. (1974). *Agric. Biol. Chem.* **38**, 1761–1762.

Kitano, K., Kintaka, K., and Nakao, Y. (1976). *J. Ferment. Technol.* **54**, 696–704.

Kitano, K., Nara, K., and Nakao, Y. (1977). *Jpn. J. Antibiot.* **30** (Suppl.), 239–245.

Koster, W. H., Cimarusti, C. M., and Sykes, R. B. (1982). *In* "The Chemistry and Biology of β-Lactam Antibiotics" (R. B. Morin and M. Gorman, eds.), Vol. 3, pp. 339–375. Academic Press, New York.

Koster, W. H., Bonner, D. P., Cimarusti, C. M., Parker, W. L., and Sykes, R. B. (1984). *Intersci. Conf. Antimicrob. Agents Chemother.* Abstract 136.

Kurita, M., Jomon, K., Komori, T., Miyairi, N., Aoki, H., Kuge, S., Kamiya, T. and Imanaka, H. (1976). *J. Antibiot.* **29**, 1243–1245.

Laskey, R. A., and Mills, A. D. (1975). *Eur. J. Biochem.* **56**, 335–341.

Lin, T. S., and Kolattukudy, P. E. (1980). *Eur. J. Biochem.* **106**, 341–351.

Liu, W. C., Parker, W. L., Wells, J. S., Principe, P. A., Trejo, W. H., Bonner, D. P., and Sykes, R. B. (1982). *Curr. Chemother. Immunother., Proc. 12th Int. Congr. Chemother., 1981* pp. 328–329.

Nagarajan, R., Boeck, L. D., Gorman, M., Hamill, R. L., Higgens, C. E., Hoehn, M. M., Stark, W. M., and Whitney, J. G. (1971). *J. Am. Chem. Soc.* **93**, 2308–2310.

Nakayama, M., Kimura, S., Tanabe, S., Mizoguchi, T., Watanabe, I., and Mori, T. (1981). *J. Antibiot.* **34**, 818–823.

Newton, G. G. F., and Abraham, E. P. (1955). *Nature (London)* **175**, 548.

Okamura, K., Hirata, S., Koki, A., Hori, K., Shibamoto, N., Okumura, Y., Okabe, M., Okamoto, R., Kouno, K., Fukagawa, Y., Shimauchi, Y., Ishikura, T., and Lein, J. (1979). *J. Antibiot.* **32**, 262–271.

Ono, H., Nozaki, Y., Katayama, N., and Okazaki, H. (1984). *J. Antibiot.* **37**, 1528–1535.

O'Sullivan, J., and Abraham, E. P. (1980). *Biochem. J.* **186**, 613–616.

O'Sullivan, J., and Aklonis, C. A. (1984). *J. Antibiot.* **37**, 804–806.

O'Sullivan, J., Gillum, A. M., Aklonis, C. A., Souser, M. L., and Sykes, R. B. (1982). *Antimicrob. Agents Chemother.* **21**, 558–564.

O'Sullivan, J., Souser, M. L., Kao, C. C., and Aklonis, C. A. (1983). *Antimicrob. Agents Chemother.* **23**, 598–602.

Otsuki, M. (1981). *J. Antibiot.* **34**, 739–752.

Parker, W. L., and Rathnum, M. L. (1982). *J. Antibiot.* **35**, 300–305.

Parker, W. L., Cimarusti, C. M., Floyd, D. M., Koster, W. H., Liu, W. C., Principe, P. A., Rathnum, M. L., and Slusarchyk, W. A. (1981). *J. Antimicrob. Chemother.* **8**(Suppl. E), 17–20.

Parker, W. L., Koster, W. H., Cimarusti, C. M., Floyd, D. M., Liu, W. C., and Rathnum, M. L. (1982a). *J. Antibiot.* **35**, 189–195.

Parker, W. L., Rathnum, M. L., Wells, J. S., Trejo, W. H., Principe, P. A., and Sykes, R. B. (1982b). *J. Antibiot.* **35**, 653–660.

Parker, W. L., Rathnum, M. L., and Liu, W. C. (1982c). *J. Antibiot.* **35**, 900–902.

Proctor, P., Gensmantel, N. P., and Page, M. I. (1982). *J. Chem. Soc., Perkin Trans.* **2**, 1185–1192.

Sagers, R. D., and Gunsalus, I. C. (1961). *J. Bacteriol.* **81**, 541–549.

Shinagawa, S., Kasahara, F., Wada, Y., Harada, S., and Asai, M. (1984). *Tetrahedron* **40**, 3465–3470.

Shoji, J., Kato, T., Sakazaki, R., Nagata, W., Terui, Y., Nakagawa, Y., Shiro, M., Matsumoto, K., Hattori, T., Yoshida, T., and Kondo, E. (1984). *J. Antibiot.* **37**, 1486–1490.

Singh, P. D., Ward, P. C., Wells, J. S., Ricca, C. M., Trejo, W. H., Principe, P. A. and Sykes, R. B. (1982). *J. Antibiot.* **35**, 1397–1399.

Singh, P. D., Johnson, J. H., Ward, P. C., Wells, J. S., Trejo, W. H., and Sykes, R. B. (1983). *J. Antibiot.* **36**, 1245–1251.

Singh, P. D., Young, M. G., Johnson, J. H., Cimarusti, C. M., and Sykes, R. B. (1984). *J. Antibiot.* **37**, 773–780.

Spratt, B. G. (1980). *Philos. Trans. R. Soc. London Ser. B* **289**, 273–283.

Stapley, E. O., Jackson, M., Hernandez, S., Zimmerman, S. B., Currie, S. A., Mochales, S., Mata, J. M., Woodruff, H. B., and Hendlin, D. (1972). *Antimicrob. Agents Chemother.* **2**, 122–131.

Stapley, E. O., Cassidy, P., Currie, S. A., Daoust, D., Goegelman, R., Hernandez, S., Jackson, M., Mata, J. M., Miller, A. K., Monaghan, R. L., Tunac, J. B., Zimmerman, S. B., and Hendlin, D. (1977). *Intersci. Conf. Antimicrob. Agents Chemother.* Abstract 80.

Stewart, W. W. (1971). *Nature (London)* **229**, 174–178.

Sweet, R. M. (1972). *In* "Cephalosporins and Penicillins" (E. H. Flynn, ed.), pp. 280–309. Academic Press, New York.

Sykes, R. B., and Wells, J. S. (1985). *J. Antibiot.* **38**, 119–121.

Sykes, R. B., Cimarusti, C. M., Bonner, D. P., Bush, K., Floyd, D. M., Georgopapadakou, N. H., Koster, W. H., Liu, W. C., Parker, W. L., Principe, P. A., Rathnum, M. L., Slusarchyk, W. A., Trejo, W. H., and Wells, J. S. (1981). *Nature (London)* **291**, 489–491.

Sykes, R. B., Parker, W. L., and Wells, J. S. (1982). *In* "Trends in Antibiotic Research: Genetics, Biosynthesis, Actions and New Substances, Proceedings of an International Conference" (H. Umezawa, A. L. Demain, T. Hata, and C. R. Hutchinson, eds.), pp. 115–124. Jpn. Antibiot. Res. Assoc., Tokyo.

Tanabe, S., Okuchi, M., Nakayama, M., Kimura, S., Iwasaki, A., Mizoguchi, T., Murakami, A., Itoh, H., and Mori, T. (1982). *J. Antibiot.* **35**, 1237–1239.

Tanaka, S. K., Bonner, D. P., Schwind, R. A., Minassian, B. F., Lalama, L. M., and Sykes, R. B. (1984). *Intersci. Conf. Antimicrob. Agents Chemother.* Abstract 137.

Tsuchiya, K., Kondo, M., Kida, M., Nakao, M., Iwahi, T., Nishi, T., Noji, Y., Takeuchi, M., and Nozaki, Y. (1981). *Antimicrob. Agents Chemother.* **19**, 56–65.

Wells, J. S., Trejo, W. H., Principe, P. A., Bush, K., Georgopapadakou, N., Bonner, D. P., and Sykes, R. B. (1982a). *J. Antibiot.* **35**, 184–188.

Wells, J. S., Trejo, W. H., Principe, P. A., Bush, K., Georgopapadakou, N., Bonner, D. P., and Sykes, R. B. (1982b). *J. Antibiot.* **35**, 295–299.

Wells, J. S., Hunter, J. C., Astle, G. L., Sherwood, J. C., Ricca, C. M., Trejo, W. H., Bonner, D. P., and Sykes, R. B. (1982c). *J. Antibiot.* **35**, 814–821.

Wells, J. S., Trejo, W. H., Principe, P. A., and Sykes, R. B. (1984). *J. Antibiot.* **37**, 802–803.

New Frontiers in Applied Sediment Microbiology

Douglas Gunnison

Environmental Laboratory (WESES-A)
USAE Waterways Experiment Station
Vicksburg, Mississippi

I. Introduction

"Mud, the essential habitat of certain essential microorganisms, is just as important as water to the economy of the planet."

"Mud is not always and everywhere the same . . . some essential microorganisms require kinds of muddy water that are inimical to other essential kinds."

Deevey (1970)

Mud and the "essential microorganisms" mentioned in Deevey's treatise "In Defense of Mud" are, indeed, vital factors in the global ecology of the earth. From the anthropocentric province of the applied sediment microbiologist, an active interest in mud (henceforth taken as meaning aquatic sediments) and its microbial inhabitants stems not from sediment as an exotic habitat nor from some set of unique properties possessed by the sediment microflora. Rather, sediment and its attendant microorganisms collectively provide an assemblage of biological, chemical, and physical

207

ADVANCES IN APPLIED MICROBIOLOGY, VOLUME 31

properties necessary for the everyday business of mineral cycling and bio-degradation of man-made products.

Sediment itself is valuable as an interface—between solid and liquid, between aerobic and anaerobic, and between sorptive and desorptive. Sediment is home for many microorganisms (cf. Zobell, 1946; Rheinheimer, 1980), a temporary refuge for others (Burton, 1985), and a hostile battleground for still others (Zhukova, 1963; Gerba and Schaiberger, 1973). Depending upon prevailing hydrodynamic conditions, sediment may serve as a source or sink of microorganisms with respect to the ambient environment (LaLiberte and Grimes, 1982; Burton et al., 1985). Alternatively, microbial activities within sediment may, in large part, determine what chemical constituents are released into or depleted from the overlying water column (see for example, Brannon et al., 1985a; Gunnison et al., 1985). Thus, the activities of microorganisms can often serve as major determinants of water quality.

This article is based largely upon the author's professional experience working with the microbiology of flooded soils, sediments, and natural waters. This experience has validated Deevey's (1970) claim that "mud is not always and everywhere the same." In like manner, it is apparent that the microorganisms associated with sediment are also not always and everywhere the same, and the problems resulting from microbial processes occurring in sediment are, likewise, often site specific. Within a short period of time, the applied sediment microbiologist may encounter questions that encompass a wide range of problems. Typical examples in this regard are the following: How to deal with the survival and possible multiplication of a microbial pathogen in the sediment of a swimming area? Will placement of a dredged sediment contaminated with sewage into an upland disposal site result in the release of pathogenic viruses? Is the potential for methylation of mercury a real problem in the basin of a reservoir having water supply and fishing as two major activities? Will clearing a new reservoir before initial filling improve water quality of the releases?

In practical terms, many of the problems encountered by the applied sediment microbiologist actually fall into only one or two broad subject areas. The first area is biochemical ecology which, as defined by Alexander (1971), deals with biochemical explanations for ecological phenomena; this area may, within the context of this article, be interpreted as (1) the process of understanding and defining the biochemical mechanisms which underlie microbiological processes of environmental significance (principally biodegradation and cycling of nutrients and metals), and (2) elucidating those environmental factors responsible for accelerating or retarding these processes. Hidden beneath the umbrella of biochemical ecology is, of course, the tacit understanding that the applied sediment microbiologist has a knowledge/command of several other fields, either personally or through his association with other scientists. Many of these fields overlap; others do not. Among the fields from which the applied sediment microbiologist often must draw are microbial physiology, geomicrobiology and sediment geochemistry, inorganic and organic chemistry, biochemistry, natural product chemistry, and aquatic ecology.

The second general subject area into which most of the remaining problems fall is aquatic microbial ecology, the science concerned with interrelationships between aquatic microorganisms and their environment. This area requires a broad knowledge of microbiology, limnology, and oceanography, but also often draws upon various aspects of microbial physiology and biochemistry, and sediment chemistry. Of course, the preceding descriptions of knowledge required are not all inclusive and information available from other fields is also required.

Applied sediment microbiology differs from its sister science of wastewater microbiology, whose concern may be summarized as the mineralization of organic matter (lowering the biochemical oxygen demand), removal of minerals through the removal of cells growing on wastes, and inactivation of pathogenic microorganisms (for a review of this subject, see Taber, 1976). In contrast, applied sediment microbiology is concerned with (1) microbial processes actively or passively occurring in sediments and the ecological phenomena influencing these processes; (2) microbiological phenomena resulting from the disturbance of a sediment (i.e., dredging and disposal operations, prop wash from passing ships, resuspension and transport by waves and currents) and from man-influenced additions to sediment of synthetic and of abnormally large amounts of natural materials; (3) the processes accompanying the transformation of soil microflora into sediment microflora upon temporary or permanent flooding of previously dry soils as, for example, in the filling of a new reservoir; (4) the processes accompanying the changes in sediment microflora associated with placing a sediment into an upland situation; and (5) the factors associated with the persistence, multiplication, or decline of population of pathogenic microorganisms in aquatic sediments.

The increasing interest in studies on interactions between aquatic microorganisms in sediment and their environment and on the processes carried out by sediment microorganisms is reflected in a number of recent papers dealing with various aspects of this subject (see, for example, Collins, 1976; Marshall, 1980; Brannon et al., 1985a; Gunnison et al., 1985). Rather than to provide an additional review of the literature, this article attempts to present a conceptual view of microbial processes as they occur in flooded soils and sediments and to point out the areas of need for future studies in applied sediment microbiology. The article will also examine the pitfalls inherent in counterbalancing the reality of dealing with microbial processes occurring in sediments with the equally important reality of requirements mandated by law, the latter two phenomena often being mutually exclusive. In presenting this information, the author believes support will be given to the working premise for this article; namely that which is of most utility in applied sediment microbiology is based on two major components: (1) what is known about sediment microorganisms and the processes they carry out, and (2) our understanding of aquatic ecosystems and the manner in which they operate. It will become apparent to the reader that applied sediment microbiology is often as concerned with the process level as with the level of the individual microorganism.

II. General Concerns

A. Microorganisms as the Driving Force in Aquatic Sediments

The role of microorganisms as agents of oxidation and reduction in flooded soils and sediment has been reviewed recently (Brannon et al., 1985a; Gunnison et al., 1985) and will be only briefly summarized here. Through their oxidation and reduction activities, sediment microorganisms translate the various chemical constituents available to them into products that determine the overall oxidation–reduction status of the sediment itself. Sediment oxidation–reduction status, in turn, has a direct influence on the processes occurring within the sediment and on sediment exchanges with the overlying water column (see Mortimer, 1941, 1942; Ponnamperuma, 1972; Brannon et al., 1985a; Gunnison et al., 1985).

Heterotrophic and chemolitho(auto)trophic microorganisms derive metabolic energy for their activities from organic and from reduced or incompletely oxidized inorganic compounds, respectively; this process actively consumes molecular oxygen in the surface layer of a flooded soil or sediment (Takai et al., 1956; Takai and Kamura, 1966; Patrick and Mikkelsen, 1971). Providing that supplies of dissolved oxygen within the overlying water column are adequate, sufficient dissolved oxygen can diffuse into the sediment surface layer to permit the surface layer to remain oxidized. The exact depth of this surface layer is extremely dependent upon the chemical and biological oxygen demand of the sediment (see Gunnison et al., 1983). However, with the presence of abundant biologically available organic matter and sources of reduced or incompletely oxidized inorganic compounds, supplies of dissolved oxygen within a flooded soil or sediment matrix are rapidly depleted resulting in a decrease of aerobic microbial activity. In addition, supplies of biologically available organic matter may also be limited, depending upon seasonal productivity and/or external sources of particulate organic matter; this, in turn, leads to a decrease in heterotrophic microbial activity. The aerobic–anaerobic layer formed at the sediment surface has also been termed the "aerobic–anaerobic double layer" (Patrick and DeLaune, 1972; DeLaune *et al.*, 1976; Reddy *et al.*, 1980) and serves as the site of several important microbially mediated processes.

Anaerobic respiration is the microbial process wherein oxidized inorganic compounds are utilized as electron accepts coupled to energy-yielding oxidation of organic or inorganic compounds (Doetsch and Clark, 1973; Yoshida, 1975). Reduction of inorganic compounds occurs in a successive series of events, generally following thermodynamic predictions. Once nearly all dissolved oxygen has been exhausted, reduction of nitrate is initiated. After nitrate has been consumed, manganese begins to be reduced. This is followed by reduction of iron, sulfate, and then carbon dioxide, respectively. The reduction products formed have different forms and solubilities resulting in a variety of consequences.

Fermentation, the anaerobic conversion of organic compounds to organic acids and alcohols, is the principal process for utilization of organic matter under oxygen-free conditions in sediment. However, accumulation of organic acids to high levels within sediment does not normally occur, presumably due to gradual utilization of these compounds in the formation of hydrogen, carbon dioxide, and methane (Lovely and Klug, 1982; Gunnison *et al.*, 1985). Organic acids may also be consumed by aerobic heterotrophs as the acids diffuse upward into aerobic zones.

B. Sources of Energy and the Concept of Biologically Available Organic Substrates

Clearly, the utilization of organic substrates is determined by two major factors. Initially, there must be present a microorganism/assemblage of microorganisms which, individually or as a group, possess the enzymatic capacity to consume the substrate. Also of importance, however, is the nature of the environment required for microbial catabolism. The second factor will be considered later in this section.

This discussion is limited to a consideration of substrates other than the organic and reduced or incompletely oxidized inorganic chemicals indicated in the preceding section; attention here is focused on those naturally occurring and man-made organic substrates regarded as difficult or impossible to degrade (also known as refractory or recalcitrant compounds—see Alexander, 1965).

Within the context of natural products, plants have a marked capacity to form large amounts of insoluble organic polymers, including aromatic macromolecules, which are unique in that only microorganisms have the required enzymatic capacity for degradation (Evans, 1977). While microorganisms have developed a vast array of enzymes for the metabolism of natural products, the lignins and tannins remain among those materials having the slowest turnover rate of all the natural products (Dickinson and Pugh, 1974; Crawford, 1981). The degradation products of these two substances, together with aromatic compounds, amino acids, proteins, etc., which result from the degradation of microbial protoplasm and associated metabolites, are combined in a complex series of processes to form humus (Gjessing, 1976; Evans, 1977). Humus in soil is persistent, but is subject to continuous modification, with the exact composition varying according to the location (Gjessing, 1976). Aquatic humus consists of both humus derived from water-soluble and particulate-sorbed components of terrestrial origin plus a component which, at least in the case of marine humus, is known to be formed *in situ* from degradation products of plankton (Gjessing, 1976). Although soluble aquatic humus has been studied extensively (see Gjessing, 1976), little appears to be known about humus in aquatic sediments. Sorption of aquatic humus to clay followed by settling of the latter material to the bottom is a likely mechanism for the import of external sources of humus into sediment. The amounts, levels, and mechanisms of formation of sedimentary humus *in situ* are not presently well understood.

The man-made xenobiotic compounds (synthetic organic chemicals that are "foreign to life") are often biotoxic, readily bioaccumulated, and frequently move through food webs (Johnson, 1982). Compounds included within the xenobiotic category are very diverse, including substances such as cyclic intermediates, organic pigments and dyes, flavor and perfume substances, plastic and resin materials, medicinals, plasticizers and elastomers, rubber processing chemicals, surface-active agents, plus pesticides and related products (International Trade Commission, 1978). Many xenobiotics, along with several of the aromatic petroleum hydrocarbons and petroleum derivatives, fall into the difficult to degrade category, depending on the environmental circumstances.

Degradation of aromatic compounds, while sometimes difficult, can be accomplished by microorganisms via one or more alternative mechanisms. Evans (1977) has reviewed both the aerobic and anaerobic catabolism of aromatic structures, with particular attention given to anaerobic dissimilation. The pathways examined include anaerobic photometabolism of benzoate by the *Athiorhodaceae*, anaerobic metabolism of benzoate through nitrate respiration, and methanogenic fermentation of benzoate and other aromatics by a microbial consortium. Many xenobiotic compounds, particularly the pesticides, are chemically structured in a manner that interferes with metabolic processes. Their structures are often similar to key biochemical intermediates and thus are theoretically susceptible to attack by appropriate microorganisms. However, addition of substituents, such as chlorine, to various positions on an aromatic structure not only increases the toxicity, but also renders the molecule more difficult to attack enzymatically, as a consequence of the hindrance imposed by the substituent group. As a result, biodegradation of many synthetic chemical compounds must be initiated by removal of the substituent group (see Alexander, 1981).

Microorganisms do have various devices for dealing with refractory materials. Often, these mechanisms involve several members of the microbial community rather than individual organisms. Among the possible strategies listed by Evans (1977) are cometabolism (Horvath, 1972), enzyme induction, transfer of metabolic plasmids, mutation resulting in production of novel enzymes, and the formation of microbial consortia. Some investigators have suggested that it is possible to stimulate biodegradation of recalcitrant molecules, either through the application of an extra carbon source to support development of a population of decomposers (Chou and Bohonas, 1979; Clark *et al.*, 1979), or through facilitating cometabolism (Horvath, 1972), or by aiding induction of appropriate enzyme systems (Chou and Bohonos, 1979; Atlas, 1981). At the same time, several of the major factors that, in addition to structural hindrances, serve to inhibit biodegradation are also known, including competitive inhibition for growth factors and nutrients (Haller and Finn, 1978) and diauxy, the preferential metabolism by microorganisms of easily degradable compounds (Chou and Bohonos, 1979; Atlas, 1981).

The effects of the degradative activities on the biological activities of xenobiotics are broad in scope and vary widely depending upon the particular xenobiotic involved. Among the effects listed by Alexander (1977) are the following: (1) *detoxication*—the conversion of a molecule from a material that is inhibitory in the concentration normally applied to the environment into a product that is nontoxic; (2) *degradation*—often the same as mineralization; the transformation of a complex molecule into simple products; (3) *conjugation, complex formation,* or *addition reactions*—the process in which the complexity of a substrate is increased or in which an organism combines the substrate with cellular metabolites; (4) *activation*—the conversion of a nontoxic substrate or potential pesticide into a toxic substrate; and (5) *altering the pattern (spectrum) of activity*—a process wherein metabolism of a pesticide or other biotoxic compound yields a change in the range of the compound from the species originally intended as the target to another dissimilar species.

The nature of the environment has a direct influence on which of the catabolic pathways, as discussed above, is utilized in the degradation of aromatic compounds. Environmental factors involved in promoting, inhibiting, or preventing microbial degradation have been discussed in detail by several authors (Matsumura and Bousch, 1971; Woodcock, 1971; Yoshida, 1975; Alexander, 1977; Evans, 1977). Several of these factors are particularly relevant to microbial metabolism of these materials in aquatic sediment. The presence of aerobic or anaerobic conditions has special importance, as indicated by the preceding discussion, and aquatic sediment has an abundance of both types of habitats, including the aerobic–anaerobic double layer where the two habitats are situated in intimate proximity. The suitability of aquatic sediment for microbial survival and growth is an important concern with respect to degradative species, as is the presence of available nutrients, appropriate pH levels, and optimum temperatures (Evans, 1977).

Another environmental factor of critical importance in aquatic sediment is the amount and nature of sorptive surfaces available. Clearly, sorptive capacity will vary from one sediment to another; this depends largely upon the textural composition, particularly with regard to the nature of the clay complex and the organic matter content (Buckman and Brady, 1969; Marshall, 1980) as well as the grain size itself, which can account for more than 80% of the variation of bacterial numbers in sediments (Dale, 1974). The materials already sorbed by the sediment are also important determinants of available surfaces. Textural composition of a given sediment may also vary with time, depending upon (1) the influence of sedimentation processes depositing new layers upon existing layers of sediment; newly deposited layers may or may not be the same as previous layers according to prevailing hydrodynamic and anthropocentric conditions; (2) the effect of resuspension of existing sediments by wind and wave action followed by removal of fine particles by currents and resettling of heavier particles (winnowing); (3) compaction— the process of settling of newly deposited materials with resulting compression of underlying layers and intergradation of new and old deposits at the interface between the two layers; and (4) variations in the oxidation status of the sediment surface layer. Sorptive surfaces, in turn, influence (1) whether a microbial cell is surface bound or mobile within pore water; (2) whether the substrate being degraded is readily available in solution or is difficult to get at as a consequence of sorption; and (3) whether enzymes required for degradation, if of an extracellular nature, are free to move around or are immediately immobilized (see Alexander, 1971).

An additional environmental concern is salinity. While the freshwater and open ocean environments are relatively stable with respect to salinity, estuarine environments are more dynamic. A number of papers consider the effect of salinity on various aspects of water and sediment microbiology, with particular reference to survival (Greenberg, 1956; Orlob, 1956; Carlucci and Pramer, 1959; Mitchell, 1968; Faust et al., 1975).

In addition to the biological and physicochemical factors mentioned above, the nature and concentration of the substrate are also important. As Evans (1977) has indicated, dilution processes and time are often sufficient to ensure that the conditions required for the factors to operate are eventually attained; however, the concentrations and hazardous properties of the substrate may be such as to require removal of the substrate-contaminated sediment.

A final area of concern in degradation is the ecology of the degrading microorganisms. The degrading microorganisms may be unable, either because of physicochemical restrictions or because of relationships with other microorganisms (due to excretion of inhibitory compounds, competition for nutrients, or the inability to keep up with the remaining members of a consortium), to attain numbers sufficient to cause appreciable degradation, and in this event, the substrate will persist. In terms of persistence, then, the cause—either biological unavailability, lack of a suitable environment, or lack of appropriate numbers of degrading microbes—is often irrelevant, unless the situation can be altered to promote degradation. Various options to bring about such changes will be explored in the last section.

C. Relationship of Sediment Oxidation–Reduction Potential to Microbiological Processes

The relationship of sediment oxidation status to the processes occurring within the sediment was examined earlier. While the oxidation–reduction potential (E_h) attained in a given aquatic sediment is a reflection of the processes occurring within the sediment matrix, the relationship of that value to what is happening within the sediment is not simple. As an example, the role of oxygen in regulating the degree of oxidation–reduction remains vague, despite a plethora of research on this subject (Brannon et al., 1978). Various mechanisms have been proposed to explain the effect of O_2 on E_h, ranging from control of E_h by the O_2/H_2O_2 couple (Breck, 1974) to the function of the platinum electrode surface as an electrode that responds to pH (Whitfield, 1974). Apparently, E_h is not responsive to varying partial pressures of O_2 (Garrels and Christ, 1956; Whitfield, 1974). The lack of response has been attributed to various causes, including the absence of electroactive surfaces in high abundance and to the presence of mixed systems that are not in equilibrium among themselves (Morris and Stumm, 1967; Stumm and Morgan, 1970). In either event, oxygen is a potent agent capable of oxidizing many naturally occurring reduced chemical species, either directly or through supporting microbial activity. Thus, the presence of oxygen results in decreased levels of reduced species while at the same time causing an increase in levels of oxidized components. The consequence is an inhibition of the ability of redox couples to exert significant exchange currents in oxygenated systems. This amounts to an indirect control of E_h by O_2, but does not explain the influence of O_2 on E_h. Minute amounts of O_2 also have a toxic effect on obligately anaerobic bacteria, either killing them outright or else preventing their proliferation (Hungate, 1968; Friedovich, 1975). In either case, production of reducing substances that accompanies the metabolism and growth of these organisms is prevented.

The example illustrated by oxygen can be repeated for other chemical species, i.e., nitrate, sulfate, plus several others, and serves to illustrate the complex interrelationship between a given chemical species, the sediment environment, the microflora, and the oxidation–reduction status of the sediment system. The oxidation–reduction status of the sediment system is not necessarily static as is indicated by the previous discussion on stepwise reduction of inorganic electron acceptors. Rather, the status changes in space and time as dictated by the availability of electron donors and electron acceptors and by the microorganisms present. The oxidation–reduction status may remain stable at various depths below the sediment surface layer, but the surface layer itself is influenced both by the overlying water column and by materials diffusing upward from the underlying sediment layers. This can be a dynamic situation, depending upon the environment involved.

D. MICROBIAL CELLS AS PACKAGED PRODUCTS

The focus of attention now shifts briefly from the activities of microorganisms in aquatic sediment to the role of sediment as a source of microorganisms that are pathogenic to man or are food for benthic animals. To the extent that microorganisms are individually capable of causing illness or death, they can be considered as discrete packages of disease. These packages become significant in areas where they are concentrated or can multiply or become mobile because it is under these circumstances that large portions of the human population are placed at risk. To the extent that microorganisms serve as food for benthic animals, the microorganisms can also be considered as packages. In this case, however, the number of organisms required to do the job is considerably greater than the number required to initiate infection.

Much of the past research on pathogens has been concerned with microbiological aspects of water quality and with the survival of indicator species microorganisms which may or may not cause disease themselves, but which by their own presence indicate the potential presence of disease-causing agents. Sediments have recently been receiving an increasing share of research attention in this regard. Several authors have examined the survival and die-off of coliforms in marine waters and have concluded that die-off occurs rapidly and is influenced by such factors as limitation of nutrient supply, toxicity resulting from salinity changes, heavy metals, pH, osmotic changes, organic matter, competition by indigenous microflora, and predation (Allen *et al.*, 1953; Greenberg, 1956; (Orlob, 1956; Carlucci and Pramer, 1959; Mitchell, 1968; Boyd *et al.*, 1969; Faust *et al.*, 1975; Verstraete and Voets, 1976; Kapuscinski and Mitchell, 1983). In contrast, others have shown that *Escherichia coli* can multiply in sea water when suitable concentrations of nutrients or sewage are added Orlob, 1956; Won and Ross, 1973), or when *E. coli* is held in a dialysis tube containing raw sewage and sea water suspended in polluted bay water (Nusbaum and Garver, 1955). In the latter case, the dialysis tube prevented access of predators and high-molecular-weight substances to the coliforms, thus also preventing these factors from adversely affecting the growth of the coliforms. In other cases, the presence of fecal coliforms in water has been questioned due to possible resuspension of viable sediment bound bacteria (La Liberte and Grimes, 1982). Much recent work has demonstrated that higher levels of coliforms and pathogens can exist in sediment than in the overlying water column (Rittenberg *et al.*, 1958; Hendrick, 1971; Van Donsel and Geldreich, 1971), that *E. coli* is readily absorbed to silts occurring in estuaries (Weiss, 1951), and that the presence of organic matter in sediment is a likely factor influencing survival of *E. coli* in sediments (Gerba and McLeod, 1976). Similar observations have been made for other bacteria (Sayler *et al.*, 1975). Die-off of bacterial pathogens can vary considerably from the pattern shown by coliform survival (Burton, 1985). For example, *Clostridium perfringens*, a spore-forming anaerobe also sometimes used as an indicator species, can persist for much longer—a fact that enables the spores of this organism to be used as a tracer in sediment studies (Brannon *et al.*, 1985b).

Other pathogenic microorganisms, in addition to bacteria, have been examined with regard to their relationship to sediment. Viruses, for example, tend to sorb strongly to sediment (LaBelle *et al.*, 1980), and bacteriophages have been found to survive for long periods of time in marine water when adsorbed to clay, as opposed to being held in free suspension (Bitton and Mitchell, 1974; Gerba and Schaiberger, 1975). Furthermore, viruses appear to persist longer than fecal coliform bacteria and enteric pathogenic bacteria (Berg, 1978; Melnick and Gerba, 1980). Factors that influence survival of viruses in sediment are similar to those affecting coliforms in sediment—i.e., temperature, presence of organics, and presence of particulate matter (Melnick and Gerba, 1980).

The occurrence of pathogenic protozoans such as various species of *Giardia*, *Naeglaria*, and *Acanthamoeba* is significant, particularly because lower numbers of these organisms are required for an infective dose than for other microorganisms and because they form resistant structures that persist for long periods. In addition, *Naegleria* is also free living, uses gram-negative bacteria and dissolved organic matter as food sources, and under appropriate circumstances, this organism can multiply and achieve significant levels in sediment microbial communities (Tyndall, 1983). *Naegleria* and *Acanthamoeba* are particularly sensitive to thermal conditions. *Naegleria*, for example, appears to occur in greater numbers at temperatures above 25°C (Tyndall, 1983). These temperatures occur often during the summer months in the southern portions of the United States, but can also be reached in sediments in cooler regions when thermal additions are made, as for example, by the effluent from a power plant.

Little research has been done on the factors that contribute to the ability of pathogenic protozoans to persist in aquatic sediment. In the past, the majority of the effort has been focused on the means to destroy these organisms in water (see discussions by Taber, 1976; Burton, 1985). Thus, little is known about the effects of pH, heavy metals, or predation upon survival in sediment. A potential inhibitory effect of *Pseudomonas fluorescens* and *Serratia marcesens* on the growth of *Naegleria fowleri* was indicated by the results of a study by Duma (1981), but these results were insufficient to make firm conclusions at that time (Duma, 1981; Tyndall, 1983).

The occurrence of any microbial pathogen in aquatic sediment is of concern in terms of the relationship of that sediment to human activities. Thus, a sediment having large numbers of pathogenic *Naegleria* is of little or no concern in many areas; but in an area where swimming occurs and where the depth is shallow enough that sediment is likely to be resuspended by swimming, the threat of human nasal contact with sediment-borne *Naegleria* becomes a real possibility. Similarly, the presence of bacterial and viral pathogens in sewage-contaminated sediment in a harbor area may pose little or no risk to the human population, but if the sediment is dredged and placed into an upland disposal site, the potential for infection may be increased, either through direct contact or through mobilization of the microorganisms into nearby waters (Grimes, 1975, 1977; Sayler *et al.*, 1975; Goyal *et al.*, 1977; Matson *et al.*, 1978).

Not a great deal is known about the relationship of microorganisms in sediment to benthic animals. It is known that animals that feed on sedimentary deposits are much more nutritionally dependent upon the attached bacteria than on the nonliving organic debris (Mann, 1973), a phenomenon that appears to be generally true of detritus as a whole (Darnell, 1958, 1961; De LaCruz and Gabriel, 1973). In addition, the distribution of deposit-feeding invertebrates is apparently closely related to sediment bacterial content (Wilson, 1955; Gray, 1966).

E. Microbial Products, Molecular
Recalcitrance, and Microbial Fallibility versus
Legislative Requirements

From the information presented in the preceding discussion, several pertinent facts concerning microbial activity in sediment become evident. First, the reduction of various elements capable of existing in more than one oxidation state is often a direct or indirect result of microbial activity occurring within and on the surface of aquatic sediment. Second, the primary processes of concern involve the consumption of dissolved oxygen, the reduction of inorganic electron acceptors, the formation of organic acids, and the synthesis of products having reducing properties. Third, the products formed have their own importances with respect to the sediment environment and the overlying water column. Some products are major nutrients, while others are micronutrients. Still others are important from a geochemical standpoint, both in terms of the aquatic sediment itself—serving to either oxidize or reduce the sediment—and in terms of the aquatic biota, particularly with regard to the bioaccumulation and/or toxicity of that element.

From the information presented in the literature, we can also say something about the metabolism of polluting and hazardous chemicals in natural ecosystems in general and in aquatic sediment in particular. It is apparent that a large assortment of chemical compounds exists in aquatic sediment. These compounds are both natural and synthetic in origin, and oftentimes, different members of the microbial community native to aquatic sediment can multiply and utilize these materials. As mentioned previously, while a microbial species having the requisite enzymes may not be present initially, there exists the possibility that a microorganism(s) having this capability may appear through one or more mechanisms, including invasion of appropriate microorganisms, and genetic modification of indigenous or invading species to permit utilization of the substrate. As Alexander (1971) has noted, many, but not all, pollutants are organic, and a large number of these are destroyed principally or only through microbial activities. Several organisms capable of complete degradation, partial conversion, or cometabolism of polluting chemicals are known (Bourquin and Pritchard, 1979; Leisinger et al., 1981). However, while some pollutants are easily degraded, others are not and may, in fact, be quite persistent. Metabolism of pesticides has been used by Alexander (1971) as an example of the ways in which microorganisms act or fail to act upon various pollutants.

For the applied sediment microbiologist, the problems that arise are often the result of the relationship of pollutants and/or microbial pathogens pre-

sent in an aquatic sediment to the use or intended use of the aquatic ecosystem the sediment is in. Problems may also arise with respect to the
intended use of another environment into which a polluted sediment may be
placed, if it is moved. In either event, a problem becomes of concern from a
legal standpoint when and if the accumulation of materials in the sediment is
high enough to cause a conflict with legislative mandates. For example, open
water disposal of contaminated dredged sediment requires consideration of
several pieces of legislation, including the Fish and Wildlife Coordination
Act, the National Environmental Protection Act, the Marine Protection,
Research and Sanctuaries Act, the Federal Water Pollution Control Act, the
National Ocean Pollution Planning Act, and the London Dumping Conven-
tion. Alternatively, fresh water concerns with aquatic sediment relate to
many of the same plus several additional pieces of legislation. These include
the Fish and Wildlife Coordination Act, the National Environmental Protection Act, the Clean Water Act, National Drinking Water Standards, plus
state and local water quality criteria.

Why should conflict with legislation be of concern to the applied sediment
microbiologist? The answer is more complex than may be apparent from the
information presented thus far. In addition to the failure of microorganisms
to degrade some pollutants or the ability of other microorganisms to form
undesirable products, consideration must also be given to the possible
changes that can occur if a sediment is moved. Movement of a sediment from
one part of an aquatic environment to another may not necessarily cause a
change in the sediment as a microbial habitat. However, placement of a
sediment into another environment can cause some severe changes. Such
changes may even be used to advantage when and if the changes enhance
the opportunity for microbial degradation. The Dutch, for example, have
used various means of environmental manipulation for the treatment of
contaminated soils (TNO, 1985) and the suitability of these procedures for
treatment of contaminated sediment will be examined later.

III. New Frontiers

The title of this section, as in the article title, suggests new areas to be
conquered in sediment microbiology. While this title may seem misleading
in that there do not appear to be vast new areas of the globe for the sediment
microbiologist to explore (there certainly are some), there are many frontiers
to conquer in terms of the problems with which the applied sediment microbiologist must deal on a day-to-day basis. There are also several frontiers to
explore in terms of alternative solutions to these problems.

The days of the one problem—one microorganism solution are gone for the
applied sediment microbiologist. Our present understanding of the rela-

tionships of microorganisms to their environment is still somewhat limited with respect to the details of certain processes and the intimacies of certain interrelationships between the microorganisms. At the same time, however, this understanding is far too broad to permit the scientist to assume the old and somewhat parochial single organism approach. Instead, problems must be addressed with the more modern understanding supplied by microbial ecology. The approach should be flexible enough to permit incorporation of new concepts as they develop. The current approach necessarily involves not only a clear understanding of microorganisms and the processes they carry out, but also requires a knowledge of the way in which the microorganisms and the processes they carry out relate to aquatic sediment and to the rest of the aquatic environment. Because the applied sediment microbiologist possesses at least some of the knowledge, he becomes responsible for the consequences of his actions, should he fail to take this knowledge into account when he makes decisions. If the applied sediment microbiologist chooses not to use this knowledge, then the legal system and the public at large may well hold him responsible.

This section will examine several situations in aquatic ecosystems wherein the problems to be faced have constituted "new frontiers" and where care has had to be given to environmental consequences of solutions to the problems. To develop a full understanding of the problems and of the environmental consequences of the solutions, it is important to give proper attention to asking the right questions. Often "the right questions" are fairly simple and directed toward making an assessment of microbial activity and its impacts, i.e., the differentiation of real, potential, and perceived concerns. Is the situation resulting from microbial activity or lack of microbial activity really a problem? Will it become a problem in the future, particularly if man's activities will result in a change in the sediment environment? The right questions may simply involve, what needs to be measured? What is the best way to go about making the measurements? Should the measurements be made in the field or are they more appropriately made in the laboratory? How does the investigator best apply his own experimental observations to solving the problem?

A. The Flooded Soil

Flooded soils have received considerable attention from several authors, both from a microbiological point of view (Yoshida, 1975; Gunnison *et al.*, 1985) and from the chemical aspect (Ponnamperuma, 1972). Flooded soils encompass a variety of habitats ranging from the field that is flooded only briefly in periods of heavy accumulation of precipitation to ephemeral pools and riverbanks that are flooded for only certain portions of the year to rice paddies that may be flooded for substantial periods of the year. In terms of environmental extremes, the flooded soil is one of the aquatic habitats exposed to the most severe of temperature and moisture changes, depending upon the presence or absence of a covering layer of water.

Flooded soils often, but not always, include a unique set of environmental conditions in marked contrast to other ecosystems. The depth of the overlying water column is normally rather shallow—ranging from a few millimeters to no more than a few centimeters, and thus the water is easily mixed by the wind, ensuring an adequate amount of dissolved oxygen. In addition, the shallowness also contributes to the possibility that a sufficient penetration of sunlight can occur to promote active growth of algae and cyanobacteria on the soil surface; this likewise promotes a high level of dissolved oxygen in the overlying water. Flooded soils often feature an abundant supply of organic matter, much of which is in readily available form owing to presence of growing plants. These factors tend to encourage the formation of a flooded soil system that has an oxidized surface layer with intensely reducing conditions in the underlying layer. In such an environment, one would expect a considerable interaction between the aerobic and anaerobic zones; processes such as nitrification–denitrification, release of iron and manganese in the reduced form followed by reoxidation and precipitation of these materials on the soil surface, and sulfate reduction to sulfide followed by reoxidation to sulfate tend to predominate (Gunnison *et al.*, 1985). While active formation of organic fermentation products and methanogenesis can occur, accumulation of the materials is retarded/prevented due to the capacity for rapid microbial oxidation to carbon dioxide.

Problems with flooded soils, from an anthropocentric point of view, tend to occur in two different situations. In soils that are periodically flooded for agronomic purposes, microbial processes associated with denitrification tend to predominate because the system is ideally suited for that portion of the nitrogen cycle (Terry and Tate, 1980), i.e., an aqueous surface overlying normally oxidized soil surface layer overlying a reduced subsurface layer. Aside from encouraging the cultivation of legumiferous plants during periods of nonflooding and N-fixing cyanabacteria during flooded periods there is, at present, little that can be done to add nitrogen to this system beyond the obvious ploy of adding external sources of N, i.e., nitrogenous fertilizers. In areas of the world where energy prices are high, the cost of sustaining large crop production by addition of fertilizers can be prohibitive.

A second problem occurs when a soil that is normally flooded during most, if not all, of the year is drained, predominantly for agricultural purposes. Most such soils are exploited because they are organic and highly fertile. However, exposure results in a loss of the anaerobic integrity that was present during the flooded state. With the loss of anaerobic conditions that permitted the development of the organic soils in the first place, comes the advent of processes inherent with aerobic conditions. These processes, both chemical and microbiological, serve to oxidize the surface of the organic soil, resulting in the loss of from one to several inches of the valuable resource per year. The specific processes involved in the aerobic destruction of these soils have been examined in detail by several authors (see, for example, Tate, 1979). Although various strategies for dealing with this situation have been examined, few areas appear to be realistic. One effective procedure involves reflooding the soil during fallow periods, minimizing atmospheric contact to only those times of the year when the soil is actually being used to support crop growth. However, this can also involve a great deal of added expense on the part of the owner–operator.

B. Flooded Soils in Transition—Marshes and New Reservoirs

The use of the word "soil" here may appear unusual. In the case of fresh and saltwater marshes, the source materials are often obtained largely from aquatic sediment plus an admixture of living and decomposed organic material derived from aquatic and terrestrial plants. If the marsh is accumulating materials, then the level of the soil tends to rise above the water level, becoming either high marsh or meadowland. In either event, the substrate is in transition from being a sediment to becoming a soil. By contrast, in new reservoirs, that which was formerly a wetland or upland soil is flooded and is in transition to becoming a sediment in the established reservoir (see Gunnison et al., 1985).

Salt water marshlands are extremely productive, microbiologically fascinating environments featuring alternating flooded and dry conditions, an abundance of organic matter, and large quantities of imported dissolved sulfate and nitrate to fuel the activities of heterotrophs, sulfate reducers, and denitrifiers, respectively. At the same time, an active source of sulfide production is, in turn, a resource for aerobic chemolithotrophs. Microbial activities in salt marshes are not normally a source of difficulties for man. However, problems may occur when man attempts to create new marshlands using dredged material as a source of sediment, if the material contains contaminants. There is the possibility for direct uptake of contaminants by marsh vegetation. Also, a possibility exists that through the movement of an aquatic sediment from an underwater situation followed by placement of the material into a wet–dry situation, environmental conditions may be altered in a manner that enhances microbial and/or direct chemical release of the contaminants; the potential then exists for entry of hazardous materials into food webs. Such considerations become more important when dealing with sediments containing xenobiotic compounds that have a high affinity for the biota; whether or not the use of a contaminated sediment will be acceptable in a particular situation depends largely upon scientific opinion, often requiring the joint efforts of the sediment microbiologist and a geochemist.

Problems resulting from microbial processes of importance in the filling and operation of new reservoirs are considered in a recent book (Gunnison *et al.*, 1985) and will be only briefly summarized here. The microbial phenomena occurring in reservoirs do not differ from those in a natural lake. The principal difference between reservoirs and lakes occurs as a result of the way in which reservoirs are operated. The operation influences the impacts that microbial processes have on the rest of the environment.

In lakes, the impacts of biological processes are largely governed by the way in which natural cycles of thermal stratification and mixing influence the formation and subsequent distribution of products resulting from microbial activities in sediments. Among the most important of these are various nutrients and metals released by microbial activity during the summer and early fall months under anaerobic conditions in the hypolimnion in lakes in temperate regions of the world.

In reservoirs, the impacts of biological processes are largely governed by the way in which water moves through the system. Unlike natural lakes, many reservoirs are equipped with bottom withdrawal facilities which means that hypolimnetic waters are released from the reservoir. If the hypolimnion happens to develop anaerobic conditions, there is a strong possibility that the reduction products formed by microbial activity may end up in the release waters. This problem is intensified in new reservoirs because standing trees, shrubs, litter, and soil surface layers are often left in place when the reservoir is filled. The decompositional activities of microorganisms acting on these substrates can be very intense with the result that not only nutrients and metals, but also sulfides and various organic decomposition products are released to the hypolimnion and from here, to the downstream environment. The consequences of the release of these materials can be quite severe both within the reservoir and downstream, particularly if water supply is one of the project purposes. Current methods for dealing with this problem are (1) remove the substrate, i.e., clear out vegetation and/or strip off the soil surface layer to eliminate these materials as fuels for microbial activities; (2) treat the products, i.e., inject air or oxygen into the hypolimnion to oxidize reduced products as they are released from the bottom. At present, both methods are expensive to apply on a basin-wide basis with the result that neither method is used very often. Aeration or oxygen injection, when used at all, is often limited to bottom waters in the immediate vicinity of the dam in order to provide releases that meet downstream water quality standards.

The construction, filling, and operation of a new reservoir can require consideration to be given to microbial processes other than decomposition of organic matter. Deposits of various materials may be subject to microbial action; for example, concern in the past has been expressed for flooding areas that contain cinnabar (mercury ore), arsenical wastes, as well as soils containing high levels of iron and/or manganese. Additional sources of microbial substrates, such as industrial plants that release large volumes of organic or metallic wastes into the reservoir basin or its tributaries, must also be examined, as must other sources such as acid mine drainage, urban, and agricultural runoff, and releases from upstream reservoirs.

C. True Aquatic Sediment

Included within the category "true" aquatic sediment are those natural materials that Ponnamperuma (1972) has characterized as generally having the following properties: (1) dark coloration; (2) low redox potential; (3) no dissolved oxygen, at least within the strata lying under the first few millimeters or centimeters of the surface; (4) reduced forms of N, S, Mn, and Fe, and CH_4 and refractory organic matter; and (5) possibly a thin, light-colored oxidized layer at the surface, depending on the aeration status of the overlying water column. Not all sediments have each of these properties, and the preceding characterization is not intended as all inclusive.

Areas of the world that have sediments heavily polluted with organic and metallic contaminants and microbial pathogens are generally associated with heavy industrial activity. However, these same areas are also often the sites of heavy sediment deposition, and many such areas require frequent dredging to keep channels open to ship traffic. For example, of the cubic meters of sediment dredged annually in the United States, an estimated 5% are considered unsuitable for open water disposal.

Problems for the applied sediment microbiologist with regard to dredging operations can occur at both the dredging and the disposal sites, although the more severe problems tend to be associated with the latter site. At the dredging end, a major source of difficulty is associated with the resuspension of settled materials along with contaminants and/or pathogens associated with the particles. This problem has been decreased in intensity somewhat by recent advances in dredging technology, i.e., the use of dustpan dredges and silt curtains, that serve to decrease the amount of resuspended sediment that leaves the dredging site.

Problems that occur at the disposal site are more varied in nature, in large part because the possibilities for varying environmental conditions at the disposal site are often rather large. Underwater disposal offers the least change in environmental conditions from the original dredging site; however, the disposal site can be very different in terms of salinity, currents, and the nature of the biota and sediments indigenous to the area. Emerging technologies for dealing with placement of contaminated sediments into comparatively clean sites are largely dependent on various ways of confining the contaminated sediment, preventing the sediment plus associated contaminants from being moved outside of the disposal area. Such technologies include capping placement of a layer of clean material over the layer of contaminated material, subaqueous burial placement of contaminated sediment beneath the surface of existing sediment, and confined aquatic disposal placement of contaminated material into a lined pit followed by capping with either a clean sediment or an impermeable substance.

Placement of contaminated dredged material into either a diked area at the water surface or into an upland situation involves a large change in environmental conditions. In particular, the material is subject to different conditions of aeration, temperature, and moisture than were present underwater. The change in environmental circumstances can result in changes in chemical and microbiological activities occurring in the contaminated sediment. Thus, a sediment laden with sulfides and organic contaminants that were not previously degraded due to anaerobic conditions now are subject to oxidation and aerobic microbial attack. Depending upon the nature of the sediment, the product resulting from this activity may be a cat clay having levels of metals toxic to plants. Alternatively, the combination of microbial activity and leaching with water may yield a fertile soil; however, this may cause undesirable effects on the water in areas adjacent to or below the disposal site. New methodologies for dealing with contaminated soils and sediments in terrestrial situations are presently being developed, but are still in their infancy. One approach currently being examined by the Dutch involves the use of microorganisms to treat soils contaminated with various organic compounds. In this approach, the contaminated material is manipulated in a manner that will enhance the microbiological degradative process (TNO, 1984). Methods do exist to provide treatment for water leaving the disposal area, thus minimizing the impact of the products of chemical and microbiological activity upon the waters in the surrounding area.

IV. Management Strategies—The Cure of Problems through the Treatment of Symptoms

Although many of the problems faced by the applied sediment microbiologist are, indeed, caused by the presence of certain microorganisms or by the activities of microorganisms, there are presently only a very limited number of ways to deal with these problems. Often, present methods for handling problems associated with microbial activity are designed to treat the symptoms—i.e., the products of the activity, rather than the cause—the presence of substrates that support activity of the microorganism. Part of the reasoning involved in this approach is based on the size of many environmental situations in which the problems occur. For example, low dissolved oxygen over the basin of a reservoir or lake is a considerably larger problem than low dissolved oxygen in a flask or even in a fermenter. The costs involved in dealing with low dissolved oxygen in a reservoir may be great enough to limit application of the treatment to only the area(s) of major concern, i.e., reservoir release waters. In this section, we will briefly examine three different management strategies for dealing with problems associated with sediment microorganisms. Many of the specific methods involved in the application of each of the strategies were described in greater detail previously.

A. Remove the Product

In this management approach, the end product is dealt with by facilitating its removal from the ecosystem containing the sediment source material. Classical water treatment methods such as coagulation and precipitation are often suitable for dealing with products that have reached the water column. Aeration of anoxic water can be used to reoxidize reduced chemical products of microbial activity in aquatic sediments, resulting in the formation of precipitates and subsequent removal of the chemical from the water column. In the case of aquatic sediments placed on land, installation of tile drains and venting systems serve to convey liquid-borne and gaseous products, respectively, from the system.

B. Restrict Product Mobility

This strategy involves retaining the product within the system or, at most, releasing it slowly from the system to minimize the environmental impact. Thus, capping of contaminated sediments or placement of a contaminated dredged material into a confined area on land with a liner and a cap are two versions of one mechanism for preventing product loss. In reservoirs having anoxic hypolimnions that accumulate reduced chemical constituents during the warm summer months, the installation of a surface withdrawal system to tap surface waters is a possible means to prevent release of reduced products during periods of hypolimnetic anoxia. Reservoirs situated in watersheds that receive acid mine drainage may be used to slow and then meter the release of acid mine drainage and associated particulate matter, thus preventing slug loads of these materials from adversely impacting downstream areas during periods of heavy runoff.

C. Remove the Substrate

This strategy involves a direct attack upon the cause of the problem, but at the same time, this strategy is rarely used. Stripping of fertile soils and clearing of vegetation from the basin are proven methods for reducing negative impacts of microbial degradation activities upon the water quality of a new reservoir. Shutting off an acid mine drainage problem by finding, filling in, and sealing abandoned mines in the watershed is an effective means for dealing with this problem. Dredging to remove a troublesome sediment laden with contaminants followed by disposal of the material into a confined facility or by treatment of the sediment to remove contaminants is rarely the primary reason for which dredging is undertaken, although this does happen.

Each of the above strategies carries with it a cost, both in terms of financial burden imposed upon the operator and in terms of the environmental burden resulting from carrying out or not carrying out the procedure. Each cost must be carefully factored in during the decision-making process.

V. Conclusions

Our present understanding of microorganisms in sediment and the way the activities these organisms carry out relate to the rest of the aquatic ecosystem still has many limitations. Nonetheless, this understanding is too great to permit us to ignore the sediment microflora when considering environmental manipulations that involve sediment. Present techniques for dealing with problems arising from microbial activities are somewhat restricted, primarily as a result of the level of approach that must be used in dealing with materials as heterogenous as aquatic sediment. In most cases, unless a specific point source can be identified, it is difficult to remove the substrate or microorganism that serve as the original source of a particular problem. Thus, it becomes necessary to deal either with the microbial product that is actively causing the problem or else with the sediment and its resident microflora as a total package. There are various measures available to handle a product-related problem, ranging from water treatment to remove the product to confinement to restrict or prevent movement of the product. Removal of a specific substrate or set of substrate or a specific microbial pathogen is oftentimes a more difficult task. Depending on the areas of flooded soil or sediment involved, the amount of substrate removal that is required may be so vast as to be cost prohibitive. Present technologies, while applicable in some situations, are still quite limited in terms of cost and general applicability. Future research in applied sediment microbiology must focus on more effective ways of dealing the presence of undesirable substrates and ways to alter or enhance their metabolism. Methods to handle problems associated with survival and mobilization of microbial pathogens in sediment must also be devised.

REFERENCES

Alexander, M. (1965). *Adv. Appl. Microbiol.* **7**, 35–80.
Alexander, M. (1971). *Annu. Rev. Microbiol.* **25**, 361–392.
Alexander, M. (1977). "Introduction to Soil Microbiology," 2nd Edition. Wiley, New York.
Alexander, M. (1981). *Science* **211**, 132–138.
Allen, L. A., Grindley, J., and Brook, E. (1953). *J. Hyg.* **51**, 185–194.
Atlas, R. M. (1981). *Microbiol. Rev.* **45**, 180–209.
Berg, G., ed. (1978). "Indicators of Viruses in Water and Food." Ann Arbor Science Press, Ann Arbor, Michigan.
Bitton, G. (1978). *In* "Water Pollution Microbiology" (R. Mitchell, ed.), Vol. 2, pp. 273–299. Wiley, New York.
Bitton, G., and Mitchell, R. (1974). *Water Res.* **8**, 227–229.
Bourquin, A. W., and Pritchard, P. H., eds. (1979). "Proceedings of the Workshop: Microbial Degradation of Pollutants in Marine Environments." U. S. Environmental Protection Agency, Report No. EPA-600/9-79-012. Gulf Breeze, Florida.
Boyd, J. W., Yoshida, T. Y., Vereen, L. E., Cada, R. L., and Morrison, S. M. (1969). "Bacterial Response to the Soil Environment." Colorado State University Sanitary Engineering Papers, No. 5 Colorado State University, Fort Collins.
Brannon, J. M., Gunnison, D., Butler, P. L., and Smith, I., Jr. (1978). "Mechanisms that Regulate the Intensity of Oxidation-Reduction in Anaerobic Sediments and Natural Water Systems." Technical Report Y-78-11. Environmental Laboratory. U. S. Army Engineer Waterways Experiment Station, Vicksburg, Mississippi.
Brannon, J. M., Chen, R. L., and Gunnison, D. (1985a). *In* "Microbial Processes in Reservoirs" (D. Gunnison, ed.). Dr. W. Junk Publisher, The Hague.
Brannon, J. M., Hoeppel, R. E., Sturgis, T. C., Smith, I., Jr., and Gunnison, D. (1985b). "Final Report. Effectiveness of Capping in Isolating Contaminated Dredged Material from the Biota and the Overlying Water." TR-E85-9. Environmental Laboratory. U. S. Army Engineer Waterways Experimental Station, Vicksburg, Mississippi.
Breck, W. G. (1974). *In* "The Sea" (E. D. Goldberg, ed.), Vol. 5. Wiley, New York.
Buckman, H. O., and Brady, N. C. (1969). "The Nature and Properties of Soils," 7th Edition. Macmillan, New York.
Burton, G. A., Jr. (1985). *In* "Microbial Processes in Reservoirs" (D. Gunnison, ed.). Dr. W. Junk, The Hague.
Burton, G. A., Jr., Gunnison, D., and Lanza, G. R. (1985). In preparation.
Carlucci, A. F., and Pramer, D. (1959). *Appl. Microbiol.* **7**, 388–392.
Chou, T. W., and Bohonos, N. (1979). *In* "Microbial Degradation of Pollutants in Marine Environments" (A. W. Bourquin and P. H. Pritchard, eds.), pp. 76–88. U. S. Environmental Protection Agency, Report No. EPA-600/9-79-012. Gulf Breeze, Florida.
Clark, R. R., Chian, E. S. K., and Griffin, R. A. (1979). *Appl. Environ. Microbiol.* **37**, 680–685.
Collins, V. G. (1976). *In* "Advances in Aquatic Microbiology" (M. R. Droop and H. W. Jannasch, eds.), Vol. 1, pp. 219–272. Academic Press, London.
Crawford, R. L. (1981). "Lignin Biodegradation and Transformation." Wiley, New York.
Dale, N. G. (1974). *Limnol. Oceanogr.* **19**, 509–518.
Darnell, R. M. (1958). *Publ. Inst. Mar. Sci., Univ. Tex.* **5**, 353–568.

Darnell, R. M. (1961). *Ecology* **42**, 553–568.

Deevey, E. S., Jr. (1970). *Bull. Ecol. Soc. Am.* **51**, 5–8.

De LaCruz, A. A., and Gabriel, B. C. (1973). Caloric, elemental, and nutritive changes in decomposing *Juncus roemerianus* leaves. *Ecology* **55**, 882–886.

De Laune, R. D., Patrick, W. H., Jr., and Brannon, J. M. (1976). "Nutrient Transformations in Louisiana Salt Marsh Soils." Sea Grant Publ. No. LSU-T-76-009. Center for Wetland Resources. Louisiana State University, Baton Rouge.

Dickinson, C. H., and Pugh, G. J. F., eds. (1974). "Biology of Plant Litter Decomposition," Vol. 1. Academic Press, London.

Doetsch, R. N., and Cook, T. M. (1973). "Introduction to Bacteria and Their Ecobiology." Univ. Park Press, Baltimore.

Duma, R. J. (1981). "Study of Pathogenic Free-Living Amebas in Fresh-Water Lakes in Virginia." Technical Report EPA-600/51-80-037. U. S. Environmental Protection Agency.

Evans, W. C. (1977). *Nature* **270**, 17–22.

Faust, M. A., Aotaky, A. E., and Hargadon, M. I. (1975). *Appl. Microbiol.* **30**, 800–806.

Friedovich, I. (1975). *Annu. Rev. Biochem.* **44**, 147–159.

Garrels, R. M., and Christ, C. L. (1956). "Solutions, Minerals, and Equilibria." Harper, New York.

Gerba, C. P., and McLeod, J. S. (1976). *Appl. Environ. Microbiol.* **32**, 114–120.

Gerba, C. P., and Schaiberger, G. E. (1975). *J. Water Pollut. Control Fed.* **47**, 93–103.

Gerba, C. P., Smith, E. M., Schaiberger, G. E., and Edmond, T. D. (1979). *In* "Methodology for Biomass Determinations and Microbial Activities in Sediment" (C. D. Litchfield and P. L. Seyfried, eds.), pp. 64–74. Amer. Soc. Testing Material, Philadelphia.

Gjessing, E. T. (1976). "Physical and Chemical Characteristics of Aquatic Humus." Ann Arbor Science Publishers, Ann Arbor Michigan.

Goyal, S. M., Gerba, C. P., and Melnick, J. L. (1977). *Appl. Environ. Microbiol.* **34**, 139–149.

Gray, J. S. (1966). *J. Mar. Biol. Assoc. U.K.* **46**, 627–646.

Greenberg, A. E. (1956). *Public Health Rep.* **71**, 77–86.

Grimes, D. J. (1975). *Appl. Microbiol.* **34**, 139–149.

Grimes, D. J. (1980). *Appl. Environ. Microbiol.* **39**, 782–789.

Gunnison, D., Chen, R. L., and Brannon, J. M. (1983). *Water Res.* **17**, 1609–1617.

Gunnison, D., Engler, R. M., and Patrick, W. H., Jr. (1985). *In* "Microbial Processes in Reservoirs" (D. Gunnison, ed.). Dr. W. Junk, The Hague.

Haller, H. D., and Finn, R. K. (1978). *Appl. Environ. Microbiol.* **35**, 890–896.

Hendricks, C. W. (1971). Increased recovery of *Salmonellae* from bottom sediments versus surface waters. *Appl. Microbiol.* **21**, 379–380.

Horvath, R. S. (1972). *Bacteriol. Rev.* **36**, 146–155.

Hungate, R. E. (1968). *In* "Methods in Microbiology" (R. Norris and D. W. Ribbons, eds.), pp. 117–132. Academic Press, London.

International Trade Commission (1978). "Synthetic Organic Chemicals: U. S. Production and Sales." U. S. Government Printing Office, Washington, D.C.

Johnson, B. T., ed. (1982). "Impact of Xenobiotic Chemicals on Microbial Ecosystems." Technical Papers of the U.S. Fish and Wildlife Service, No. 107. U. S. Department of the Interior, Fish and Wildlife Service, Washington, D.C.

Kapuscinski, R. B., and Mitchell, R. (1983). *Environ. Sci. Technol.* **17**, 1–6.

LaBelle, R. L., Gerba, C. P., Goyal, S. M., Melnick, J. L., Ceck, I., and Bogdan, G. F. (1980). *Appl. Environ. Microbiol.* **39**, 588–596.

LaLiberte, P., and Grimes, D. J. (1982). *Appl. Environ. Microbiol.* **43**, 623–628.

Leisinger, T., Cook, A. M., Hütter, M., and Nüesch, J., eds. (1981). "Microbial Degradation of Xenobiotics and Recalcitrant Compounds." Academic Press, London.

Lovely, D. R., and Klug, M. J. (1982). *Appl. Environ. Microbiol.* **43,** 552–560.

Mann, K. H. (1973). *Mem. 1st Ital. Idrobiol.* **29** (Suppl.), 353–383.

Marshall, K. G. (1980). *In* "Adsorption of Microorganisms to Surfaces" (G. Bitton and K. C. Marshall, eds.), pp. 317–329. Wiley, New York.

Marshall, N. (1970). *In* "Marine Food Chains" (J. H. Steele, ed.), pp. 52–66. Univ. of California, Los Angeles.

Matson, E. A., Horner, S. G., and Buck, J. D. (1978). *J. Water Pollut. Control. Fed.* **50,** 13–20.

Matsummura, F., and Boush, G. M. (1971). *In* "Soil Biochemistry" (A. D. McLaren and J. J. Skujins, eds.), Vol. 2, pp. 320–336. Dekker, New York.

Melnick, J. L., and Gerba, C. P. (1980). *Crit. Rev. Environ. Control* **10,** 65–93.

Mitchell, R. (1968). *Water Res.* **2,** 535–543.

Morris, J. C., and Stumm, W. (1967). *In* "Equilibrium Concepts in Natural Water Systems" (R. F. Gould, ed.). Advances in Chemistry Series, No. 67. American Chemical Society, Washington, D.C.

Mortimer, C. H. (1941). *J. Ecol.* **29,** 280–329.

Mortimer, C. H. (1942). *J. Ecol.* **30,** 147–201.

Mortimer, C. H. (1971). *Limnol. Oceanogr.* **16,** 387–404.

Nussbaum, I., and R. M. Garver. 1955. Survival of Coliform organisms in Pacific Ocean coastal waters. *Sewage Ind. Wastes* **27,** 1383–1390.

Orlob, G. T. (1956). *Sewage Ind. Wastes* **28,** 1147–1167.

Patrick, W. H., Jr., and DeLaune, R. D. (1977). *Geosci. Man.* **18,** 131–137.

Patrick, W. H., Jr., and Mikkelson, D. S. (1971). *In* "Fertilizer Technology and Use." Soil Soc. Amer., Madison, Wisconsin.

Ponnamperuma, F. N. (1972). *Adv. Agron.* **24,** 29–88.

Reddy, K. R., and Patrick, W. H., Jr. (1975). *Soil Biol. Biochem.* **7,** 87–94.

Reddy, K. R., Patrick, W. H., Jr., and Phillips, R. E. (1980). *J. Soil Sci. Soc. Am.* **44,** 1241–1246.

Reinheimer, G. (1980). "Aquatic Microbiology," 2nd Edition. Wiley, Chichester.

Rittenberg, S. C., Mittiver, T., and Ivler, O. (1958). Coliform bacteria in sediments around three marine sewage outfalls. *Limnol. Oceanogr.* **3,** 101–108.

Sayler, G. S., Nelson, J. D., Jr., Justice, A., and Colwell, R. R. (1975). *Appl. Microbiol.* **30,** 625–638.

Stumm, W., and Morgan, J. J. (1970). "Aquatic Chemistry." Wiley, New York.

Taber, W. A. (1976). *Annu. Rev. Microbiol.* **30,** 263–277.

Takai, Y., and Kamura, T. (1966). *Folia Microbiol.* **11,** 304–313.

Takai, Y., Koyama, T., and Kamura, T. (1956). *Soil Sci. Plant Nutr.* **2,** 63–66.

Tate, R. L., III. (1979). *Soil Sci.* **128,** 267–273.

Terry, R. E., and Tate, R. L., III. (1980). *Soil Sci.* **129,** 88–91.

TNO (1984). "Literature Study on the Feasibility of Microbiological Decontamination of Polluted Soils." Netherlands Organization for Applied Scientific Research (TNO), Groningen.

Tyndall, R. L. (1983). *Crit. Rev. Environ. Control* **13,** 195–226.

Van Donsel, D. J., and Geldreich, E. E. (1971). *Water Res.* **5,** 1079–1087.

Verstraete, W., and Voets, J. P. (1976). *Water Res.* **10,** 129–136.

Weiss, C. M. (1951). *Sewage Ind. Wastes* **23,** 227–237.

Whitfield, M. (1974). *Limnol. Oceanogr.* **19,** 857–865.

Wilson, D. P. (1955). *J. Mar. Biol. Assoc. U.K.* **34,** 531–543.

Won, W. D., and Ross, H. (1973). *J. Environ. Eng. Div. Am. Soc. Civ. Eng.* **99,** 205–211.

Woodcock, D. (1971). *In* "Soil Biochemistry" (A. D. McLaren and J. J. Skujins, eds.), Vol. 2, pp. 337–360. Dekker, New York.

Yoshida, T. (1975). *In* "Soil Biochemistry" (E. A. Paul and A. D. McLaren, eds.), Vol. 3, pp. 83–122. Dekker, New York.

Zhukova, A. I. (1963). *In* "Symposium on Marine Microbiology" (C. H. Oppenheimer, ed.), pp. 699–710. Thomas, Springfield, Illinois.

Zobell, C. E. (1946). *Sci. Mon.* **55,** 320–330.

Ecology and Metabolism of *Thermothrix thiopara*

Daniel K. Brannan* and Douglas E. Caldwell[†]

*The Procter and Gamble Company
Cincinnati, Ohio
†Department of Applied Microbiology and Food Science
The University of Saskatchewan
Saskatoon, Canada

I. Introduction

For over two centuries, biologists have been intrigued by thermobiosis (Sonnerat, 1774). They have speculated about the origin of thermophilic bacteria since the initial discovery of a thermophilic bacillus by Miguel (1888). Origin hypotheses range from exotic Venusian beginnings (Arrhenius, 1927) to a more likely assumption that bacterial life began when the earth's environment was hot and reducing (Brock, 1967; Gaughran, 1947). If this latter assumption is correct, then studies of autotrophic thermophiles may allow a better understanding of both the early evolution of life and the chemical evolution of Earth (Corliss *et al.*, 1981).

Descriptions of thermophilic autotrophs growing at neutral pH are sparse (Brierley and LeRoux, 1977; Brock *et al.*, 1971; Tansey and Brock, 1978). *Thermothrix thiopara*, the subject of this article, is a filamentous sulfur-oxidizing bacterium growing at neutral pH and high temperature (Brannan and Caldwell, 1980, 1982, 1983; Caldwell *et al.*, 1976, 1983, 1984; Kieft and Caldwell, 1984a,b). It is found in sulfide-containing hot springs where it grows at discrete sulfide–oxygen interfaces and contributes to mineral weathering. It is distributed throughout the Mammoth Hot Springs of Yellowstone National Park and the Jemez Hot Springs of New Mexico, occur-

ADVANCES IN APPLIED MICROBIOLOGY, VOLUME 31

ring in springs varying in temperature from 45 to 75°C. Electron acceptor-limited growth results in the induction of cell filaments from flagellated bacilli. This has been observed *in situ* and confirmed in continuous culture studies (Caldwell *et al.*, 1984). Filament formation is an adaptation necessary to position cells within the oxic–anoxic interface of geothermal springs. The growth rate of *T. thiopara* is rapid compared with mesophilic sulfur-oxidizing *Thiobacillus* species despite higher maintenance requirements. Its growth yields exceed those of many mesophilic sulfur oxidizers. It is both facultatively chemolithotrophic and facultatively anaerobic. In contrast, thiobacillus-like thermophiles are obligately aerobic, facultatively chemolithotrophic, and rod shaped (Brierley *et al.*, 1978; Le Roux *et al.*, 1977; Williams and Hoare, 1972) while *Sulfolobus* species are lobed, facultatively chemolithotrophic, and aerobic (Brock *et al.*, 1972). *T. thiopara* is thus unique among the extremely thermophilic sulfur oxidizers.

II. Isolation and Cultivation

Difficulties encountered in the isolation of *Thermothrix thiopara* using heterotrophic media are inherent in the cultivation of all extremely thermophilic autotrophs. They result from the stress of metabolic transition at high temperature. The maintenance rate of *T. thiopara*, like other thermophiles, is high compared to those of mesophiles. Any added maintenance costs, such as those incurred during the transition between autotrophic and heterotrophic metabolism, may prevent successful adaptation from one metabolic state to another. As a result, it is difficult to isolate cells growing autotrophically *in situ* if heterotrophic laboratory media are used. Cells are more readily cultivated autotrophically and then adapted to heterotrophic growth conditions.

A knowledge of potential difficulties is important in successful cultivation. Due to the heat resistance of thermophilic spores (Zeikus, 1979), prolonged autoclave periods may be required to ensure sterility. Pure cultures can be obtained by end-point dilution of inocula in liquid media. Streak plates, however, are generally more useful because contaminants can easily be detected. However, syneresis and extreme drying occurs at 70°C. Gellan Gum (K9A40, Kelco, Division of Merck & Co.) can be used as an agar substitute to reduce syneresis. Plates should be sealed with plastic tape to prevent drying, and incubated in a circulated water bath to control temperature. Five hours or more are required for forced air incubators to warm a sleeve of 20 petri plates to 70 from 4°C. The temperature of plates also varies depending upon positioning within the incubator. The lower shelves may be 10°C or more below the temperature reading at the top. Plates should be cooled to room temperature before opening to prevent excessive evaporation. Colonies should be subcultured immediately when visible. Large colonies are frequently nonviable or in death phase. Although stationary phase suspensions quickly become nonviable at 70°C, they remain viable for up to 7 days when stored at 4°C. Lyophilized cells are difficult to recover; however, cells maintain viability when frozen in 20% glycerol (w/v) at −50°C.

When *T. thiopara* is grown in liquid culture, vigorous aeration is required to prevent oxygen limitation. The solubility of oxygen at 70°C is approximately 3 mg/liter. The concentration of the oxidant is thus generally three orders of magnitude below that of the energy source. Consequently, vigorous aeration is necessary to prevent oxygen limtation of cell suspensions.

III. Metabolism

A. Autotrophic Growth

Thermothrix thiopara is capable of growth on a variety of inorganic reduced sulfur sources. *In situ*, its primary source of energy is sulfide (Caldwell *et al.*, 1976, 1984). It also utilizes thiosulfate, elemental sulfur, sulphite, and pyrite but not sulfur-containing copper concentrates (Brannan and Caldwell, 1980). When grown on thiosulfate, *T. thiopara* has a maximum growth rate of 0.56 generations per hour at 73°C. Its optimum pH for growth is between 6.7 and 7.1. Oxidation of thiosulfate to sulfate occurs with the formation of sulfate, sulfur, and polythionate intermediates (Fig. 1). One mole of thiosulfate is oxidized to 2 mol of sulfate. Sulfide is not detected as an intermediate of thiosulfate oxidation by *T. thiopara*. However, at high temperature and neutral pH, it may be chemically converted to sulfur (Roy and Trudinger, 1970). Sulfite, sulfur, and polythionate are formed from thiosulfate, and subsequently oxidized to sulfate. Although sulfite and elemental sulfur are intermediates in the formation of sulfate, the significance of polythionates as intermediates is unclear (Kelly, 1982).

1. Sulfur Deposition

T. thiopara deposits elemental sulfur as extracellular globules when grown using thiosulfate as the sole energy source in batch culture (Fig. 2). It also deposits sulfur during transfer from heterotrophic to autotrophic growth (Brannan and Caldwell, 1980). In this case, the cell filaments become encased in a coating of elemental sulfur (Fig. 3). In continuous culture on thiosulfate, elemental suflur does not form during steady state growth. However, it is deposited during the transition from low to high dilution rates (Brannan and Caldwell, 1983). The source of extracellular sulfur may be the sulfane group of thiosulfate (Roy and Trudinger, 1970). In contrast, the

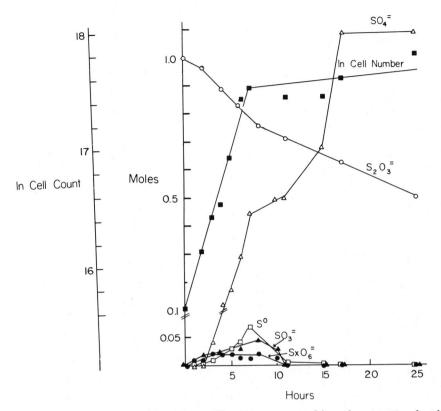

FIG. 1. Oxidation of thiosulfate (○) to sulfate (△) accompanied by a drop in pH and with formation of the intermediates sulfur (□), sulfite (●), and polythionates (▲) during growth of *T. thiopara* (■) on TXB at 73°C. Taken from Brannan and Caldwell (1980) with permission of the American Society for Microbiology.

Beggiatoaceae and *Chromatiaceae (Thiorhodaceae)* catalyze intracellular sulfur deposition biologically when grown on sulfide (Vishniac and Santer, 1953).

2. Leaching of Metal Sulfides

T. thiopara oxidizes elemental sulfur and pyrite chemosynthetically. When cultured on precipitated sulfur, growth is accompanied by a drop in pH and the formation of sulfate (Brannan and Caldwell, 1980). Growth on pyrite is slow and cell densities are lower than on sulfur or thiosulfate. A small amount of sulfate is produced from pyrite during the death phase and could be due to oxidation of polythionates, polysulfide, or other sulfur impurities. This observation and the absence of growth on copper sulfides

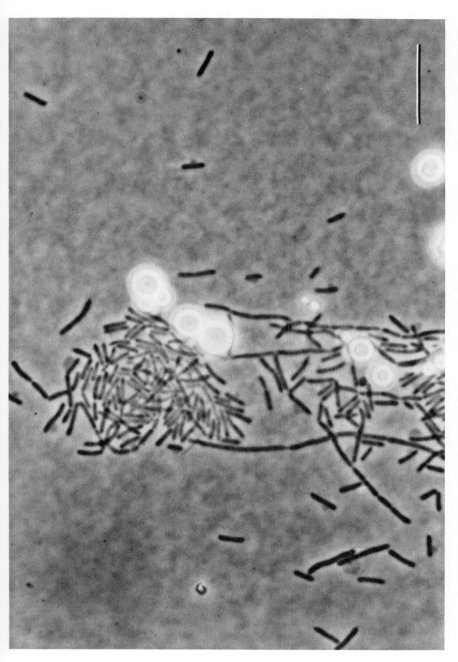

Fig. 2. Extracellular deposition of elemental sulfur (spherical granules) by *Thermothrix thiopara* during oxidation of thiosulfate. The culture was highly aerated, thus producing rod-shaped cells and cell chains but no cell filaments. Bar equals 10 μm.

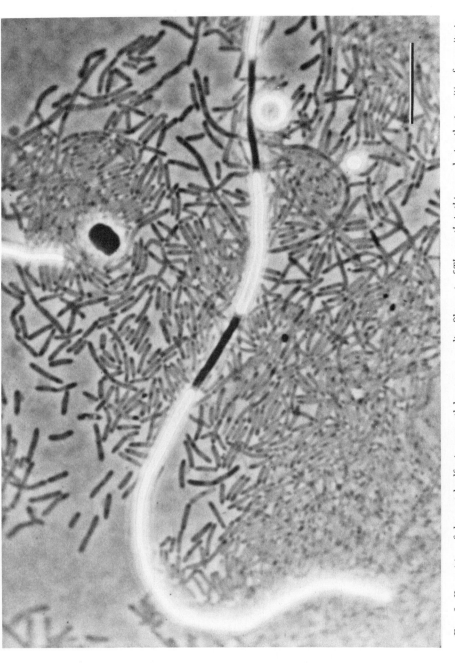

FIG. 3. Deposition of elemental sulfur in a smooth layer surrounding filaments of *Thermothrix thiopara* during the transition from nitrate broth to a synthetic medium with thiosulfate as the sole energy source. Phase micrograph. Bar equals 10 μm.

suggest that *T. thiopara* does not directly catalyze metal–sulfide bond breakage.

Thiobacillus species play a significant role in metal sulfide leaching (Brierley, 1978; Wainwright, 1978). The activity of thermophilic *Thiobacillus* species is usually limited to temperatures below 55°C and organic supplements are required for growth (Brierley and LeRoux, 1977; Fliermans and Brock, 1972; LeRoux *et al.*, 1977; Williams and Hoare, 1972). In contrast, *Thermothrix thiopara* oxidizes sulfur and sulfur compounds at 73°C and does not require organic supplements. *T. thiopara* may thus be necessary for the succession of *Sulfolobus* species which are both thermophilic and acidophilic. In this way it may indirectly accelerate thermophilic leaching processes.

3. RuBP Carboxylase

Thermothrix thiopara fixes carbon dioxide when grown autotrophically on thiosulfate (Brannan and Caldwell, 1980). The average fixation rate of $^{14}CO_2$ is 3.4×10^{-1} μmol min^{-1} g^{-1} of log-phase cells (wet weight). It contains a cold-labile form of ribulose-1,5-bisphosphate (RuBP) carboxylase (Brannan, 1981). Energy-dependent ($S_2O_3^{2-}$) CO_2 fixation is greatest during late log phase and falls off rapidly once the stationary phase of growth is reached. RuBP-dependent CO_2 fixation is greatest during mid-log phase. The enzyme activity from cells harvested beyond this growth period is low. Therefore, cultures having a high cell density have poor enzyme activity and it is difficult to obtain a large quantity of cells with high activity. This, coupled with the lability of the enzyme, has made purification difficult (Table I).

Crude extracts of RuBP carboxylase are cold labile. The enzyme is completely inactivated from *T. thiopara* within 18 hours when stored at -20 or 5°C. The crude extract can be stored in 0.01% azide at 25°C with a 16% loss in activity after 24 hours. Nearly 70% of the activity is lost after 48 hours at 25°C.

The effect of temperature on enzyme activity is shown in Fig. 4. The activity is highest at 83°C (1.07 μmol CO_2 fixed min^{-1} mg^{-1} protein (Fig. 4). This temperature is twice that for the RuBP carboxylase isolated from *Cyanidium caldarum* (Ford, 1979). Studies of neutral thermal springs (70 to 90°C) have demonstrated autotrophic carbon dioxide fixation *in situ* (Brock *et al.*, 1971). *Thermothrix thiopara* is thus far the only isolate that can account for this fixation. Based on its molecular weight and quaternary structure, RuBP carboxylase has been dichotomized into T- and O-type enzymes (Lawlis *et al.*, 1979). It has been assumed that the gene for RuBP carboxylase was first established in anaerobic bacteria (McFadden and Purohit, 1978; McFadden and Tabita, 1974). The O-type carboxylase for *Rhodospirillum rubrum*, a facultatively anaerobic facultative autotroph, is the simplest yet

TABLE I

Ribulose-1,5-Bisphosphate Carboxylase Activity in Crude Extracts
of *Thermothrix thiopara*[a]

		Units of enzyme activity (μmol CO_2 fixed minute^{-1})				
Step	Volume (ml)	Total activity (units)	Total protein (mg)	Specific activity (units mg^{-1})	Purification (fold)	Yield
1. 12,000 g supernatant	2.5	43.5	132.5	0.33	1	100%
2. Crude extract (44,000 g supernatant)	2.5	35.0	43.8	0.8	2.4	80%

[a] *T. thiopara* was grown at 70°C in 30 liters of thiosulfate-mineral salts medium (Brannan and Caldwell, 1980), harvested by centrifugation at 9000 g for 20 minutes and washed in TEMB buffer (Lawlis *et al.*, 1979). The pellet was resuspended in TEMB with 5 mM 2-mercapto-ethanol (TEMMB) and sonicated. Cell-free crude extracts were prepared by centrifugation at 12,000 g for 20 minutes. The supernatant was centrifuged at 44,000 g for 2 hours. The pellet was resuspended in TEMMB, sonicated, centrifuged again at 44,000 g, and added to the super-natant. Due to the cold lability of the enzyme, the purification steps were performed at 25°C.

Ribulose-1,5-bisphosphate carboxylase was assayed by the method of McFadden (McFadden *et al.*, 1974; Tabita and McFadden, 1974).

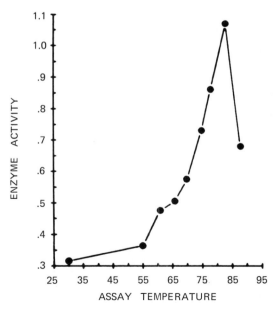

FIG. 4. Effect of assay temperature on the activity of ribulose-1,5-bisphosphate carboxylase in crude extracts from *T. thiopara*. One unit equals 1 μmol $^{14}CO_2$ per minute.

described and may represent an ancestral form of the enzyme (McFadden *et al.*, 1974). It is significant in this respect that *Thiobacillus denitrificans*, a facultatively anaerobic chemolithotroph, contains RuBP carboxylase of intermediate size between *R. rubrum* and obligately aerobic chemolithotrophs (McFadden and Purohit, 1978; McFadden and Tabita, 1974). However, there is disagreement as to which represents the ancestral form (Schloss *et al.*, 1979). Based on matrix analyses of four proteins, a phylogenetic tree has been proposed which places organisms with the complex octameric T-type enzyme at an earlier stage than organisms with the similar O-type RuBP carboxylase (Fitch and Margoliash, 1967; Schwartz and Dayhoff, 1978). Further information from extremely thermophilic autotrophs is needed to develop an accurate phylogenetic tree for autotrophic prokaryotes.

Thermodynamic arguments suggest that the earth's ancient atmosphere was thermal and reducing (Urey, 1952). Thermophilic autotrophs, which may represent primitive life forms, have been discovered in remanent thermal environments that are today considered extreme (Brannan and Caldwell, 1980; Brock *et al.*, 1972; Caldwell *et al.*, 1976). Thermophily thus provides an environmental barrier between primitive bacteria which presumably evolved in thermal environments and more modern mesophilic bacteria. Studies of thermophilic autotrophs may thus provide insights concerning the early biochemical evolution of RuBP carboxylase.

B. HETEROTROPHIC GROWTH

Thermothrix thiopara is capable of heterotrophic growth if a reduced form of sulfur, either organic or inorganic, is provided. It was previously reported (Caldwell *et al.*, 1976) that heterotrophic growth occurred in defined media on organic compounds without a reduced sulfur source. In a subsequent study (Brannan and Caldwell, 1980) growth on organic compounds in defined media did not occur unless the culture was supplied with glutathione, methionine, or thiosulfate. The discrepancy was traced to nutrient carryover from the seed inoculum (grown on nitrate broth). After three successive transfers, growth occurred in defined organic media supplemented with reduced sulfur compounds. However, growth was poor compared to that obtained on complex organic media or on synthetic inorganic media. *T. thiopara*, like *Thiobacillus intermedius* (Roy and Trudinger, 1970), cannot synthesize the necessary enzymes for heterotrophic growth if sulfate is the sole sulfur source.

1. Heterotrophic–Autotrophic Transition

Thermothrix thiopara can be adapted from autotrophic to heterotrophic growth and vice versa, but transfer is difficult, often resulting in loss of the culture. A 2- to 5-day lag period occurs when transferring from one to the other. This lag period may be a result of the time needed for induction, repression, or derepression of enzymes characteristic of heterotrophic or autotrophic metabolism (Matin, 1978). During this transitional period the obsolescence of enzymes is assumed to result in an increased rate of maintenance. This increased rate is added to maintenance requirements that are already extremely high due to high temperature. *Thiobacillus* species also need an adaptation period between heterotrophic and autotrophic growth (McCarthy and Charles, 1974). However, in *T. thiopara* the effects of high temperature and metabolic transition may be synergistic, making transition from autotrophy and heterotrophy unusually difficult.

2. Glucose Uptake and Metabolism

T. thiopara can be grown in a mineral salts medium containing either glucose (0.5%) or amino acids (20 mM each of methionine, glutamate, aspartate, and serine) and supplemented with small amounts of yeast extract (0.05%) to stimulate otherwise poor growth on defined organic media. Cells from these cultures accumulate and respire glucose (Brannan and Caldwell, 1980). Both amino acid and glucose grown cells respire glucose at nearly the same rate. Uptake of glucose-grown cells is twice that of amino acid grown cells; this may be due to an inducible transport system specific for glucose. Autotrophically grown cells accumulate glucose at one-third the rate of heterotrophically grown cells; glucose respiration to CO_2 is one-eighth that of heterotrophically adapted cells. This decreased rate is not unusual and, as for the thiobacilli, is presumably due to repression of glucose catabolism when grown autotrophically on thiosulfate (Matin and Rittenberg, 1970a). Unlike *Thiobacillus intermedius* (Matin and Rittenberg, 1970b), complete repression does not occur since autotrophically grown cells are capable of low rates of glucose respiration. The ability to assimilate preformed organic compounds from the environment gives the organism a selective advantage over autotrophs without this ability (London and Rittenberg, 1966).

Although *T. thiopara* is capable of glucose respiration, glucose alone does not support growth unless supplemented with reduced sulfur compounds. *Thiobacillus thioparus*, *T. neapolitanus*, and *T. thiooxidans* also do not grow on glucose. These lack the Entner–Doudoroff and Embden–Meyerhoff pathways but do have the pentose-phosphate pathway. Thus they are capable of dissimilating glucose to pyruvate as evidenced by the presence of glucokinase, glucose-6-P dehydrogenase, 6-phosphogluconate dehydrogenase activity, and by the presence of the pentose-phosphate shunt (Campbell *et al.*, 1966; Johnson and Abraham, 1969).

3. Reduced Sulfur Requirements

Inability to grow on glucose has been related to the requirement for a reduced sulfur source. *Thiobacillus novellus* is a facultative chemolithotroph that can utilize glucose only if sulfur-containing organics or a reduced inorganic sulfur source is available (Matin, 1978). This, as with *T. thiopara*, is presumably due to an inability to assimilate SO_4^{2-} from the medium (Smith and Rittenburg 1974). The reduced sulfur sources added to glucose-containing media, however, are also potential energy sources; thus it is difficult to determine whether growth occurs due to the supply of a usable sulfur source or an alternative energy source.

4. Heterotrophic Denitrification

T. thiopara uses nitrate as its terminal electron acceptor when grown anaerobically in nitrate broth (Caldwell *et al.*, 1976; Brannan and Caldwell, 1980). Anaerobic oxidation occurs with production of dinitrogen and nitrous oxide; nitrite is also detected. Ammonia is not detected even after addition of 2 *N* NaOH. When grown on yeast extract (5 g/liter) or nutrient broth, each supplemented with 1 g/liter KNO_3, nitrate is reduced to nitrite. Growth rates on yeast extract and nutrient broth with or without 1.0 g/liter KNO_3 are identical when grown aerobically. No growth occurs anaerobically on these media without added nitrate and the amount of growth obtained is proportional to the amount of added nitrate up to 1 g/liter. Under aerobic conditions, little or no effect from added nitrate is seen.

Growth does not occur on anaerobic, nitrate-containing mineral salts medium supplemented with either thiosulfate, sulfide, or elemental sulfur. *T. thiopara* is capable of anaerobic respiration on heterotrophic media but, like *Thiobacillus* A2 (*Thiobacillus versutus*) (Taylor and Hoare, 1969), it does not use NO_3^- when grown autotrophically. This might be due to the requirement for elemental oxygen in the metabolism of sulfur compounds as for many *Thiobacillus* species (Charles and Suzuki, 1966; Suzuki, 1965a,b; Suzuki and Silver, 1966). However, such a requirement would be inconsistent with the view that oxygen in production of sulfate is derived from water rather than elemental oxygen (Lu and Kelly, 1984). No such requirement occurs during autotrophic sulfur oxidation and simultaneous denitrification by *Thiobacillus denitrificans* (Justin and Kelly, 1978).

IV. Growth Kinetics

In contrast to mesophilic bacteria (Monod, 1942; Ng, 1969; Senez, 1962), thermophiles generally attain maximal yields at temperatures below their optimum for growth rate (Coultate and Sunderam, 1975). This is due to high maintenance requirements in thermophiles which usually result in lower yields compared to metabolically similar mesophiles (Sukatch and Johnson, 1972; Zeikus, 1979). *Thermothrix thiopara* also has high maintenance requirements; but despite this, its high overall metabolic efficiency (reflected by higher actual growth yields: Y_{max}) is greater than any known mesophilic sulfur oxidizer. It also attains its maximal yield at temperatures that are optimal for growth rate rather than at temperatures below this optimum as for other thermophiles. As a result, *T. thiopara* appears to be uniquely adapted to growth and metabolism of sulfur compounds at high temperature.

High maintenance requirements do not necessarily dictate low yields. Maintenance can be represented as the consumption of potential biomass (Pirt, 1975) and results in a lower growth efficiency (E_g) and yield. Growth efficiency can be defined as the fraction of potential growth obtained and is equal to $\mu/(\mu + a)$ where a is the specific maintenance rate and μ is the growth rate (Brannan and Caldwell, 1982). Alternatively, E_g can be determined from the ratio of Y_{max}/Y_G where Y_{max} is the actual yield observed at μ_{max} and Y_G is the potential theoretical yield in the absence of maintenance. A low E_g indicates poor growth efficiency and high maintenance energy requirements. Despite poor growth efficiency, actual yields (Y_{max}) can still be high if the potential theoretical yield in the absence of maintenance (Y_G) is high. Final yields are not only a function of maintenance but also depend on energy available from substrate and how well the organism conserves substrate energy. Therefore, low yield due to high maintenance may be overcome in thermophiles if metabolism is thermodynamically more favorable at the higher temperature or if it has evolved a more efficient mechanism of energy conservation.

A. GROWTH IN CONTINUOUS CULTURE

Thermothrix thiopara appears to have either a more thermodynamically favorable metabolism at high temperature (increased free energy yield from substrate oxidation) or a more efficient mechanism of energy conservation than mesophilic sulfur-oxidizers. In thiosulfate-limited continuous culture, the maximum specific growth rate (0.57 hour^{-1}), specific maintenance rate (0.11 hour^{-1}), actual molar growth yield at μ_{max} ($Y_{max} = 16$ g mol^{-1}), and theoretical molar growth yield ($Y_G = 24$ g mol^{-1}) are all higher for *T. thiopara* (72°C) than for mesophilic (25–30°C) *Thiobacillus* species (Brannan and Caldwell, 1983). Table II shows the kinetic parameters of *T. thiopara* and Table III shows comparative parameters for mesophilic bacteria. The growth efficiency of *T. thiopara* at μ_{max} (0.84) is lower than that for *Thiobacillus ferrooxidans* (0.94) and *Thiobacillus denitrificans* but greater than that of *Thiobacillus neapolitanus* (0.60).

The high maintenance rates of *T. thiopara* do not prevent high yields. Although the losses are severe in thermophiles, they can be compensated for if the potential biomass (Y_G) is higher. Thus, although *T. thiopara* has higher maintenance expenditures than most *Thiobacillus* species, it still has a higher overall metabolic (energy) efficiency reflected by higher observed yields.

TABLE II

KINETIC PARAMETERS OF *T. thiopara* GROWN IN CONTINUOUS CULTURE
AT 65, 70, AND 75°C[a]

Temperature (°C)	a (hour^{-1})	μ_{max} (hr^{-1})	Y_{max} (g mol^{-1})[b]	Y_G (g mol^{-1})	$E_g{}^c$
65	0.42 ± 0.14	0.36 ± 0.004	19.03 ± 2.2	41.4 ± 12	0.47 ± 0.08
70	0.11 ± 0.04	0.57 ± 0.009	20.27 ± 2.7	24.3 ± 4	0.84 ± 0.04
75	0.15 ± 0.09	0.54 ± 0.01	14.29 ± 1.1	18.8 ± 4	0.78 ± 0.1

[a] Values are reported as the mean with 95% confidence limits ($n = 6$).
[b] Y_{max} is the actual yield at μ_{max} obtained from linear regression equations for plots of $1/Y$ versus $1/D$ and of q vs D. This value can be calculated using the formula $Y_{max} = (E_g)(Y_G)$.
[c] Calculated with the formula $E_g = \mu_{max}/(\mu_{max} + a)$. The growth efficiency, E_g, is the fraction of potential biomass (Y_G) conserved at μ_{max}. Taken from Brannan and Caldwell (1983) with permission from the American Society for Microbiology.

Actual cell yields (Y_{max}) for *T. thiopara* at 65 or 70°C are not significantly different. In general, however, thermophiles show higher yields at temperatures below their thermal optimum for growth (Coultate and Sundaram, 1975). The effect is due to increased maintenance requirements at the higher temperatures (Zeikis, 1979). Higher maintenance requirement for *T. thiopara* at 65°C (0.42 hour^{-1}), compared with the requirement at 70 or 75°C (0.11 or 0.15 hour^{-1}, respectively), contradicts this generalization.

In the case of *T. thiopara* metabolic efficiency is higher than in other mesophilic sulfur oxidizers. The Y_{max} $S_2O_3{}^{2-}$ for *T. thiopara* at 65, 70, or 75°C is greater than that of any known mesophilic sulfur chemoautotroph despite its higher maintenance requirements. The high yield despite high maintenance is likely to be a result of a different, and more efficient, mechanism of energy conservation that more than compensates for the losses due to maintenance. This is also seen in other thermophiles. For example, growth yields of *Methanobacterium thermoautotrophicum* grown at 55°C on methane are of the same order as those of the mesophile *Methanobacterium barkeri* (Schonheit *et al.*, 1980; Taylor and Pirt, 1977). *Thermoactinomyces* species (Lee and Humphrey, 1979) grown at 55°C on glucose have higher molar growth yields than *Microbacterium thermosphactum* (Hitchener *et al.*, 1979) grown at 25°C with limiting glucose (Table III). Therefore, thermophiles are not necessarily limited to lower growth yields simply because they have a higher maintenance requirement. In fact, some thermophiles appear to have evolved more efficient mechanisms of energy conservation to compensate for the increased maintenance requirements at high temperature. This has not occurred in all thermophiles, however. For example,

TABLE III

KINETIC PARAMETERS OF FIVE BACTERIA GROWN IN CONTINUOUS CULTURE UNDER VARIOUS CONDITIONS OF TEMPERATURE AND LIMITING SUBSTRATE

Organism	Reference	Growth conditions	a (hr^{-1})	μ_{max} (hr^{-1})	Y_{max} $(g\ mol^{-1})$[a]	Y_G $(g\ mol^{-1})$	E_g[b]
Thiobacillus ferrooxidans	Eccleston and Kelly (1978)	Aerobic; thiosulfate limiting; 30°C	0.0073	0.129	7.0	7.48	0.94
Thiobacillus denitrificans	Justin and Kelly (1978)	Aerobic; thiosulfate limiting; 25°C	0.007	0.13	13.8	14.69	0.94
		Anaerobic; thiosulfate limiting; 25°C	0.015	0.08	9.6	11.37	0.84
Thiobacillus neapolitanus	Hempfling and Vishniac (1967)	Aerobic; thiosulfate limiting; 25°C	0.303	0.48	8.3	13.9	0.60
Thiomicrospira denitrificans	Hoor (1981)	Anaerobic; thiosulfate limiting; 25°C	0.0079	0.06	1.9	5.65	0.88
Microbacterium thermosphactum (psychrotroph)	Hitchener et al. (1979)	Aerobic; glucose limiting; 25°C	0.015	0.495	70.8	73.0	0.97
		Anaerobic; glucose limiting; 25°C	0.018	0.46	44.2	46.0	0.96
Thermoactinomyces sp.	Lee and Humphrey (1979)	Aerobic; glucose limiting; 55°C	0.01	0.36	73.7	76.0	0.97

[a] Y_{max} is the actual yield at μ_{max}, calculated with the formula $Y_{max} = (E_g)(Y_G)$. This value could also be obtained by reading the values of Y at μ_{max} from a plot of $1/Y$ versus $1/\mu$.

[b] See footnote c, Table II. Taken from Brannan and Caldwell (1983) with permission from the American Society for Microbiology.

some thermophilic fermenters have lower yields than their mesophilic counterparts (BenBassat and Zeikus, 1981).

Reasons for optimization of sulfur autotrophy at higher temperature might include not only more efficient mechanisms of energy conservation but also increased energy yields from sulfur compounds, a relative reduction in the oxygenase function of RuBP carboxylase, and increased solubility of sulfur at the high temperatures.

Higher yields despite high maintenance may be a result of thermodynamically favorable metabolism at high temperature. The implication for industrial microbiology is that many microbial processes are not necessarily optimized to the conditions under which the organisms evolved. If evolution has occurred at both moderate and extreme temperatures, determination of Y_{max}, μ_{max}, a, and Y_G for additional extreme thermophiles may reveal that some metabolic processes are more favorable at higher temperatures.

B. Growth within Surface Microenvironments

Microorganisms growing on surfaces are a significant component of the microbiota in nearly all natural ecosystems (Geesey *et al.*, 1978; Fletcher, 1979). Attachment to solid surfaces may confer an increased growth capability in oligotrophic environments (Heukelian and Heller, 1940; Zobell, 1943; Jannasch, 1958; Marshall, 1976) since organic nutrients may adsorb to surfaces to create a nutrient-rich microenvironment at the solid–liquid interface. However, only a few studies have quantitated the attachment and growth of autotrophic microorganisms on surfaces (Staley, 1971; Brannan and Caldwell, 1982). In studies of *Thermothrix thiopara*, attachment rates and growth rates were determined on a variety of surfaces including nutrient-enriched membranes and crystals of both calcite and pyrite (naturally occurring minerals *in situ*). Attachment of *T. thiopara* to surfaces is necessary in fast-flowing spring environments where planktonic cells are rapidly washed out.

Several mathematical models have been used to quantitate microbial surface growth rates in nature including the exponential growth equation (Bott and Brock, 1970a,b; Brock, 1971), surface colonization equation (Caldwell *et al.*, 1981; Brannan and Caldwell, 1982), and surface growth rate equation (Caldwell *et al.*, 1983a). The exponential growth equation considers only growth, whereas the surface colonization equation includes simultaneous microbial growth and attachment. Thus the colonization rate is equal to the sum of the instantaneous growth rate (μN) and attachment rate (A) as described by the differential

$$dN/dt = \mu N + A \tag{1}$$

where N = number of cells on the surface (cells field^{-1})
   t = incubation period (hours)
   μ = specific growth rate (hour^{-1})
   A = attachment rate (cells field^{-1} hour^{-1}).
Integration gives the colonization equation [Eq. (2)]. It predicts the number of cells, N, present on the surface at any point in time, t:

$$N = (A/\mu)e^{\mu t} - A/\mu \qquad (2)$$

The effect of integration is to continuously sum the number of attaching cells and their progeny from the time of each attachment. This predicts the number of cells (N) present on the surface at any point in time. The growth rate equation [Eq. (3)] gives the specific growth rate as a function of cell distribution (Caldwell et al., 1983a):

$$\mu = \ln(N/C_4 + 1)/t \qquad (3)$$

where C_4 = number of colonies containing three to six cells.

1. Mineral Colonization in Continuous Culture

Kieft and Caldwell (1984b) studied colonization of calcite and pyrite by $T.$ $thiopara$ both in the spring environment and in laboratory chemostat studies where growth rates of cells colonizing the surfaces could be compared with those of suspended cells. They found that cells colonizing calcite and pyrite in the chemostat grew at about one-third the rate of suspended cells and attached to pyrite faster than to calcite. $T.$ $thiopara$ thus grew more slowly when colonizing mineral surfaces than when growing as a cell suspension. The lower growth rate on surfaces may be due to a reduced cell surface area for nutrient uptake, an increased specific maintenance rate, or the depletion of substrate within the hydrodynamic boundary layers of surface microenvironments. This contradicts the view that bacteria grow at faster rates when attached to solid surfaces. In studies of $T.$ $thiopara$ the growth-limiting nutrient was an inorganic ion, which may not be concentrated at solid–liquid interfaces as in the case of many organic molecules. For attached cells, there could be a reduction in available cell surface area for nutrient transport. The original hypothesis of increased growth rates on surfaces (Zobell, 1943) may thus apply primarily to heterotrophs. Studies of surface colonization by $Pseu$-$domonas$ $aeruginosa$, on surfaces in continuous culture, confirm that heterotrophic growth is as rapid or more rapid than that for suspended cells in the same culture (Malone and Caldwell, 1983).

Although the chemical environment in continuous culture is constant, the hydrodynamics of surfaces are not. As a result, the laminar flow velocities and/or surface turbulence remain undefined, and it is not possible to determine whether these factors might have affected the outcome of surface colonization studies. Studies in capillaries with controlled laminar flow would be useful to quantitate those effects.

Thermothrix thiopara is able to use pyrite as its sole energy source (Caldwell *et al.*, 1983); it does not grow faster on pyrite than on other surfaces, even in thiosulfate-limited chemostats. Thus, the rate of oxidation for the sulfide moiety of pyrite is not significant when compared with the rate of thiosulfate oxidation. *Thiobacillus ferrooxidans* and other bacteria capable of directly oxidizing minerals (Duncan *et al.*, 1967; Beck and Brown, 1968; Arkensteyn, 1979), have not been studied on mineral surfaces in chemostat culture.

2. Mineral Colonization in Situ

Although *T. thiopara* does not grow faster on pyrite than on calcite, it does attach to pyrite at a significantly higher rate (Kieft, 1984a). The surface charge and hydrophobicity characteristics of pyrite may favor nonspecific bacterial adsorption, or the sulfide moiety of the mineral could elicit a specific chemoadherent response. There is a lack of significant differences between *in situ* rates of attachment to calcite and pyrite, however, because attachment rates *in situ* are low due to rapid dilution of planktonic cells. The lack of significant differences between *in situ* growth rates on calcite and pyrite surfaces can be explained by the excess of sulfide (1.0 ppm) *in situ* (Caldwell *et al.*, 1983). As in continuous culture, pyrite oxidation appears to be insignificant (Kieft, 1982; Kieft and Caldwell, 1984a,b).

V. Cell Morphology and Fine Structure

Morphology of *Thermothrix thiopara* is variable both in the laboratory and *in situ* depending on growth conditions. When grown under oxygen-limited conditions, filamentous cells are produced. Under highly aerated conditions, the filaments fragment to form rod-shaped cells ($0.5-1 \times 3-5$ μm) that possess a single polar flagellum. This variable morphology is an adaptive response that permits *T. thiopara* to locate at sulfide–oxygen interfaces that are optimal for growth in its natural environment (see Section VI,A).

T. thiopara stains gram negative; its cell wall has a typical gram-negative appearance (Fig. 5) and it is lysozyme and penicillin sensitive (Brannan and Caldwell, 1980). The filamentous cells are composed of a series of protoplasts separated by peptidoglycan septa. The filament is surrounded by an outer membrane which invaginates slightly between individual protoplasts but does not completely separate them.

Flagellated unicells of *T. thiopara* colonize pyrite and calcite surfaces of neutral geothermal springs. These develop into filaments and form nets composed of a mixture of filaments and unicells that intertwine to form

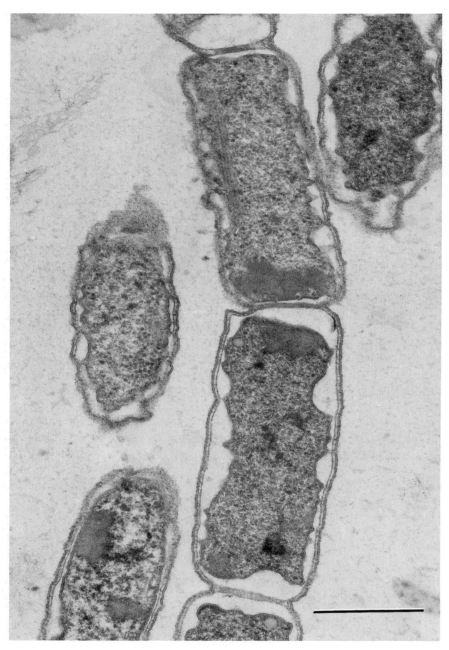

Fig. 5. Electron micrograph of *Thermothrix thiopara* taken from Jemez Spring. The membrane is separated from the cell wall. Bar equals 0.5 μm. The cells possess a gram-negative wall, septa, and unit membrane. The cell wall is composed of three electron-dense layers. Taken from Caldwell *et al.* (1976) with permission of the *Canadian Journal of Microbiology.*

macroscopically visible "streamers" up to 10 cm or more in length (Figs. 6, 7, and 8). This variable morphology and its relationship to positioning of *T. thiopara* within sulfide–oxygen interfaces is discussed further in Section VI.

Induction of the filamentous growth form is dependent on electron accep-tor limitations. *In situ*, filaments are found only on the reducing side of the sulfide–oxygen interface. Once these streamers grow further downstream to oxidizing waters, they break into flagellated unicells which are quickly washed out unless they enter the hydrodynamic boundary layers of tufa, and back-growth occurs allowing cells to reach more favorable reducing environ-ments. Attachment occurs preferentially upstream and on pyrite surfaces where electron acceptor-limiting conditions exist and result in filamentous cell growth. Downstream, under oxidizing conditions, mats of unicells re-sult. Cell filaments can also be induced in the laboratory by limiting the amount of oxygen available. In batch culture, the filament length of *T. thiopara* cells grown in nutrient broth with only a 50-ml headspace of air is typically longer (33.5 ± 5.1 µm) than cells grown where the headspace (and available oxygen) is increased to 150 ml (4 ± 0.5 µm). Formation of "stream-ers" composed of filament bundles like those found *in situ* can be obtained only in continuous culture under oxygen-limiting conditions. In continuous culture, as in the spring, unicells and unattached bacteria are quickly wash-ed out. The selective pressure for formation of streamers is greater under continuous culture conditions than in batch culture due both to the physical conditions and precise control of the rate-limiting electron acceptor.

VI. Ecology

A. Adaptation to Sulfide–Oxygen Interfaces

Thermothrix thiopara is found *in situ* within sulfide–oxygen interfaces of neutral geothermal springs located at Mammoth Hot Springs (Yellowstone National Park, Wyoming) and Jemez Hot Springs (Jemez Springs, New Mexico).

1. Geochemistry

The primary water source of the Jemez Springs is located at the top of a porous tufa mound. These mounds form as a result of carbonate precipitation when carbon dioxide outgasses from the spring and the temperature drops (Jennings, 1971). The sulfide present in the spring serves as the energy source for *T. thiopara*, which grows in well-defined regions of the spring as macroscopic bundles of filaments (streamers). The pH varies from 6.3 at the source to 6.7 at the outflow, because of carbon dioxide outgassing. The temperature is 74°C at the source. The dilution rate of the spring is rapid,

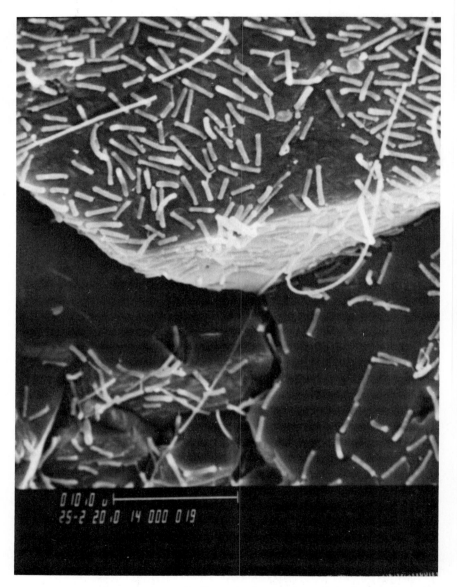

FIG 6. Scanning electron micrograph of natural calcium carbonate samples from Jemez Hot Spring showing *T. thiopara* unicells colonizing freshly precipitated calcium carbonate (bar = 10 μm). Taken from Kieft and Caldwell (1984a) with permission of the *Geomicrobiology Journal*.

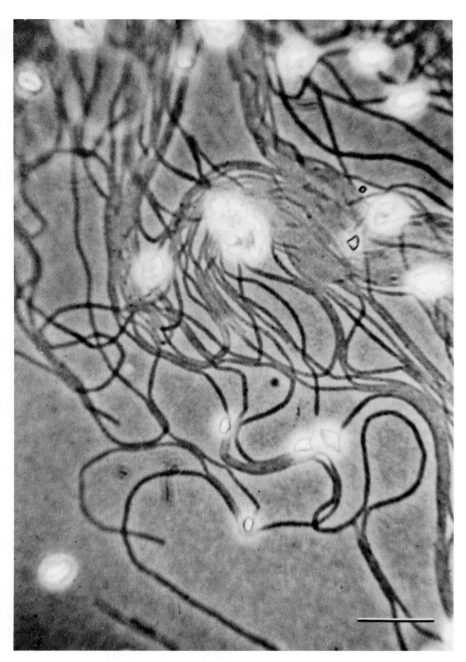

FIG. 7. Filaments of *Thermothrix thiopara* collected from the sulfide–oxygen interface of a thermal spring (Jemez Springs, NM). The filaments form streamers which are visible macroscopically as shown in Fig. 8. Phase micrograph. Bar equals 10 μm.

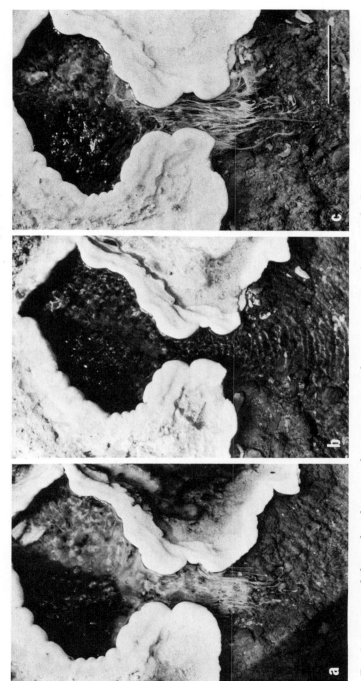

FIG. 8. Streamers of *Thermothrix thiopara* as they appear *in situ*. The streamers consist of intertwined cell filaments. Bar equals 10 cm. (a) Original appearance of spring. (b) Appearance of spring after displacing the sulfide–oxygen (anoxic–oxic) interface for 8 days using a plastic covering. (c) Return of streamers 8 days following the restoration of the chemical interface. Taken from Caldwell *et al.* (1984) and reproduced with the permission of the *Geomicrobiology Journal*.

and typically varies from 100 to 150 hour^{-1}. These variations are due to continued changes in the morphology of the mound. Water at the source contains 0.96 ppm (29 μmol/liter) sulfide and 0.3 ppm (9.4 μmol/liter) oxygen, indicating that small amounts of oxygen enter during the passage of water through the interior of the porous tufa.

In addition to sulfide, other sulfur compounds are present at constant levels: 1.8 ppm thiosulfate (15.9 μmol/liter), 1.8 ppm polythionates (7.9 μmol/liter), 54 ppm sulfate (564 μmol/liter), and 1.0 ppm elemental sulfur (32.5 μmol/liter). As the source water moves from the mouth further downstream, the concentration of oxygen increases and the concentration of sulfide decreases. Within the sulfide–oxygen interface, *T. thiopara* forms macroscopic bundles of filaments (streamers) up to 100 mm in length. Streamers are visible at locations within the sulfide–oxygen interface where the molar ratio of sulfide to oxygen (HS^-/O_2) is between 0.3 and 0.8, the optimum for metabolism of sulfur autotrophs being 0.5. Macroscopically visible "filaments" do not occur at ratios above 0.8 or below 0.3.

2. Positioning within Sulfide–Oxygen Interfaces

The mechanism by which *Thermothrix thiopara* localizes at these sulfide–oxygen interfaces is unique. In most environments, bacteria position themselves along chemical gradients using either flagella, gas vesicles, or gliding motility (Baas-Becking, 1925; Caldwell, 1977; Jorgenson, 1982). For *T. thiopara*, the induction of a filamentous growth form by oxygen-limited growth conditions results in the precise positioning of cell streamers at sulfide–oxygen interfaces that are optimal for growth. The high temperature and continual outgassing of carbon dioxide purges the spring of sulfide after spring waters travel only a few decimeters from the source. *Thermothrix thiopara* requires both sulfide and oxygen for growth and forms streamers only in a relatively narrow zone where atmospheric oxygen mixes with the sulfide of the spring. Filamentous cells are induced only under oxygen (electron acceptor)-limited conditions and flagellated unicells only under sulfide (electron donor)-limited conditions. This mechanism results in the formation of streamers positioned precisely at the interface (Caldwell *et al.*, 1984). Flagellated unicells colonize surfaces of tufa, primarily pyrite and calcite (Kieft and Caldwell, 1984a,b). Pyrite is colonized more rapidly than calcite due to a more rapid rate of attachment (Kieft and Caldwell, 1984a) which may be due to a chemoadherent response. As a result, the cells preferentially attach upstream from the interface. Once attachment occurs on the reducing side of the interface, growth becomes oxygen limited. These conditions induce filamentous growth which results in streamers that move the cells downstream to regions where conditions are more oxidizing (Fig. 9). The elongation of filamentous cells continues to move the streamers beyond the interface where sulfide is limiting. Then the filaments fragment to form

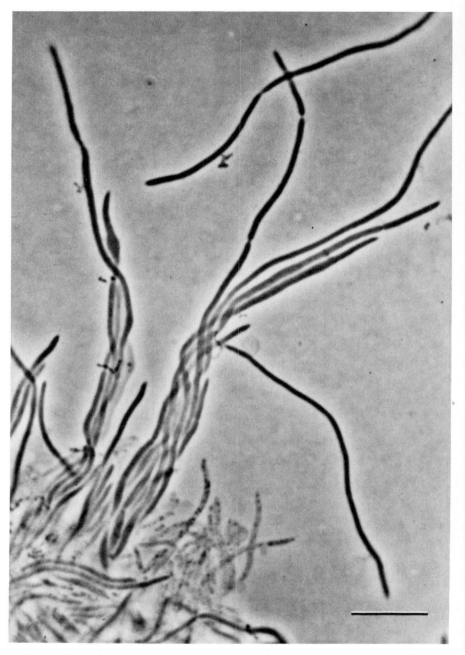

FIG. 9. Filaments of *Thermothrix thiopara* grown using nutrient broth under oxygen-limited conditions, required for the induction of filaments. Phase micrograph. Bar equals 10 μm. Taken from Caldwell *et al.* (1984) and reproduced with the permission of the *Geomicrobiological Journal*.

flagellated unicells that colonize other sulfide–oxygen interfaces on the tufa and result in backgrowth to areas upstream. If the unicells colonize tufa on the oxidizing side of the interface, they form colonies of unicells instead of cell filaments. These eventually form a mat of cells. The calcite beneath the mat is weathered away by the sulfuric acid produced until the mat becomes a net suspended in the liquid phase.

The positioning of streamers has been studied by shifting the interface with a plastic covering (Caldwell *et al.*, 1984). Due to carbon dioxide outgassing, the covered portion of the spring becomes reducing and turns black in color from deposition of pyrite. The streamers elongate and eventually disappear from the original interface and reposition at the new interface that forms at the edge of the covering. Upon removing the cover, the original interface and streamers are restored, confirming that the streamers actively reposition at new sulfide–oxygen interfaces. This also confirms that sulfide and oxygen serve as electron donor and acceptor for *T. thiopara in situ* (Caldwell *et al.*, 1984). When membrane enrichment vials (Brannan and Caldwell, 1982) containing either oxygen or thiosulfate are positioned on the oxidizing and reducing sides of the interface, thiosulfate stimulates colonization on the oxidizing side while oxygen does not. On the reducing side of the interface, colonization is stimulated by oxygen but not by thiosulfate.

T. thiopara forms filaments only when oxygen is limiting and rapid dilution selects for attachment of filaments. Similar adaptations may be found in many organisms that become oxygen limited within surface biofilms located in flowing waters. Investigations of the thiobacilli and other organisms in oxygen-limited continuous culture are needed to determine whether they are also capable of producing filaments and/or streamers. The use of high dilution rate culture systems may be important in the selection and morphogenesis of organisms that attach to mineral and other surfaces (Yoshikama and Takiguchi, 1979). If the dilution rate is zero, as in batch culture, or less than μ_{max}, as in continuous culture, this selects for planktonic rather than for attached bacteria and requires that the cells grow in the presence of metabolic waste. This explains, in part, the easy isolation of acidophilic thiobacilli and the difficulty in isolation of attached filamentous sulfur autotrophs requiring a neutral pH. As a result, additional continuous culture studies and continuous enrichments will be needed to understand adequately the effects of dilution rate, surface chemistry, surface hydrodynamics, and nutrition on the selection of sulfur autotrophs and other bacteria associated with surfaces.

B. Mineral Weathering

Thermothrix thiopara colonizes the carbonate and sulfur minerals of tufa mounds. It weathers calcium carbonate by production of sulfuric acid from hydrogen sulfide and contributes to the porosity of tufa.

Kieft and Caldwell (1984a) used SEM and energy-dispersive X-ray analysis to observe bacterial colonization and weathering of sulfur, pyrite, and calcite incubated within the sulfide–oxygen interfaces of Jemez Spring (see Section VI,A). Cleaved calcite crystals were rapidly colonized by bacteria that pitted surfaces and rounded edges. Biofilms of *T. thiopara* developed on crystal surfaces and deposited sulfur. Cleaved calcite crystals incubated *in situ*, but exposed to constant UV irradiation, showed dissolution though they were not colonized by bacteria. Pyrite crystals were also colonized but were not noticeably altered even after 6–8 days' exposure. The crystals became coated with a calcium carbonate precipitate which was subsequently colonized and, as described above, solubilized by bacterial action. UV-irradiated controls were not colonized or weathered but became coated with calcium carbonate. Sulfur was colonized and became coated with carbonate and iron precipates.

Carbonate minerals are particularly susceptible to solubilization by sulfur-oxidizing bacteria (Ehrlich, 1981; Krumbein, 1968; Wagner and Schwartz, 1965). Sulfur-oxidizing organisms have been isolated from the surfaces of decaying limestone buildings (Paine *et al.*, 1933), concrete (Parker, 1947; Fjerdingstad, 1969), and have been implicated as agents resonsible for the formation of caverns in karst environments (Pohl and White, 1965). Metal sulfides serve as energy sources for sulfur-oxidizing bacteria. They are solubilized by direct oxidation of the sulfide moiety (Duncan *et al.*, 1967; Beck and Brown, 1968; Arkensteyn, 1979) or through indirect oxidation by ferric iron produced by *T. ferrooxidans* (Silverman and Ehrlich, 1964). The oxidation of sulfide minerals is exploited in the solubilization of metals during microbial leaching processes (Brierley, 1978), and causes acidic drainage from coal mines (Paciorek *et al.*, 1981; Olson *et al.*, 1981).

The dissolution of calcium carbonate by *T. thiopara* occurs concurrently with calcium carbonate precipitation, resulting in formation of a porous tufa mound. Although the pH appears to be neutral throughout the spring, microenvironmental differences in pH control the dissolution and precipitation of calcium carbonate. The primary site of dissolution is beneath biofilms of *T. thiopara*. Carbonate dissolution in the absence of direct bacterial colonization (in UV-irradiated calcite controls) suggests that physical erosion is also a factor.

Although *T. thiopara* oxidizes pyrite and sulfur in the laboratory, these minerals are not noticeably weathered by bacterial action *in situ*. The spring contains an excess of sulfide which is the preferred substrate for growth. Instead of being solubilized, these minerals become coated with a calcium carbonate precipitate that is colonized and subsequently solubilized. Sulfur is also deposited on mineral surfaces due to both chemical and biological activity.

VII. Comparison of *T. thiopara*
to Other Extreme Thermophiles

Tables IV, V, and VI compare the characteristics of *Thermothrix thiopara* to other extreme thermophiles (growing above 70°C) and pyrophiles (growing over 100°C). Both the metabolic diversity and the rapid growth rates of these organisms indicate that 70°C is not an extreme temperature for life. Thus it is not reasonable to assume that 37°C, the body temperature of *Homo sapiens*, is "normal" and optimal for the majority of prokaryotic organisms. The first prokaryotes are more likely to have evolved when the earth was at a temperature exceeding 70°C (Corlis *et al.*, 1981). Study of extreme thermophiles have shown that they are unique not only because they grow at high temperature but also because of their cell structure, biochemistry, and metabolism. This has required that several new kingdoms be proposed and indicates that the diversity of thermophiles could eventually exceed that of mesophiles.

Thermophilic life forms may become increasingly important in biotechnology (Sonnleitner, 1983). They greatly increase the range of conditions under which chemical reactions can be biologically catalyzed. In the case of *Sulfolobus*, this range has been extended to include a pH of 1 at a temperature of 80°C and has potential applications in microbial leaching of metal sulfide ores. In the fermentation industry, thermophiles offer the possibility of thermal self-sufficiency. Large-scale industrial fermentations require expensive cooling systems to remove metabolic heat. Heating is less expensive and may not be required if thermophiles are grown in insulated reactors. Thermophiles may be less susceptible to microbial contamination problems because extreme thermophiles are not as abundant in the biosphere as are mesophiles and the possibility of contamination at extreme temperatures is greatly reduced. Anaerobic fermentations by thermophiles offer the possibility of continuous distillation during fermentation. Metabolic heat may provide a portion of the energy necessary for the first distillation step in ethanol production at 78.2°C (optimal for the distillation of ethanol from ethanol–water mixtures). Continuous distillation would also prevent end-product inhibition.

The industrial potential for utilization of extreme thermophiles remains to be determined. However, if successfully applied, thermophiles could have a major impact on fermentation technologies. Further isolation and characterization will be necessary to define the range of conditions under which they can be utilized and the range of chemical conversions which they carry out. If the next decade is as fruitful in this respect as the last, numerous new applications are in the offing.

TABLE IV

Comparison of Extremely Thermophilic Prokaryotes—Heterotrophs[a]

	Organisms						
	Thermo-coccus celer	Thermofilium pendens	Desulfuro-coccus mucosus	Thermus aquaticus	Thermo-microbium thermophila	Thermo-anaerobacter ethanolicus	Bacillus flavothermus
Cardinal temperatures (°C)							
Minimum				40	47	37	30
Optimum	92	90	85	70	70	69	65
Maximum		100		79	85	78	70
Substrates respired (K_s)							
Glucose						−	+
Pectin						−	
Cellobiose						−	
Maltose						−	+
Xylose						−	
Cellulose						−	
Substrates fermented							
Glucose				−		+	+
Pectin				−			
Cellobiose				−		+	
Maltose				−		+	+
Xylose				−		+	
Cellulose				−		−	
Fermentation end products							
Acetate				−		+	
Lactate				−		+	
Gas				−		+	
Acid				−		+	
Butanol				−			
Ethanol				−		+	
Butanediol (acetoin)				−			+
Flagellation				−		+	
Cytochromes						−	
Spores				−			+
Fermentation		+		−		+	+
Aerobic respiration		−		+		−	+
Heterotrophic denitrification		+					+
Respiration using S°	+	+	+	−	−	−	−
Autotrophic denitrification							−
Facultatively anaerobic		−				−	+
Facultatively autotrophic						−	
Sulfide production (from thiosulfate)							−

			Organisms				
Thermobacteroides acetoethylicus	Clostridium thermosulforogenes	Thermoleophilum album	Thermoplasma	Bacillus stearothermophilus	Clostridium thermocellum	Desulfovibrio thermophilus	Clostridium thermohydrosulfuricum
40	35	45	37	40			
65	60	60	59	55	55		
80	75	70	65	70		70	70
−		13–20C al-	−	+	−		
−		kanes			−		
−					−		
−				+	−		
−					−		
+	+		−	+			
	+		−				
+	+		−		+		
+	+		−				
	+		−	+			
	−		−		+		
+	+		−	+	+		
−	+		−	+	+		
+	+		−	−			
+	+		−	+	+		
			−				
+	+		−		+		
			−	+			
+	+		−				
−				+	−		
−	+	+	−	+	+		
+	+	−	−	+	+		
−	−	+	+	+	−		
				+			
−	−	−	−	−	−	−	
				−	−		
				+	−		
				−	−		
+	S° from S$_2$O$_3^{2-}$						

(*continued*)

TABLE IV (Continued)

	Thermococcus celer	Thermofilium pendens	Desulfurococcus mucosus	Thermus aquaticus	Thermomicrobium thermophila	Thermoanaerobacter ethanolicus	Bacillus flavothermus
				Organisms			
pH optimum	6.0	5.0	5.8	7.6	7.0	7.0	
GC%		57		66	69	35	
Tryptophanase (indol)						–	
Amylase							
Cellulase							–
Oxidase						–	
Catalase						–	+
Gelatin liquefication						+	–
Motility							
Color				Yellow orange	Yellow orange	White	Dark yellow
μ_{max} (hour^{-1})	0.23	0.07	0.87	0.83	2.1	0.29	
Cell yields (grams cells per mole glucose)							
Maintenance rates							
Peptidoglycan (diamino acid as constituent)		No muramic acid		ORN			
Cell wall structure				G–		Atypical (G+)	
Cell morphology		Rods to filaments, 0.2 × 1 μm, rarely branched		Rods to filaments, 0.6 × 5 μm	Rods, 3 μm	Rods to filaments, unequal division	Rods, 3 μm
Gram stain	G–	G–	G–	G–	G–	G±	G+
Lysozyme sensitivity	–	–	–				
Antibiotic susceptibilities							
Penicillin						–	
Chloramphenicol						+	
Polymycin B						+	
Tetracycline						–	
Erythromycin						–	
Streptomycin							
References	Stetter and Zellig (1985)	Zillig et al. (1983), Stetter and Zellig (1985)	Stetter and Zellig (1985)	Brock and Freeze (1969)	Phillips and Perry (1976)	Wiegel and Ljungdahl (1981)	Heinen et al. (1982)

[a] See Sonnleitner (1983) for additional information on strains of potential industrial value.

	Organisms						
Thermo-bacteroides aceto-ethylicus	Clostridium thermo-sulforogenes	Thermo-leophilum album	Thermo-plasma	Bacillus stearo-thermophilus	Clostridium thermocellum	Desulfovibrio thermophilus	Clostridium thermo-hydro-sulfuricum
7.0	6.0	7.0	2.0	7.0			
31	32	69		50			
				−			
				−	+		
−	−						
				+	+		
		White					
1.4		0.11					
18.3							
		DAP ORN LYS					
No outer membrane Rods, 0.6 × 2 μm	No outer membrane Rods to filaments, 0.5 × 72 μm	Rods, 0.4 × 1.0 μm	No cell wall Pleiomorphic, 0.3–2 μm	No outer membrane Rods, 0.7 × 3 μm	No outer membrane Oval, 0.6 × 3 μm		
G−	G−	G−	G±	G±	G−		
				−			
+	+						
−	+						
	+						
+	+						
BenBassat and Zeikus (1981)	Schink and Zeikus (1983)	Zarilla and Perry (1984)	Darland et al. (1970)	Coultate and Sundaram (1975)	Viljoen et al. (1926)	Wiegel et al. (1979), Roza-nova and Khudyakova (1974)	

TABLE V

COMPARISON OF EXTREMELY THERMOPHILIC PROKARYOTES—AUTOTROPHS

	Organisms							
	Pyrodictium occultum	*Thermoproteus tenax*	*Thermodiscus maritimus*	*Methanococcus jannaschii*	*Sulfolobus acidocaldarius*	*Thermothrix thiopara*	*Hydrogenobacter thermophilus*	*Methanobacterium thermoautotrophicum*
Cardinal temperatures (°C)								
Minimum	—	—	—	50	55		37	40
Optimum	105	88	87	85	73	72	72	67
Maximum	—	—	—	85	85	80	80	70
Electron donor	H_2	H_2	H_2	H_2	$S°$	H_2S	H_2	H_2
Electron acceptor	$S°$	$S°$	$S°$	CO_2	O_2	O_2	O_2	CO_2
Autotrophs	Obligate	Facultative	Obligate	Obligate		Facultative	Obligate	
Flagellation				Two bundles near one pole				
Cytochromes								
Spores								
Fermentation					+	+	+	
Aerobic respiration								
Heterotrophic denitrification						+		
Respiration using $S°$								
Autotrophic denitrification						+		
Facultatively anaerobic						+		
Facultatively autotrophic					+			
Sulfide production								
pH optimum	6.5	5.5	6.5	6.0	2.5		7.0	
GC%	63		53	31	64		43	52

	Fischer et al. (1983), Stetter and Zellig (1985)	Fischer et al. (1983), Stetter and Zellig (1985)	Fischer et al. (1983), Stetter and Zellig (1985)	Jones et al. (1983)	Brierly and Brierly (1982)	Braman and Caldwell (1980, 1982, 1983), Caldwell et al. (1976)	Kawasumi et al. (1984)
Tryptophanase							
Amylase							
Cellulase							
Oxidase							
Catalase							
Gelatin liquidation							
Motility							
Color				Yellow			Tan
μ_{max} (hour^{-1})			0.39	1.6		0.57	
Cell yields							
Maintenance rates							
Peptidoglycan (di-amino acid as constituent)						+	
Cell wall structure				Single layer, surface covered by hexagonal structure	No peptidoglycan		
Cell morphology	Hyphal network	Rods to filaments (highly branched)	Flat disk	Irregular cocci, 1.5-μm diameter	Lobed spheres, 1-μm diameter	Rod shaped, 0.4 × 2.5 μm	Rods to filaments, 0.5 × 3 μm
Gram stain				G−	G−	G−	G+
Lysozyme sensitivity							
Antibiotic susceptibilities							
Penicillin							
Chloramphenicol							
Polymycin B							
Tetracycline							
Erythromycin							
Streptomycin							
References	Fischer et al. (1983), Stetter and Zellig (1985)	Fischer et al. (1983), Stetter and Zellig (1985)	Fischer et al. (1983), Stetter and Zellig (1985)	Jones et al. (1983)	Brierly and Brierly (1982)	Braman and Caldwell (1980, 1982, 1983), Caldwell et al. (1976)	Kawasumi et al. (1984)

TABLE VI

AMERICAN TYPE CULTURE COLLECTION STRAINS
WITH A RECOMMENDED INCUBATION TEMPERATURE
OF 50°C OR ABOVE[a]

Culture collection number (ATCC)	Organism	Temperature
33488	*Acetogenium kivui*	60°C
27009	*Bacillus acidocaldarius*	55°C
7953	*Bacillus stearothermophilus*	55°C
8005	*Bacillus kaustophilus*	55°C
10149	*Bacillus calidolactis*	55°C
12016	*Bacillus* sp.	55°C
12976	*Bacillus* sp.	55°C
12977	*Bacillus* sp.	55°C
12978	*Bacillus* sp.	55°C
12979	*Bacillus* sp.	55°C
12980	*Bacillus* sp.	55°C
29492	*Bacillus thermodenitrificans*	60°C
29493	*Bacillus thermodenitrificans*	60°C
27405	*Clostridium thermocellum*	55°C
31549	*Clostridium thermocellum*	55°C
31924	*Clostridium thermocellum*	60°C
33223	*Clostridium thermohydrosulfuricum*	65°C
25773	*Clostridium thermosaccharolyticum*	50°C
31925	*Clostridium thermosaccharolyticum*	60°C
19858	*Desulfotomaculum nigrificans*	55°C
33518	*Excellospora viridinigra*	55°C
19998	*Desulfotomaculum nigrificans*	55°C
29096	*Methanobacterium thermautotrophicum*	60°C
29183	*Methanobacterium thermautotrophicum*	60°C
29034	*Micropolyspora faeni*	55°C
33515	*Micropolyspora rectivirgula*	55°C
33909	*Sulfolobus acidocaldarius*	70°C
35091	*Sulfolobus solfataricus*	75°C
31558	*Thermus aquaticus*	65°C
31557	*Thermus lacteus*	65°C
31556	*Thermus rubens*	55°C
27868	*Thermoactinomyces antibodicus*	55°C
31550	*Thermoanaerobacter ethanolicus*	55°C
33075	*Thermoanaerobium brockii*	60°C
29033	*Thermomicrobium fosteri*	60°C
27502	*Thermomicrobium roseum*	70°C
25905	*Thermoplasma acidophilium*	60°C
27656	*Thermoplasma acidophilium*	55°C
27657	*Thermoplasma acidophilium*	55°C

TABLE VI (*Continued*)

Culture collection number (ATCC)	Organism	Temperature
27658	*Thermoplasma acidophilium*	55°C
29244	*Thermothrix thiopara*	70°C
25104	*Thermus aquaticus*	70°C
27737	*Thermus* sp.	70°C
27978	*Thermus* sp.	74°C
27634	*Thermus thermophilus*	75°C
23841	*Thiobacillus thermophilica imschenetskii*	55°C
27599	Unidentified bacteria	60°C
27979	Unidentified bacteria	64°C

[a] From the 15th edition, *American Type Culture Collection Catalogue of Strains* (1982) and March 1984 update.

ACKNOWLEDGMENTS

J. A. Malone and T. W. Trumper are acknowledged for helpful discussions. D. P. Kelly is acknowledged for review of the manuscript.

REFERENCES

Arkensteyn, G. J. M. W. (1979). *Antonie van Leeuwenhoek* **45**, 423–435.
Arrhenius, S. (1927). *Z. Phys. Chem.* (Frankfurt am Main) **130**, 516–519.
Baas-Becking, L. G. M. (1925). *Ann. Bot.* **39**, 613–651.
Baross, J. A. (1983). *Nature* **303**, 423–426.
Baross, J. A., Lilley, M. D., and Gordon, L. I. (1982). *Nature* **298**, 366–368.
Beck, J. V., and Brown, D. G. (1968). *J. Bacteriol.* **96**, 1433–1434.
BenBassat, A., and Zeikus, J. G. (1981). *Arch. Microbiol.* **128**, 365–370.
Bott, T. L., and Brock, T. D. (1970a). *Limnol. Oceanogr.* **15**, 333–342.
Bott, T. L., and Brock, T. D. (1970b). *Appl. Microbiol.* **19**, 100–102.
Brannan, D. K. (1981). Ph.D. dissertation, Univ. of New Mexico.
Brannan, D. K., and Caldwell, D. E. (1980). *Appl. Environ. Microbiol.* **40**, 211–216.
Brannan, D. K., and Caldwell, D. E. (1982). *Microb. Ecol.* **8**, 15–21.
Brannan, D. K., and Caldwell, D. E. (1983). *Appl. Environ. Microbiol.* **45**, 169–173.
Brierley, C. L. (1978). *CRC Crit. Rev. Microbiol.* **6**, 207–262.
Brierley, C. L., and Brierley, J. A. (1982). *Zbl. Bakt. Hyg.* **3**, 289–294.

Brierley, J. A., and LeRoux, N. W., (1977). *Gesellsch. Biotechnol. Frosch. Mongor. Ser.* **4,** 55–56.

Brierley, J. A., Norris, P. R., Kelley, D. P., and LeRoux, N. W. (1978). *Eur. J. Appl. Microbiol. Biotechnol.* **5,** 291–299.

Brock, T. D. (1967). *Science* **158,** 1012–1018.

Brock, T. D. (1971). *Bacteriol. Rev.* **35,** 39–58.

Brock, T. D., and Freeze, H. (1969). *J. Bacteriol.* **98,** 289–297.

Brock, T. D., Brock, M. L., Bott, T. L., and Edwards, M. R. (1971). *J. Bacteriol.* **107,** 303–314.

Brock, T. D., Brock, K. M., Belly, R. T., and Weiss, R. L. (1972). *Arch. Mikrobiol.* **84,** 54–68.

Caldwell, D. E. (1977). *CRC Crit. Rev. Microbiol.* **5,** 305–370.

Caldwell, D. E., Caldwell, S. J., and Laycock, J. P. (1976). *Can. J. Microbiol.* **22,** 1509–1517.

Caldwell, D. E., Brannan, D. K., and Morris, M. E. (1981). *Microbiol. Ecol.* **7,** 1–11.

Caldwell, D. E., Malone, J. A., and Kieft, T. L. (1983a). *Microbiol. Ecol.* **9,** 1–6.

Caldwell, D. E., Brannan, D. K., and Kieft, T. L. (1983b). *In* "Environmental Biogeochemistry" (R. Halberg, ed.), Vol. 35, pp. 129–134. Ecological Bulletin, Stockholm.

Caldwell, D. E., Kieft, T. L., and Brannan, D. K. (1984). *Geomicrobiol. J.* **3,** 181–200.

Campbell, A. E., Hellebust, J. A., and Watson, S. W. (1966). *J. Bacteriol.* **91,** 1178–1185.

Charles, A. M., and Suzuki, I. (1966). *Biochim. Biophys. Acta* **128,** 510–521.

Corliss, J. B., Baross, J. A., and Hoffman, S. E. (1981). *Oceanol. Acta* 59–69.

Coultate, T. P., and Sundaram, T. K. (1975). *J. Bacteriol.* **121,** 55–64.

Darland, G., Brock, T. D., Samsonoff, W., and Conti, S. F. (1970). *Science* **170,** 1416–1418.

Doelle, H. W. (1975). "Bacterial Metabolism," 2nd Ed. Academic Press, New York.

Duncan, D. W., Landesman, J., and Walden, C. C. (1967). *Can. J. Microbiol.* **13,** 397–403.

Eccleston, M., and Kelly, D. P. (1978). *J. Bacteriol.* **134,** 718–727.

Ehrlich, H. L. (1981). *In* "Geomicrobiology." Dekker, New York.

Ferguson, T. J., and Mah, R. A. (1982). *Appl. Environ. Microbiol.* **45,** 265–274.

Fischer, F., Zillig, W., Stetter, K. O., and Schreiber, G. (1983). *Nature* **301,** 511–513.

Fitch, W. M., and Margoliash, E. (1967). *Science* **155,** 279–284.

Fjerdingstad, E. (1969). *Water Res.* **3,** 21–30.

Fletcher, M. (1979). *Arch. Microbiol.* **122,** 271–274.

Fliermans, C. B., and Brock, T. D. (1972). *J. Bacteriol.* **111,** 343–350.

Ford, T. W. (1979). *Biochim. Biophys. Acta* **569,** 239–248.

Gaughran, E. R. L. (1947). *Bacteriol. Rev.* **11,** 189–225.

Geesey, G. G., Mutch, R., Costerton, J. W., and Green, R. B. (1978). *Limnol. Oceanogr.* **23,** 1214–1223.

Gibson, T., and Gordon, R. E. (1974). "Bergey's Manual of Determinative Bacteriology" (R. E. Buchanan and N. E. Gibbons, eds.), Vol. 8, pp. 539–540. Williams & Wilkins, Baltimore.

Heinen, W., Lauwers, A. M., and Mulders, J. W. M. (1982). *Antonie van Leeuwenhoek* **48,** 265–272.

Hempfling, W. P., and Vishniac, W. (1967). *J. Bacteriol.* **93,** 874–878.

Heukelakian, H., and Heller, A. (1940). *J. Bacteriol.* **40,** 547–558.

Hitchener, B. J., Egan, A. F., and Rogers, P. J. (1979). *Appl. Environ. Microbiol.* **37,** 1047–1052.

Hoor, T. A. (1981). *Antonie van Leeuwenhoek J. Microbiol. Serol.* **47,** 231–243.

Hutchinson, M., Johnstone, K. I., and White, D. (1967). *J. Gen. Microbiol.* **47,** 17–23.

Hyon, H. H., Zeikus, J. G., Longin, R., Millet, J., and Ryter, A. (1983). *J. Bacteriol.* **156,** 1332–1337.

Jannasch, H. W. (1958). *J. Gen. Microbiol.* **18,** 609–620.

Jennings, J. N. (1971). "Kurst." M.I.T. Press, Cambridge, Mass.

Johnson, E. J., and Abraham, S. (1969). *J. Bacteriol.* **97,** 1198–1208.

Jones, W. J., Leigh, J. A., Mayer, F., Woese, C. R., and Wolfe, R. S. (1983). *Arch. Microbiol.* **136,** 254–261.

Jorgensen, B. B. (1982). *Philos. Trans. R. Soc. London Ser. B* **298**, 543–561.

Justin, P., and Kelly, D. P. (1978). *J. Gen. Microbiol.* **107**, 123–130.

Kawasumi, T., Igarashi, Y., Kodama, T., and Minoda, Y. (1984). *Int. J. Syst. Bacteriol.* **34**, 5–10.

Kelly, D. P. (1967). *Arch. Mikrobiol.* **58**, 99–116.

Kelly, D. P. (1978). *In* "Companion to Microbiology" (A. T. Bull and P. M. Meadow, eds.), Ch. 5. Academic Press, New York.

Kelly, D. P. (1982). *Phil. Trans. R. Soc. Lond.* **298**, 499–528.

Kieft, T. L. (1982). Ph.D. dissertation, University of New Mexico.

Kieft, T. L., and Caldwell, D. E. (1984a). *Geomicrobiol. J.* **3**, 201–216.

Kieft, T. L., and Caldwell, D. E. (1984b). *Geomicrobiol. J.* **3**, 217–219.

Krumbein, W. E. (1968). *Z. Allg. Mikrobiol.* **8**, 107–117.

Lawlis, V. B., Gordon, G. L. R., and McFadden, B. A. (1979). *J. Bacteriol.* **139**, 287–298.

Lee, S. E., and Humphrey, A. E. (1979). *Biotechnol. Bioeng.* **21**, 1277–1288.

LeRoux, N. W., Wakerley, D. S., and Hunt, S. D. (1977). *J. Gen. Microbiol.* **100**, 197–201.

London, J., and Rittenberg, S. C. (1966). *J. Bacteriol.* **91**, 1062–1069.

Lowry, O. H., Rosebrough, N. J., Farr, A. L., and Randall, R. J. (1951). *J. Biol. Chem.* **193**, 265–275.

Lu, W. P., and Kelly, D. P. (1984). *In* "Microbial Growth on C1 Compounds. Proceedings of the 4th International Symposium" (American Society for Microbiology), pp. 34–41.

McCarthy, J. T., and Charles, A. M. (1974). *Can. J. Microbiol.* **20**, 1577–1584.

McFadden, B. A., and Purohit, K. (1978). *In* "Photosynthetic Carbon Assimilation, Basic Life Sciences" (H. W. Siegelman and G. Hind, eds.), Vol. II. Plenum, New York.

McFadden, B. A., and Tabita, F. R. (1974). *Biosystems* **6**, 93–112.

McFadden, B. A., Tabita, F. R., and Kuehn, G. D. (1974). *In* "Methods in Enzymology" (S. Colowick and P. Kaplan, eds.). Academic Press, London.

Malone, J. A., and Caldwell, D. E. (1983). *Microbiol. Ecol.* **9**, 299–305.

Marshall, K. C. (1976). *In* "Interfaces in Microbial Ecology." Harvard Univ. Press, Cambridge, Mass.

Matin, A., and Rittenberg, S. C. (1970a). *J. Bacteriol.* **104**, 234–238.

Matin, A., and Rittenberg, S. C. (1970b). *J. Bacteriol.* **104**, 239–246.

Matin, A., and Rittenberg, S. C. (1971). *J. Bacteriol.* **107**, 179–186.

Matin, A. (1978). *Annu. Rev. Microbiol.* **32**, 433–468.

Matin, A., Konings, W. N., Kuenen, J., and Emmens, M. (1974). *J. Gen. Microbiol.* **83**, 311–318.

Miquel, P. (1888). *Ann. Microgr.* **1**, 3–10.

Monod, J. (1942). "Recherches sur La Croissance des Cultures Bacteriennes." Hermann, Paris.

Ng, H. (1969). *J. Bacteriol.* **98**, 232–237.

Olsen, G. J., McFeters, G. A., and Temple, K. L. (1981). *Microb. Ecol.* **7**, 39–50.

Paciorek, K. J. L., Kratzer, R. H., Kimble, P. F., Toben, W. A., and Vatasescu, A. L. (1981). *Geomicrobiol. J.* **2**, 363–374.

Paine, S. G., Linggood, F. V., Schimmer, F., and Thruppl, T. C. (1933). *Philos. Trans. R. Soc. London* **222B**, 97–127.

Parker, C. D. (1947). *Nature (London)* 439–441.

Phillips, W. E., Jr., and Perry, J. J. (1976). *Int. J. Syst. Bacteriol.* **26**, 220–225.

Pirie, N. W. (1983). *Nature* **305**, 8.

Pirt, S. J. (1975). *In* "Principles of Microbe and Cell Cultivation." Wiley, New York.

Pohl, E. R., and White, W. B. (1965). *Am. Mineralogist.*

Roy, A. B., and Trudinger, P. A. (1970). *In* "The Biochemistry of Inorganic Compounds of Sulfur." Cambridge Univ. Press, London and New York.

Rozanova, E., and Khydyakova, A. I. (1974). *Microbiologiya* **43**, 1069–1075.

Schink, B., and Zeikus, J. G. (1983). *J. Gen. Microbiol.* **129**, 1149–1158.
Schloss, J. V., Phares, E. I., Long, M. V., Norton, I. L., Stringer, C. D., and Hartman, I. C. (1979). *J. Bacteriol.* **137**, 490–501.
Schonheit, P., Moll, J., and Thauer, R. K. (1980). *Arch. Microbiol.* **127**, 59–65.
Schwartz, R. M., and Dayhoff, M. O. (1978). *Science* **199**, 395–403.
Senez, J. C. (1962). *Bacteriol. Rev.* **26**, 95–107.
Silverman, M. P., and Ehrlich, H. L. (1964). *Adv. Appl. Microbiol.* **6**, 153–206.
Smith, D. W., and Rittenberg, S. C. (1974). *Arch. Microbiol.* **100**, 65–71.
Sonnerat, (1774). *J. Phys.* **3**, 256–257.
Sonnleitner, B. (1983). *In* "Advances in Biochemical Engineering/Biotechnology" (A. Fiechter, ed.), Vol. 28, pp. 69–138. Springer-Verlag, New York.
Staley, J. T. (1971). *J. Phycol.* **7**, 13–17.
Stetter, K. O. (1982). *Nature* **300**, 258–260.
Stetter, K. O., and Zillig (1985). *In* "The Bacteria," Vol. VIII, pp. 85–170. Academic Press, Orlando, Florida.
Sukatch, D. A., and Johnson, M. J. (1972). *Appl. Microbiol.* **23**, 543–546.
Suzuki, I. (1965a). *Biochim. Biophys. Acta* **110**, 97–101.
Suzuki, J. (1965b). *Biochim. Biophys. Acta* **104**, 359–371.
Suzuki, I., and Silver, M. (1966). *Biochim. Biophys. Acta* **122**, 22–33.
Tabita, F. R., and McFadden, B. A. (1974). *J. Biol. Chem.* **249**, 3454–3458.
Tansey, M. R., and Brock, T. D. (1978). *In* "Microbial Life in Extreme Environments" (D. J. Kushner, ed.), pp. 159–216. Academic Press, London.
Taylor, B. F., and Hoare, D. S. (1969). *J. Bacteriol.* **100**, 487–497.
Taylor, B. F., and Hoare, D. S. (1971). *Arch. Mikrobiol.* **80**, 262–276.
Taylor, G. T., and Pirt, S. J. (1977). *Arch. Microbiol.* **113**, 17–22.
Trent, J. D., Chastain, R. A., and Yayanos, A. A. (1984). *Nature* **307**, 737–740.
Trudinger, P. A. (1969). *Adv. Microb. Physiol.* **3**, 111–158.
Urey, H. C. (1952). *Proc. Natl. Acad. Sci. U.S.A.* **38**, 351–363.
Viljoen, J. A., Fred, E. B., Peterson, W. H. (1926). *J. Agric. Sci.* **16**, 1–17.
Vishniac, W., and Santer, M. (1953). *Bacteriol. Rev.* **21**, 195–213.
Wagner, E., and Schwartz, W. (1965). *Z. Allg. Mikrobiol.* **7**, 33–52.
Wainwright, M. (1978). *Sci. Prog. (Oxford)* **65**, 459–475.
Walsby, A. E. (1983). *Nature* **303**, 381.
Watson, S. W., Graham, L. B., Remsen, C. C., and Valois, F. W. (1971). *Arch. Mikrobiol.* **76**, 183–203.
Weimer, P. J., Wagner, L. W., Knowlton, S., and Ng, T. K. (1984). *Arch. Microbiol.* **138**, 31–36.
White, R. H. (1984). *Nature* **310**, 430–432.
Wiegel, J., and Dykstra, M. (1984). *Appl. Microbiol. Biotechnol.* **20**, 59–65.
Wiegel, J., and Ljungdahl, L. G. (1981). *Arch. Microbiol.* **128**, 343–348.
Wiegel, J., Ljungdahl, L. G., and Rawson, J. R. (1979). *J. Bacteriol.* **Sept.**, 800–810.
Williams, R. A. D., and Hoare, D. S. (1972). *J. Gen. Microbiol.* **70**, 555–566.
Yoshikawa, H., and Takiguchi, Y. (1979). *Appl. Environ. Microbiol.* **38**, 200–204.
Zarilla, K. A., and Perry, J. J. (1984). *Arch. Microbiol.* **137**, 286–290.
Zeikus, J. G. (1979). *Enzyme Microb. Technol.* **1**, 243–252.
Zeikus, J. G., Hegge, P. W., and Anderson, M. A. (1979). *Arch. Microbiol.* **48**, 41–48.
Zillig, W., Gierl, A., Schreibner, G., Wunderl, S., Janekovik, D., Stetter, K. O., and Klenk, H. P. (1983). *Syst. Appl. Microbiol.* **4**, 79–87.
Zobell, C. E. (1943). *J. Bacteriol.* **46**, 34–56.

Enzyme-Linked Immunoassays for the Detection of Microbial Antigens and Their Antibodies

JOHN E. HERRMANN

Division of Infectious Diseases
University of Massachusetts Medical School
Worcester, Massachusetts

I. Introduction

Enzyme-immunoassay (EIA) evolved as a result of the findings by Nakane and Pierce (1966) that antibodies could be labeled with enzymes for use in histochemical staining procedures, and by Catt and Tregear (1967) who described solid-phase radioimmunoassays (RIA). The substitution of enzyme labels for radioactive ones in the solid-phase RIA resulted in solid-phase EIA tests for human chorionic gonadotropin (Van Weeman and Schuurs, 1971) and for IgG detection (Engvall and Perlmann, 1971). The latter authors coined the term "enzyme-linked immunosorbent assay (ELISA)" for solid-phase EIA tests.

Initial solid-phase EIA tests were not as sensitive as the corresponding RIA, but improvement in enzyme-labeling techniques have made the two types of assays comparable for detecting a number of antigens and antibodies. In some systems where RIA and EIA have been directly compared,

271

ADVANCES IN APPLIED MICROBIOLOGY, VOLUME 31

there is little difference in the sensitivity or specificity of the two assays (Sarkkinen *et al.*, 1981c).

The advantages of enzyme labels over radioactive ones are mainly convenience in use, in that the labeled immunoreagents are stable for long periods, and the precautions and disposal procedures required for radioisotopes are unnecessary. In addition, the use of chromogenic substrates for the enzyme labels permits visual interpretation of test results in some cases. The only real disadvantages of EIA tests are the loss of antibody reactivity that may result from conjugation to enzymes, and the limits of substrate detection. For example, use of enzymes that have molecular weights higher than that of IgG molecules such as β-D-galactosidase (MW 540,000 Da) can cause steric hindrance of antibody activity (Herrmann and Morse, 1974). With regard to limits of substrate detection, improvement of enzyme detection by use of fluorogenic, luminescent, or radioactive substrates (reviewed by Yolken, 1982) has been proposed.

The general principles of EIA tests and details of earlier studies have been reviewed a number of times (Yolken, 1980, 1982; Hildebrand, 1979; Voller *et al.*, 1981) and will not be repeated in detail here. Rather, the major emphasis will be on current developments in EIA methodology and the application of EIA to diagnosis of infectious diseases. This will include tests for both antigen and antibody detection in viral, rickettsial, bacterial, and mycotic infections. EIA tests for diagnosis of parasitic agents, hormones, and other antigens have been described but will not be discussed here.

II. Design of Enzyme Immunoassays

A. SOLID-PHASE ASSAYS

The solid-phase or heterogeneous EIA requires immobilization of antigens or antibodies on a solid surface as a means of separating antigen–antibody complexes. Solid-phase surfaces used to immobilize antigens or antibodies have for most applications been polystyrene beads, tubes, and wells of microtiter plates, or wells of polyvinyl chloride microtiter plates. Coupling of proteins to these surfaces is usually done by passive adsorption. More recently, adsorption of antigens or antibodies to nitrocellulose membranes has been adapted to detection of viruses by EIA (Bode *et al.*, 1984). Covalent linkage of antibodies or antigens to a variety of surfaces, including porous glass (Lynn, 1975), nylon (Hendry and Herrmann, 1980), cellulose (Ferrua *et al.*, 1979), agarose (Streefkerk and Deelder, 1975), and polyacrylamide (Avrameas and Guilbert, 1971) has been described.

The majority of solid-phase EIA tests that have been found to be clinically useful utilize plastic microtiter plates or beads, with antigen or antibody passively adsorbed to the solid phase. In situations where a given antigen does not readily attach, antibody to the antigen is applied first. For some antigens, nonspecific adsorbents such as poly-L-lysine have been used to enhance antigen adsorption, or the Clq component of complement to capture antigen–antibody complexes (Yolken, 1982). Most of the assays described are of the noncompetitive type, although a number of competitive assays have been described. The disadvantage of many competitive assays for antigen detection is that they use labeled antigen, which is usually more difficult to prepare than labeled antibody. With the advent of monoclonal antibodies, competitive assays for specific antibodies are becoming more common. A number of different formats of EIA tests for antigens and antibodies are possible. These are discussed in Section II,C and D.

B. Homogeneous Assays

To avoid the need for separation of antigen–antibody complexes, homogeneous EIA tests were developed (Rubinstein *et al.*, 1972). Homogeneous assays are based on the reaction of antigen with an antibody–enzyme complex. This results in steric hindrance of the enzyme, which causes a decrease in product after reaction with enzyme substrate. The major advantage of homogeneous assays is that they do not require the separation and washing steps required in heterogeneous assays. The major disadvantage of this type of EIA is that it has been difficult to apply it to detection of high-molecular-weight antigens with the degree of sensitivity required. Thus, the homogeneous assay has been used mainly for detection of hormones, drugs, and other low-molecular-weight substances. Means to improve the sensitivity of homogeneous assays have been devised (review by Yolken, 1982) but have not been widely used. Thus, the applications discussed below will be limited to EIA tests of the solid-phase type.

C. Types of Assays for Antibodies

Enzyme immunoassays for antibodies to microbial agents have been utilized for almost all of the common infectious diseases, because the sensitivity required is well within the range of EIA. A summary of the procedures most often used is given in Table I. The choice of assay depends on the sensitivity required, the availability of reagents, and whether a class-specific test is desired. For detection of IgM, either the noncompetitive EIA or class capture methods can be used. In the noncompetitive EIA, enzyme-labeled antiglobulin (step 3) specific for IgM is used. The advantage of IgM capture

TABLE I

METHODS FOR ANTIBODY EIA

Noncompetitive EIA	Competitive EIA
1. Specific antigen is attached by passive adsorption or with specific antibody	1. Specific antigen is attached by passive adsorption or with specific antibody
2. Test serum or solution containing specific antibody is added	2. Test serum, or solution plus enzyme-labeled specific antibody, is added
3. Enzyme-labled antiglobulin is added	3. Enzyme substrate is added
4. Enzyme substrate is added	4. Substrate hydrolysis product is inversely proportional to the amount of antibody present
5. Substrate hydrolysis product is proportional to the amount of antibody present	

Class-specific capture EIA	
Labeled antigen EIA	Labeled antibody EIA
1. Solid phase is coated with IgM-specific anti-IgM antibody	1. Solid phase is coated with IgM-specific anti-IgM antibody
2. Test serum or solution containing IgM is added	2. Test serum or solution containing IgM is added
3. Enzyme-labeled specific antigen is added	3. Specific antigen is added
4. Enzyme substrate is added	4. Antisera specific for antigen are added (unlabeled for indirect test, enzyme-labeled for direct test)
5. Substrate hydrolysis product is proportional to the amount of IgM present	5. If indirect test, enzyme-labeled antiglobulin is added
	6. Enzyme substrate is added
	7. Substrate hydrolysis product is proportional to the amount of IgM present

methods is that there is less problem with sera containing rheumatoid factor (RF), although precautions must be taken in designing the test to avoid weakly reactive sera containing RF (Parry, 1984; Briantais *et al.*, 1984).

The use of enzyme-labeled antigen in tests for IgM has been useful for diagnosis of a number viral infections, e.g., cytomegalovirus (Schmitz *et al.*, 1980), Epstein–Barr virus (Schmitz, 1982), and flaviviruses (Schmitz and Emmerich, 1984) but requires purified antigen for labeling. Thus, it is limited to those agents where production and purification of antigen is relatively simple and offers improved diagnosis as well.

The majority of class-specific EIA tests are for IgM (sometimes referred to as MAC-ELISA for M antibody capture-ELISA), because IgM is the most important serological marker indicating recent infection when only single serum specimens are available. However, substitution of other class-specific markers can be used.

D. Assays for Antigen Detection

Methods for detecting antigen by EIA can be done by competitive and noncompetitive formats, as with assays for antibodies. The lower limit of sensitivity for detecting antigen by EIA in most studies is approximately 100 pg to 1 ng, although lower limits have been described (Kato *et al.*, 1975). This level of sensitivity is sufficient to detect virtually all culture-propagated infectious agents, but is not always sensitive enough to detect antigen directly in clinical specimens.

The type of formats most often used are shown in Table II.

III. Factors in Sensitivity and Specificity

A number of factors determine how efficient an assay is in detecting antigens and antibodies. Some are inherent and cannot be controlled, e.g., the amount of antigen that is usually present in a positive clinical specimen, and others, such as test design, can be controlled. Some of the more important variables that can be controlled are discussed here.

A. Enzyme, Substrates, and Labeling Procedures

Most EIA formats require covalent coupling of enzymes to antibody or antigen. A number of enzymes and coupling techniques have been tried (reviewed by Yolken, 1982). The most consistent results have been obtained with horseradish peroxidase coupled by use of periodate (Nakane and Kawaoi, 1974) and alkaline phosphatase coupled by use of glutaraldehyde (Avrameas, 1969). Most of the assays found to be clinically useful in diagnosing infectious diseases use chromogenic substrates, although fluorogenic and radioactive substrates have been described (Yolken, 1982; Avrameas and Guesdon, 1982), as have luminescent ones (review by Seitz, 1984).

A more recent development in EIA which has found application in diagnostic microbiology is the use of avidin and biotin (Guesdon *et al.*, 1979). The test is based on the high affinity constant for binding biotin to avidin. The most common method is to use specific antibody labeled with biotin; the indicator system is enzyme-labeled avidin. A recent adaptation of this method using biotinillated beta-lactamase in combination with avidin was effective for detecting rotavirus antigen (Yolken and Wee, 1984).

TABLE II

Assays for Antigen Detection

Noncompetitive EIA

Direct EIA	Indirect EIA
1. Antigen-specific antibody is attached to solid phase	1. Antigen-specific antibody is attached to solid phase
2. Test specimen containing antigen is added	2. Test specimen containing antigen is added
3. Enzyme-labeled specific antibody is added	3. Specific antibody (prepared in species different from that used in step 1) is added
4. Enzyme substrate is added	4. Enzyme-labeled antiglobulin specific for antibody used in step 3 is added
5. Substrate hydrolysis product is proportional to the amount of antigen present	5. Enzyme substrate is added
	6. Substrate hydrolysis product is proportional to the amount of antigen present

Competitive EIA

Enzyme-labeled antigen method	Enzyme-labeled antibody method
1. Antigen-specific antibody is attached to solid phase	1. Antigen is attached to solid phase (may require use of specific antibody)
2. Test specimen containing antigen is added	2. Test specimen containing antigen plus enzyme-labeled specific antibody is added
3. Enzyme-labeled antigen is added	3. Enzyme substrate is added
4. Enzyme substrate is added	4. Substrate hydrolysis product is inversely proportional to the amount of antigen present
5. Substrate hydrolysis product is inversely proportional to the amount of antigen present	

B. Antibody Immobilization

In addition to antibody affinity and the sensitivity of the indicator system used, the sensitivity of many antigen detection systems depends on the amount of antibody that can be immobilized effectively on a solid phase. Methods for immobilization of antibody on plastic surfaces are usually based on simple adsorption, although convalent-linking methods have been utilized. The amount of immunoglobulin that can be immobilized on various plastics is given in Table III (Herrmann, 1981). Increasing the amount of antibody bound to a solid-phase surface should result in increased sensitivity

TABLE III

QUANTITATION OF IMMUNOGLOBULIN IMMOBILIZATION ON PLASTICS[a]

Immunoglobulin (Ig) bound	Solid-phase support	Immobilization technique	Maximum Ig bound (ng/mm^2)
Sheep IgG	Polystyrene (latex)	Covalent linkage	100
Rabbit IgG	Nylon beads	Covalent linkage	590
Rabbit IgG	Polystyrene (latex)	Adsorption	5.7
Rat Ig	Polymethylmetoacrylic beads	Adsorption	0.9
Bovine IgG1	Polystyrene tubes	Adsorption	3.2
Bovine IgM	Polystyrene tubes	Adsorption	2.9
Human IgG	Polystyrene (latex)	Adsorption	3.6

[a] Adapted from Herrmann (1981).

of the EIA. However, it has been noted that increasing the concentration of antibody for coating surfaces beyond 10 μg/ml does not give an increase in immunoassay sensitivity. This is apparently due to desorption of antibody from the plastic surface, and steric hindrance of antibody that is adsorbed.

Solid-phase surfaces other than plastic tubes, beads, or particles, such as porous glass (Lynn, 1975) have been used to immobilize antibody. However, the majority of EIA tests for microbial infections use either plastic plates or beads. A more recent development mentioned above that may be applicable to a variety of microbial antigens and antibodies is the use of nitrocellulose membrane disks as a solid phase. This was successfully developed as a visual readout method for detecting adenovirus antigens and antibodies (Bode *et al.*, 1984).

C. IMMUNOREAGENTS

The use of antibodies that are highly specific and have high affinity is the most critical aspect of most EIA techniques. An example of how the reagents used determines EIA effectiveness was shown in the two papers discussed below on detection of *Clostridium dificile* toxin, where the sensitivity was increased from 58.6 to 95% by changing the immunoreagents used (Laughon *et al.*, 1984). The diluents used for antigen preparation can also alter the sensitivity. Disrupting microbial agents with detergents or other chemicals may increase the sensitivity of some assays but decrease others (Yolken, 1982). A number of diluents not usually used in EIA tests was tested by Conroy and Esen (1984) for adsorbing a plant protein to polystyrene. These included detergents, acids, alcohols, and urea. Use of alcohols or urea increased the EIA substantially. Whether this would be applicable to microbial antigens remains to be tested.

IV. Application of Assays for Microbial Antibodies

The diagnosis of infectious disease has been accomplished in the past by either isolation of the infectious agent, or by measuring serological conversion to a given agent. Serological conversion can be measured by a high level of IgM, or by an increase in total antibody in convalescent sera compared with acute sera. The use of EIA is an extension of previously used serological tests, using enzyme-labeled antibody or antigen to determine antibody content. Direct detection of antigen by EIA represents a more dramatic departure from previous methods based on culture. Also, the method has enabled detection of infectious agents that are difficult to cultivate, such as hepatitis A virus and rotavirus, or agents that cannot be cultivated, such as hepatitis B virus (Section V).

A. BACTERIAL AND MYCOTIC INFECTIONS

A summary of the bacterial and mycotic infections for which EIA serological tests have been devised is given in Table IV,A. Because EIA techniques are not difficult to develop for antibody detection, there is an ever-increasing number of tests reported. However, developing tests for antibody which have consistent diagnostic accuracy can be difficult, and the tests cited in Table IVA report various degrees of sensitivity and specificity. In a review by Hill and Matsen (1983) a sensitivity greater than 95% was reported for some assays, but the sensitivity was as low as 50% in many others. Thus, each test must be carefully examined to determine how useful it is for a specific infection. In addition to problems with low sensitivity, the major drawback to many assays is the lack of standardization. Without the availability of standard serum samples for evaluation of assays, new assays require testing by a number of investigators before their validity can be assessed. In many instances, however, serological diagnosis is the only means available to many laboratories for diagnosing some of the more exotic diseases. For example, for diagnosis of Lyme disease, culture of the causative spirochete is possible, but is often ineffective. Use of an EIA test for specific IgM and IgG response in patients with proven Lyme disease was diagnostic in 11 of 12, and the EIA gave no false positive results in 40 control subjects (Craft et al., 1984). Early recognition of disease is another area where EIA tests for specific IgM and IgG may be the best or only method available to some, such as diagnosis of tuberculous meningitis (Hernandez et al., 1984), although new-

TABLE IV

Enzyme Immunoassay for Serodiagnosis:
Bacterial and Mycotic Infections (A)
and Viral and Rickettsial Infections (B)

(A) Microbial agent	Primary reference
Bacteria	
Bacillus anthracis	Johnson-Winegar (1984)
Bacteroides fragilis	Rissing *et al.* (1979)
Bordetella pertussis	Vijanen *et al.* (1982)
Borrelia burgdorferi	Craft *et al.* (1984)
Brucells abortus	Magee (1980)
Chlamydia trachomatis	Rai *et al.* (1983); Duc-Goiren (1983)
Clostridium tetani (toxin)	Haberman and Heller (1976)
Corynebacterium diphtheriae (toxin)	Svenson and Larsen (1977)
Escherichia coli (toxin)	Jodal *et al.* (1974)
Francisella tularensis	Carlsson *et al.* (1979)
Legionella pneumophila	Farshy *et al.* (1978)
Leptospira icterohaemorrhagiae	Adler *et al.* (1980)
Mycobacterium tuberculosis	Kalish *et al.* (1984)
Mycobacterium leprae	Douglas and Worth (1984)
Mycoplasma hominis	Miettinen *et al.* (1983)
Mycoplasma pneumoniae	Raisanen *et al.* (1980)
Neisseria gonorrhoeae	Ison *et al.* (1981)
Neisseria meningitidis	Sippel *et al.* (1980)
Salmonella sp.	Carlsson *et al.* (1972)
Shigella dysenteriae	Lindberg *et al.* (1984)
Staphylococcus aureus	Mackowiak and Smith (1978)
Streptococcus group A	Russel *et al.* (1976)
Treponema pallidum	Veldkamp and Visser (1975)
Ureaplasma urealyticium	Wiley and Quinn (1984)
Vibrio cholerae (toxin)	Majumbar *et al.* (1981)
Yersinia enterocolitica	Granfors *et al.* (1981)
Yersinia pestis	Cavanaugh *et al.* (1979)
Fungi and actinomycetes	
Candida albicans	Hommel *et al.* (1976)
Aspergillus fumigatus	Hommel *et al.* (1976)
Nocardia brasiliensis	Zlotnick *et al.* (1984)
Paracoccidioides brasiliensis	Mendes-Giannini *et al.* (1984)

(B) Virus or rickettsia	Primary reference
Adenovirus	Voller *et al.* (1976)
Coxsackievirus	Voller *et al.* (1976)
Cytomegalovirus	Voller and Bidwell (1976)

(*continued*)

TABLE IV (*Continued*)

(A) Microbial agent	Primary reference
Dengue virus	Dittmar *et al.* (1979)
Epstein–Barr virus	Hopkins *et al.* (1982)
Hepatitis A virus	Mathiesen *et al.* (1978)
Hepatitis B virus	Feinstone *et al.* (1979)
Herpes simplex virus	Gilman and Docherty (1977)
Influenza A virus	Leinikki and Passila (1977)
Japanese encephalitis virus	Burke *et al.* (1982)
Lassa fever virus	Niklasson *et al.* (1984)
Measles virus	Voller and Bidwell (1976)
Mumps virus	Ukkonen *et al.* (1980)
Parainfluenzae b	Drow *et al.* (1979); Pepple *et al.* (1980); Sippel *et al.* (1984)
Rabies virus	Atanasiu *et al.* (1977)
Rickettsia typhi	Halle *et al.* (1977)
Rochalimia quintana	Herrmann *et al.* (1977)
Rift Valley fever virus	Niklasson *et al.* (1984)
Ross River virus	Oseni *et al.* (1983)
Rotavirus	Yolken *et al.* (1978)
Rubella virus	Gravell *et al.* (1977)
St. Louis encephalitis virus	Monath *et al.* (1984)
Tick-borne encephalitis virus	Hofmann *et al.* (1979)
Varicella–zoster virus	Forghani *et al.* (1978)

er developments in use of latex particle agglutination for diagnosis of this disease may prove more useful than serology (Krambovitis *et al.*, 1984).

B. Viral and Rickettsial Infections

The applications of antibody EIA for diagnosis of viral and rickettsial infections are given in Table IV,B. As discussed above for bacterial and mycotic infections, serological diagnosis by EIA for most of the agents listed is still largely experimental. The most frequently used applications are in screening for immune status, such as rubella testing, for cytomegalovirus antibody, and for antibodies to hepatitis B antigens. Serological diagnosis of infectious mononucleosis is also the method of choice. The heterophile antibody test used is not specific for Epstein–Barr virus, and the test is usually done by agglutination, but EIA tests may be more sensitive (Halbert, 1982). Specific EIA tests for antibody to Epstein–Barr virus components have also been described (Hopkins *et al.*, 1982) but are not yet widely used.

Because many viruses are difficult to isolate, or haven't yet been cultivated, serological tests are often the most useful for diagnosis. The use of IgM capture EIA for determining recent viral infection is becoming more common, and may provide aid in diagnosis where antigen detection methods are not available.

V. Application of Assays for Microbial Antigens

The use of EIA tests for detection of microbial antigens provides an alternative to culture as a means for direct identification of a specific microbial agent. It also provides a means to detect microbial agents which have not been successfully propagated. The detection of circulating antigen or detection of antigen in other body fluids by EIA is more difficult than detection of antibody because of the sensitivity required, and because of interfering substances in specimens such as feces and respiratory secretions. For this reason, very few antigen detection assays have the sensitivity and specificity required to be used as a primary diagnostic test. The number of tests that have been developed, however, is impressive and because of the possibilities for rapid, specific diagnosis, the interest in antigen detection by EIA remains high.

A. BACTERIAL AND MYCOTIC INFECTIONS

The tests developed for bacterial infections are primarily for diseases which have causative agents difficult to culture, or where rapid diagnosis will permit prompt treatment. As can be noted by comparing Tables IV,A and V,A, there are far fewer EIA antigen detection tests than antibody tests, for reasons cited above. The efficiencies of the assays reported are variable, but none is as sensitive as the corresponding culture technique. Only one test is commercially available at this writing, an EIA for detecting *Neisseria gonorrhoeae* antigens. The EIA has been evaluated by anumber of laboratories (Table V,A). In general, almost all reports have found that the EIA was equivalent to culture for detection of gonorrhoeae in males. In females, sensitivities have ranged from 74.4 (Papasian *et al.*, 1984) to 90.9% (Danielson *et al.*, 1983) and specificities from 86.5 (Manis *et al.*, 1984) to 100% (Danielson *et al.*, 1983). Two other extensive studies reported specificities of 98% (Stamm *et al.*, 1984; Demetriou *et al.*, 1984). Thus, the reliability of EIA appears to depend on the efficiency of the culture method used, and perhaps variability in performing the test itself.

EIA tests have also been developed for another agent of sexually transmitted disease, *Chlamydia trachomatis*. This agent is more difficult to cultivate than *N. gonorrhoeae*, in that cell cultures are required. Only one EIA has

TABLE V

Enzyme Immunoassays for the Detection of Microbial Antigens
in Clinical Specimens: Bacteria and Fungi (A) and Viruses (B)

(A) Microbial agent	References
Bacteria	
Bacterioides fragilis	Rissing *et al.* (1984)
Chlamydia trachomatis	Herrmann *et al.* (1983); Jones *et al.* (1984); Stokes and Khan (1984)
Clostridium botulinum (toxins)	Dezfulian *et al.* (1984)
Clostridium dificile (toxins)	Lyerly *et al.* (1983); Laughon *et al.* (1984)
Hemophilus influenzae b	Drow *et al.* (1979); Pepple *et al.* (1980); Sippel *et al.* (1984)
Legionella pneumophila	Sathapatayavongs *et al.* (1982); Berdel *et al.* (1979); Bibb *et al.* (1984)
Mycobacterium sp.	Sada *et al.* (1983)
Mycoplasma hominis	Miettinen *et al.* (1984)
Neisseria gonorrhoeae	Aardoom *et al.* (1982); Burns *et al.* (1983); Danielsson *et al.* (1983); Papasian *et al.* (1984a,b); Martin *et al.* (1984); Schacter *et al.* (1984); Stamm *et al.* (1984); Nachamkin *et al.* (1984); Demetriou *et al.* (1984); Manis *et al.* (1984)
Neisseria meningitidis	Sippel and Voller (1980); Sippel *et al.* (1984); Sugasawara *et al.* (1984)
Streptococcus group A	Knigge *et al.* (1984)
Streptococcus pneumoniae	Sippel *et al.* (1984); Yolken *et al.* (1984)
Yersinia pestis	Williams *et al.* (1984)
Fungi	
Candida albicans	Segal *et al.* (1979)

(B) Virus	References
Adenovirus	Sarkkinen *et al.* (1980); Johansson *et al.* (1980); Sarkkinen *et al.* (1981c); Harmon and Pawlick (1982)
Coronavirus	Macnaughton *et al.* (1983)
Coxsackievirus	Yolken and Torsch (1980, 1981)
Cytomegalovirus	Pronovost *et al.* (1982)
Hepatitis A virus	Mathiesen *et al.* (1978)
Hepatitis B virus	Wolters *et al.* (1976)
Herpes simplex virus	Miranda *et al.* (1977); Grillner and Landqvist (1983); Land *et al.* (1984); Lawrence *et al.* (1984); Morgan and Smith (1984); Nerukar *et al.* (1984); Warford *et al.* (1984)
Influenza A virus	Berg *et al.* (1980); Harmon *et al.* (1983); Sarkkinen *et al.* (1981b)
Parainfluenza virus	Sarkkinen *et al.* (1981a,c)
Respiratory syncytial virus	Chao *et al.* (1979); Sarkkinen *et al.* (1981c); Hornsleth *et al.* (1981); McIntosh *et al.* (1982); Meurman *et al.* (1984)
Rotavirus	Yolken *et al.* (1977, 1980); Sarkkinen *et al.* (1980)
Varicella–zoster virus	Ziegler (1984)

been examined with a significant number of samples, a commercially produced test under development (Chlamydiazyme, Abbott Laboratories). Premarket evaluation of this test on 416 patients showed the EIA had a sensitivity of 83% (63/76) and a specificity of 94% (Jones *et al.*, 1984). In a larger study involving 2384 specimens the EIA had a sensitivity of 83% and a specificity of 94% (Herrmann *et al.*, 1983).

Mycoplasma hominis may also be involved in sexually transmitted disease, and diagnosis by culture requires expertise. An antigen EIA has been developed (Miettinen *et al.*, 1984) and was positive for six specimens positive by culture. More extensive evaluation is required to determine the utility of the assay as a screening procedure.

Detection by EIA of bacterial antigens in cerebrospinal fluid (CSF) and respiratory tract secretions has also been attempted, with good results for some antigens. Yolken *et al.* (1984) reported 100% sensitivity for detecting pneumococcal antigen in 25 CSF specimens, but others have found difficulty in differentiating pneumococcal antigens from meningococcal antigens in CSF by EIA (Sippel *et al.*, 1984). Use of monoclonal antibody for detecting group A meningococcal antigens has been described (Sugasawara *et al.*, 1984) but was 84% as sensitive as polyclonal serum for detecting antigen in the same CSF samples. Detection of *Haemophilous influenzae* B by EIA has been shown to be effective in limited clinical trials. Drow *et al.* (1979) developed an EIA which was 100% sensitive on 11 positive CSF samples, and Sippel *et al.* (1984), using a similar EIA system, detected 17 of 20 samples that were positive for *Haemophilus* by counterimmunoelectrophoresis or coagulation. There were 17 positive by culture. Antigen detection by EIA for respiratory infections has been described for *Legionella pneumophila*, tuberculosis, and streptococcus group A infection. The most extensive study for detecting *L. pneumophila* antigens in urine was done by Sathapatayavongs *et al.* (1982), who obtained a 82.9% (39/47) sensitivity and a specificity of 100% in 178 urines from patients with other diseases. An inhibition EIA for detection of streptococcus group A antigen is throat swabs was also effective, giving a sensitivity of 97.0% and a specificity of 97.9% (Knigge *et al.*, 1984). Confirmed diagnosis of tuberculosis is difficult because of the long period required to culture the causative organism. Preliminary results of an EIA developed by Sada *et al.* (1983), utilizing rabbit antibody to BCG, showed a sensitivity of 81.2% in 16 samples from patients with tuberculosis meningitis. Because antibodies to BGG may cross react with other mycobacteria, as well as with species of *Nocardia* and *Corynebacterium*, the test needs further evaluation for specificity.

There have also been EIA tests developed for detection of bacterial toxins in clinical samples, most notably assays for *Clostridium difficile* toxins. Lyerly *et al.* (1983) developed an EIA for *C. difficile* A toxin which was 100% specific in 31 samples, but only 58.6% (17/29) sensitive. An improved assay for this toxin and for B toxin was reported by Laughon *et al.* (1984). Of 79 tissue-culture-positive specimens, 91% were positive for toxin A and 80% were positive for toxin B. Combined, 95% were positive for either A or B toxin. Thus, this is one EIA test that appears to be a marked improvement over the difficult tissue culture toxin assay. Detection by EIA in stool of toxins from other *Clostridium* species, *C. perfringens* A (McClane and Strouse, 1984) and *C. botulinum* A and B (Dezfulian *et al.*, 1984) has also been reported.

Two other unrelated infections have been diagnosed by antigen EIA, *Bacteroides fragilis* and *Yersinia pestis* infections. A test for *B. fragilis* in urine was 100% specific, and detected antigen in 73% (11/15) of individuals shown to be infected with *B. fragilis* (Rissing *et al.*, 1984). Use of monoclonal antibody against the F1 antigen of *Y. pestis* was insensitive in an EIA, detecting antigen in 20% (2/10) sera from patients with acute bubonic plague (Williams *et al.*, 1984).

B. VIRAL INFECTIONS

The interest in EIA methods for rapid diagnosis of viral infections has been high, because of the time and expense required for isolation of the agents in cell culture. Further, some viruses cannot be cultivated or are difficult to cultivate. The latter includes hepatitis A and B viruses, and rotavirus. Tests for hepatitis B surface antigen and e antigen have been commercially available for some time, have been extensively evaluated, and need not be elaborated on here. Detection of hepatitis A antigen by EIA has also been reported (Mathiesen *et al.*, 1978; Locarini *et al.*, 1978) but is not at this time commercially available. The sensitivity of the EIA developed by Mathiesen *et al.* was 77% (10/13) compared with immune electron microscopy (IEM). Locarini *et al.* were able to detect hepatitis A in 85% (17/20) of samples positive by IEM, and found no false positive EIA reactions in fecal samples from patients with hepatitis B or non-A-non-B hepatitis virus infections.

In addition to detection by EIA of hepatitis viruses, the EIA tests most frequently developed have been for respiratory viruses, herpesviruses, and gastroenteritis viruses (Table V,B).

1. Respiratory Viruses

Because antiviral agents are becoming available for some of the respiratory virus infections, rapid methods of diagnosis are essential for prompt treatment. Rapid diagnosis by EIA has been proposed for a number of respiratory viruses. Several have been described for diagnosis of respiratory syncytial virus. Specificity does not appear to be a problem with any EIA reported, but the sensitivity is less than that found by culture. Compared with culture, sensitivities have been found to be 79.3% (23/29) (Chao *et al.*, 1979), 60.9% (25/41) (Hornsleth *et al.*, 1981), 78.7% (37/47), (Hornsleth *et al.*, 1982), and 82.8% (77/93) (McIntosh *et al.*, 1982).

Diagnosis of viral influenza by EIA has also been reported, with variable results. Compared with culture, Harmon and Pawlik (1982) reported a sensitivity of 53% (21/40). A later report by Harmon *et al.* (1983) on an EIA using fluorogenic substrates gave a sensitivity of 87% (27/31). By use of a radioactive substrate, Coonrod *et al.* (1984) were able to detect influenza virus in nasal washes, but the maximum sensitivity at any given day of infection was 48% (12/25). A similar assay described by Yolken (1980) on samples from 12 volunteers gave sensitivities of 78 to 100%, depending on the day tested. Tests for adenovirus in respiratory secretions have also been developed. Harmon and Pawlik (1982) compared an EIA with tissue culture isolation and were able to detect by EIA 62% (13/21) of adenovirus-positive specimens.

2. Herpesviruses

The interest in sexually transmitted herpesvirus and the availability of treatment have led to development of a number of EIA tests for rapid diagnosis of herpes infection. Most lack sufficient sensitivity to be used as a substitute for culture. Two evaluations of a commercial EIA for herpes genital infection (Ortho Diagnostic Systems, Inc.) have been reported. Morgan and Smith (1984) found the test to be 71.9% (105/146) sensitive and 100% specific in 366 control specimens. Warford *et al.* (1984), however, found the test to be only 52.5% (155/295) sensitive and 96.9% (834/860) specific. Some of the other EIA tests developed have given similar results. Lawrence *et al.* (1984) developed an EIA which was 50.5% (94/186) sensitive and 99.1% (423/427) specific. An EIA reported by Grillner and Landquist (1983) was 75.9% (44/58) sensitive and 100% specific. Two assays reported appear to have higher efficiencies. Nerurkar *et al.* (1984), using a biotin–avidin EIA obtained a 95.6% sensitivity and a 91.4% specificity; Land *et al.* (1984), using a detergent-treated specimen, obtained a sensitivity and specificity of 94%.

Development of EIA tests for other viruses in the herpesvirus group have also been reported. Ziegler (1984) developed an EIA for varicella–zoster viral antigens which detected 8/8 culture positive specimens, and Pronovost *et al.* (1982) developed a chemiluminescent EIA for cytomegalovirus antigen which detected 9/11 culture-positive specimens.

3. Gastroenteritis Viruses

The two most important gastroenteritis viral agents for which EIA tests have been developed are rotavirus and enteric adenoviruses. An EIA test for rotavirus was first developed by Yolken *et al.* (1977), and commercial assays are now available. A recent evaluation of two commercial products (Rotazyme, Abbott Laboratories; Enzygnost, Behring) showed the sensitivity of Rotazyme to be 88% and Enzygnost, 98% (Morinet *et al.*, 1984). The standard for comparison was electron microscopy (EM). Both EIA methods appear suitable for use if EM is not available, although Rotazyme is known to cause false positive results in samples from neonates (Krause *et al.*, 1983; Chrystie *et al.*, 1983) and is insensitive in samples from adults (Herrmann *et al.*, 1985).

Enteric adenoviruses (types 40 and 41) are difficult to isolate; therefore EIA methods would be preferable if the sensitivity was satisfactory. Preliminary results from Johansson *et al.* (1980) suggest that development of an EIA specific for enteric adenoviruses is possible.

VI. Use of Monoclonal Antibodies

The use of monoclonal antibodies in EIA tests offers two potential advantages: (1) improved specificity due to the nature of monoclonal antibodies, and (2) improved sensitivity by allowing for clearer EIA cut-off values. Sensitivity could also be increased by increasing the amount of detector antibody used in an EIA. However, because monoclonal antibodies react with only one epitope of a given antigen, more than one monoclonal antibody may be needed to achieve the desired sensitivity. In practice, monoclonal antibodies have been used successfully in latex agglutination tests and also in immunofluorescence techniques (Nowinkski *et al.*, 1983). Their use in EIA has been limited to date, but the number described for microbial antigens to date suggest that applications in clinical diagnosis will be increasing. The EIA tests that have been developed look promising. For EIA detection of adenovirus group antigens in stools, monoclonal antibodies were as sensitive as polyclonal ones, and were more sensitive for detecting noncultivatable adenoviruses (presumably enteric serotypes). All 12 stool samples positive by EM were monoclonal EIA positive (Anderson *et al.*, 1983).

Use of monoclonal antibodies for detection by EIA of microbial antigens in cerebrospinal fluid (CSF) also look promising. In a comparison of polyclonal and monoclonal EIA tests for group A meningococcal antigens in CSF, 21 of 25 CSF specimens positive by polyclonal EIA were positive by monoclonal EIA (Sugasawara *et al.*, 1984). In a preliminary study, 5/5 CSF specimens positive for group B streptococcal antigen reacted in a monoclonal EIA (Morrow *et al.*, 1984).

For diagnosis of rotavirus infection, we have found that a monoclonal EIA was 100% sensitive and specific for samples from adults and neonates as well as young children (Herrmann *et al.*, 1985). This was possible due to the high affinity and broad group specificity of the monoclonal antibody used (Cukor *et al.*, 1984).

Preliminary results of a monoclonal EIA for diagnoses of legionellosis showed positive correlations in 3/3 cases (Bibb *et al.*, 1984). However, in a preliminary study on using a monoclonal EIA for diagnosis of bubonic plague, only 2 of 10 were positive (Laughon *et al.*, 1984). In this situation, where the sensitivity is low, use of polyclonal sera in a control EIA would be desirable. This would help determine if the problem was the monoclonal antibody or the amount of antigen present in the clinical sample.

VII. Future Prospects

Although many of the current EIA tests for microbial antigens and antibodies have not realized their potential, there are reasons to believe this situation will improve. For detection of antibodies, the major problems are standardization of reagents and EIA methodology. This should improve when more reagents become commercially available, and when a standard method is selected from the variety of procedures now available. For detection of antigen, which offers a rapid and direct means of diagnosing microbial infections, the major problems has been lack of sensitivity. Increasing the sensitivity of polyclonal EIA tests by using more concentrated immunoreagents or more sensitive enzyme substrates has often resulted in a loss of specificity. From the reports available to date, it appears that use of the appropriate monoclonal antibodies may solve the problem of sensitivity for detecting many infectious agents in clinical samples. With the increasing number of monoclonal antibodies available for varius microbial antigens, we can expect that more of them will be utilized for EIA detection systems. If the affinities of monoclonal antibodies can be increased, the EIA tests may be sufficiently sensitive.

Another approach which is being taken for rapid diagnosis is the use of nucleic acid probes in hybridization techniques (review by Richman *et al.*, 1984). To date, most of the probes have used radioactive labels (^{32}P) and require 1 or 2 days for assay, which makes them impractical for clinical laboratories. The use of biotin labels coupled with enzyme markers may improve both the speed of the assay and the sensitivity of the assay (Richman *et al.*, 1984). Whether this technique will become useful and practical for direct detection of microorganisms in clinical specimens remains to be determined.

ACKNOWLEDGMENTS

This work was supported by Contract DAMD 17-83-C3087 from the U.S. Army Medical Research and Development Command, by Cooperative Agreement CR 8-10803-01 from the U.S. Environmental Protection Agency, and by a grant from the World Health Organization.

REFERENCES

Aardoom, H. A., Hoop, D. D., Iserief, C. O. A., Michel, M. F. and Stolz, E. (1982). *Br. J. Vener. Dis.* **58**, 359–362.

Adler, B., Murphy, A. M., Locarini, S. A., and Faine, S. (1980). *J. Clin. Microbiol.* **11**, 452–457.

Anderson, L. J., Godfrey, E., McIntosh, K., and Hierholzer, J. C. (1983). *J. Clin. Microbiol.* **18**, 463–468.

Atanasiu, P., Savy, V., and Perrin, P. (1977). *Ann. Microbiol. (Paris)* **128A**, 489–498.

Avrameas, S. (1969). *Immunochemistry* **6**, 43–50.

Avrameas, S., and Guesdon, J. L. (1982). *In* "Medical Virology" (L. de la Maza and E. M. Peterson, eds.), pp. 33–54. Elsevier, New York.

Avrameas, S., and Guilbert, B. (1971). *C. R. Acad. Sci. (Paris)* **273**, 2705–2707.

Berdal, B. P., Farshy, C. E., and Feeley, J. C. (1979). *J. Clin. Microbiol.* **9**, 575–578.

Berg, R. A., Yolken, R. H., Rennard, S. I., Dolin, R., Murphy, B. R., and Straus, S. E. (1980). *Lancet* **1**, 851–853.

Bibb, W. F., Arnow, P. M., Thacker, L., and McKinney, R. M. (1984). *J. Clin. Microbiol.* **20**, 478–482.

Bode, L., Beutin, L., and Kohler, H. (1984). *J. Virol. Methods* **8**, 111–121.

Briantais, M., Grangeot-Keros, L., and Pillot, J. (1984). *J. Virol. Methods* **9**, 15–26.

Burke, D. S., and Nisalak, A. (1982). *J. Clin. Microbiol.* **15**, 353–361.

Burns, M., Rossi, P. H., Cox, D. W., Edwards, T., Kramer, M., and Krause, S. J. (1983). *Sex. Transm. Dis.* **10**, 180–183.

Carlsson, H. E., Lindberg, A. A., and Hammarstrom, S. (1972). *Infect. Immun.* **6**, 703–708.

Carlsson, H. E., Lindberg, A. A., Hederstedt, B., Karlsson, K. A., and Agell, B. D. (1979). *J. Clin. Microbiol.* **10**, 615–621.

Catt, K., and Tregear, G. W. (1967). *Science* **158**, 1570–1572.

Cavanaugh, D. C., Fortier, M. K., Robinson, D. M., Williams, J. E., and Rust, J. H., Jr. (1979). *Bull. Pan. Am. Health Org.* **13**, 393–402.

Chao, R. K., Fishaut, M., Schwartzmann, J. D., and McIntosh, K. (1979). *J. Infect. Dis.* **139**, 483–486.

Chrystie, I. L., Totterdell, B. M., and Banatvala, J. E. (1983). *Lancet* **2**, 1028.

Conroy, J. M., and Esen, A. (1984). *Anal. Biochem.* **137**, 182–187.

Coonrod, J. D., Betts, R. F., Linnemann, C. C., Jr., and Hsu, L. C. (1984). *J. Clin. Microbiol.* **19**, 361–365.

Craft, J. E., Grodzicki, R. L., and Steere, A. C. (1984). *J. Infect. Dis.* **149**, 789–795.

Cukor, G., Perron, D. M., Hudson, R., and Blacklow, N. R. (1984). *J. Clin. Microbiol.* **19**, 888–892.

Danielson, D., Moi, H., and Forslin, L. (1983). *J. Clin. Pathol.* **36**, 674–677.

Delia, S., Russo, V., Vullo, V., Aceti, A., and Ferone, U. (1977). *Lancet* **1**, 1364.

Demetriou, E., Sackett, R., Welch, D. F., and Kaplan, D. W. (1984). *J. Am. Med. Assoc.* **252**, 247–252.

Dezfulian, M., Hatheway, C. L., Yolken, R. H., and Bartlett, J. G. (1984). *J. Clin. Microbiol.* **20**, 379–383.

Dittmar, D., Cleary, T. J., and Castro, A. (1979). *J. Clin. Microbiol.* **9**, 498–502.

Douglas, J. T., and Worth, R. M. (1984). *Int. J. Lepro.* **52**, 26–33.

Drow, D. L., Maki, D. G., and Manning, D. D. (1979). *J. Clin. Microbiol.* **10**, 442–450.

Duc-Goiren, P., Raymond, J., Leaute, J. B., and Orfiler, J. (1983). *Eur. J. Clin. Microbiol.* **2**, 32–38.

Engvall, E., and Perlmann, P. (1971). *Immunochemistry* **8**, 871–874.

Farshy, C. E., Klein, G. C., and Feeley, J. C. (1978). *J. Clin. Microbiol.* **7**, 327–331.
Feinstone, S. M., Barker, L. F., and Purcell, R. H. (1979). In "Diagnostic Procedures for Viral, Rickettsial, and Chlamydial Infections" (E. H. Lennette and N. J. Schmidt, eds.), pp. 879–925. Amer. Public Health Assoc., Washington, D.C.
Ferrua, B., Maiolini, R., and Masseyeff, R. (1979). *J. Immunol. Methods* **25**, 49–53.
Forghani, B., Schmidt, N. J., and Dennis, J. (1978). *J. Clin. Microbiol.* **8**, 545–552.
Gilman, S. C., and Docherty, J. J. (1977). *J. Infect. Dis. (Suppl.)* **136**, S286–S293.
Granfors, K., Viljaner, M. K., and Toivanen, A. (1981). *J. Clin. Microbiol.* **14**, 6–14.
Gravell, M., Dorsett, P., Gutenson, O., and Ley, A. C. (1977). *J. Infect. Dis. (Suppl).* **136**, S300–S303.
Grillner, L., and Landqvist, M. (1983). *Eur. J. Clin. Microbiol.* **2**, 39–42.
Guesdon, J. L., Ternynck, T., and S. Avrameas. (1979). *J. Histochem. Cytochem.* **27**, 1131–1139.
Haberman, E., and Heller, I. (1976). In "Protides of the Biological Fluids" (H. Peeters, ed.), Vol. 24. Pergamon, Oxford.
Halbert, S. P., Anken, M., Henle, W., and Golubjatnikov, R. (1982). *J. Clin. Microbiol.* **15**, 610–616.
Halle, S., Dasch, G. A., and Weiss, E. (1977). *J. Clin. Microbiol.* **6**, 101–110.
Harmon, M. W., and Pawlik, K. M. (1982). *J. Clin. Microbiol.* **17**, 305–311.
Harmon, M. W., Russo, L. L., and Wilson, S. Z. (1983). *J. Clin. Microbiol.* **17**, 305–311.
Hendry, R. M., and Herrmann, J. E. (1980). *J. Immunol. Methods* **35**, 285–296.
Hernandez, R., Munoz, O., and Guiscafre, H. (1984). *J. Clin. Microbiol.* **20**, 533–535.
Herrmann, J. E. (1981). In "Methods in Enzymology" (J. J. Langone and H. Van Vainakis, eds.), Vol. 73. Academic Press, New York.
Herrmann, J. E., and Morse, S. A. (1974). *Immunochemistry* **11**, 79–82.
Herrmann, J. E., Hollingdale, M. R., Collins, M. F., and Vinson, J. W. (1977). *Proc. Soc. Exp. Biol. Med.* **154**, 285–288.
Herrmann, J. E., Howard, L. V., Kurpiewski, G., and Craine, M. C. (1983). *Eur. Congr. Clin. Microbiol., 1st* (Abstr.) A311.
Herrmann, J. E., Blacklow, N. R., Perron, D. M., Cukor, G., Krause, P. J., Hyams, J. S., Barrett, H. J., and Ogra, P. L. (1985). *J. Infect. Dis.* **152**, 830–832.
Hildebrand, R. L. (1979). In "Rapid Diagnosis in Infectious Disease" (M. W. Rytel, ed.), pp. 71–88. CRC Press, Boca Raton, Florida.
Hill, H. R., and Matsen, J. M., 1983). *J. Infect. Dis.* **147**, 258–263.
Hofmann, H., Frisch-Niggemeyer, W., and Heinz, F. (1979). *J. Gen. Virol.* **42**, 305–511.
Hommel, M., Truong, T. K., and Bidwell, D. E. (1976). *Nouv. Press. Med.* **5**, 2789–2791.
Hopkins, R. F., III, Neubauer, R. H., and Rabin, H. (1982). *J. Infect. Dis.* **146**, 734–740.
Hornsleth, A., Grauballe, P. C., Genner, J., and Pedersen, I. R. (1981). *J. Clin. Microbiol.* **14**, 510–515.
Hornsleth, A., Friis, B., Andersen, P., and Brenoe, E. (1982). *J. Med. Virol.* **10**, 273–281.
Ison, C. A., Hadfield, S. G., and Glynn, A. A. (1981). *J. Clin. Pathol.* **34**, 1040–1043.
Jodal, U., Ahlstedt, S., Carlsson, B., Hanson, L. A., Lindberg, U., and Sohl, A. (1974). *Int. Arch. Allergy Appl. Immunol.* **47**, 537–546.
Johansson, M. E., Unoo, I., Kidd, A. H., Madely, C. R., and Wadell, G. (1980). *J. Clin. Microbiol.* **12**, 95–100.
Johnson-Winegar, A. (1984). *J. Clin. Microbiol.* **20**, 357–361.
Jones, M. F., Smith, T. F., Houglum, A. J., and Herrmann, J. E. (1984). *J. Clin. Microbiol.* **20**, 465–467.
Kalish, S. B., Radin, R. C., Phair, J. P., Levitz, D., Zeiss, C. R., and Metzger, E. (1983). *J. Infect. Dis.* **147**, 523–530.
Kato, K., Yamaguchi, Y., Fukui, M., and Ishikawa, E. (1975). *J. Biochem.* **78**, 235–237.

Knigge, K. M., Babb, J. L., Firca, J. R., Ancell, K., Bloomster, T. G., and Marchlewicz, B. A. (1984). *J. Clin. Microbiol.* **20**, 735–741.

Krambovitis, E., Lock, P. E., McIllmurray, M. B., Hendrickse, W., and Holzel, H. (1984). *Lancet* **2**, 1229–1231.

Krause, P. J., Hyams, J. S., Middleton, P. J., Herson, V. C., and Flores, J. (1983). *J. Pediatr.* **10**, 259–262.

Land, S. A., Skurrie, I. J., and Gilbert, G. (1984). *J. Clin. Microbiol.* **19**, 865–869.

Laughon, B. E., Viscidi, R. P., Gdovin, S. L., Yolken, R. H., and Bartlett, J. G. (1984). *J. Infect. Dis.* **149**, 781–788.

Lawrence, T. G., Budzko, D. B., and Wilcke, B. W. (1984). *Am. J. Clin. Pathol.* **81**, 339–341.

Leinikki, P. O., and Passila, S. (1977). *J. Infect. Dis. (Suppl.)* **136**, S294–S299.

Lindberg, A. A., Haeggeman, S., Karlsson, K., DacCam, P., and DucTrach, D. (1984). *Bull. WHO* **62**, 597–606.

Locarini, S. A., Garland, S. M., Lehmann, N. I., Pringle, R. C., and Gust, I. D. (1978). *J. Clin. Microbiol.* **8**, 277–282.

Lyerly, D. M., Sullivan, N. M., and Wilkins, T. D. (1983). *J. Clin. Microbiol.* **17**, 72–78.

Lynn, M. (1975). *In* "Immobilized Enzymes, Antigens, Antibody, and Peptides" (H. H. Weetall, ed.), pp. 1–48. Dekker, New York.

McClane, B. A., and Strouse, R. J. (1984). *J. Clin. Microbiol.* **19**, 112–115.

McIntosh, K., Hendry, R. M., Fahnestock, M. L., and Pierik, L. T. (1982). *J. Clin. Microbiol.* **16**, 329–333.

Mackowiak, P. A., and Smith, J. W. (1978). *Ann. Intern. Med.* **89**, 494–496.

Macnaughton, M. R., Flowers, D., and Isaacs, D. (1983). *J. Med. Virol.* **11**, 319–325.

Magee, J. T. (1980). *J. Med. Microbiol.* **13**, 167–172.

Majumbar, A. S., Dutta, P., Dutta, D., and Gose, A. C. (1981). *Infect. Immun.* **32**, 1–8.

Manis, R. D., Harris, B., and Geiseler, P. J. (1984). *J. Clin. Microbiol.* **20**, 742–746.

Martin, R., Wentworth, B., Coopes, S., and Larson, E. H. (1984). *J. Clin. Microbiol.* **19**, 893–895.

Mathiesen, L. R., Feinstone, S. M., Wong, D. C., Skinhoej, P., and Purcell, R. H. (1978). *J. Clin. Microbiol.* **7**, 184–193.

Mendes-Giannini, M. J. S., Camargo, M. E., Lacaz, C. S., and Ferreira, A. W. (1984). *J. Clin. Microbiol.* **20**, 103–108.

Meurman, O., Sarkkinen, H., Ruuskanen, O., Hanninen, P., and Halonen, P. (1984). *J. Med. Virol.* **14**, 61–65.

Miettinen, A., Paavonen, J., Jansson, E., and Leinikki, P. (1983). *Sex. Transm. Dis.* **10**, 289–293.

Miettinen, A., Turunen, H. J., Paavonen, J., Jansson, E., and Leinikki, P. (1984). *J. Immunol. Methods* **69**, 267–275.

Miranda, Q. R., Bailey, G. D., Fraser, A. S., and Tenoso, H. J. (1977). *J. Infect. Dis. (Suppl.)* **136**, S304–S310.

Monath, T. P., Nystrom, R. R., Bailey, R. E., Calisher, C. H., and Muth, D. J. (1984). *J. Clin. Microbiol.* **20**, 784–790.

Morgan, M. A., and Smith, T. F. (1984). *J. Clin. Microbiol.* **19**, 730–732.

Morinet, F., Ferchal, F., Colimon, R., and Perol, Y. (1984). *Eur. J. Clin. Microbiol.* **3**, 136–140.

Nachamkin, I., Sondheimer, S. J., Barbagallo, S., and Barth, S. (1984). *Am. J. Clin. Pathol.* **82**, 461–465.

Nakane, P. K., and Kawaoi, A. (1974). *J. Histochem. Cytochem.* **22**, 1084–1091.

Nakane, P. K., and Pierce, G. B. (1966). *J. Histochem. Cytochem.* **14**, 929–931.

Nerurkar, L. S., Namba, M., Brashears, G., Jacob, A. J., Lee, Y. J., and Sever, J. L. (1984). *J. Clin. Microbiol.* **20**, 109–114.

Niklasson, B. S., Jahrling, P. B., and Peters, C. J. (1984a). *J. Clin. Microbiol.* **20**, 239–244.
Niklasson, B. S., Peters, C. J., Grandien, M., and Wood, O. (1984b). *J. Clin. Microbiol.* **19**, 225–229.
Nowinski, R. C., Tam, M. R., Goldstein, L. C., Stong, L., Kuo, C. C., Corey, L., Stamm, W. E., Handsfield, H. H., Knapp, J. S., and Holmes, K. K. (1983). *Science* **219**, 637–644.
Oseni, R. A., Donaldson, M. D., Dalglish, D. A., and Aaskov, J. G. (1983). *Bul!. WHO* **61**, 703–708.
Papasian, C. J., Bartholomew, W. R., and Amsterdam, D. (1984a). *J. Clin. Microbiol.* **19**, 347–350.
Papasian, C. J., Bartholomew, W. R., and Amsterdam, D. (1984b). *J. Clin. Microbiol.* **20**, 641–643.
Parry, J. V. (1984). *J. Virol. Methods* **9**, 35–44.
Pepple, J. M., Moxon, E. R., and Yolken, R. H. (1980). *J. Pediatr.* **97**, 233–237.
Pronovost, A. D., Baumgarten, A., and Andiman, W. A. (1982). *J. Clin. Microbiol.* **16**, 345–349.
Rai, A., and Mahajan, V. M. (1983). *Eur. J. Clin. Microbiol.* **2**, 129–134.
Raisanen, S., Suni, J., and Leinkki, P. (1980). *J. Clin. Pathol.* **33**, 836–840.
Richman, D. D., Cleveland, P. H., Redfield, D. C., Oxman, M. N., and Wahl, G. M. (1984). *J. Infect. Dis.* **149**, 298–310.
Rissing, J. P., Buxton, T. B., and Edmondson, H. T. (1979). *J. Infect. Dis.* **140**, 994–998.
Rissing, J. P., Buxton, T. B., Harris, R. W., and Shockley, R. K. (1984). *J. Infect. Dis.* **149**, 929–934.
Rote, N. S., Taylor, N. L., Shigeoka, A. O., Scott, J. R., and Hill, H. R. (1980). *Infect. Immun.* **27**, 118–123.
Rubinstein, K. E., Schneider, R. S., and Ullman, E. F. (1972). *Biochem. Biophys. Res. Commun.* **47**, 846–851.
Russel, H., Facklam, R. R., and Edwards, L. R. (1976). *J. Clin. Microbiol.* **3**, 501–505.
Sada, E., Ruiz-Palacios, G. M., Lopez-Vidal, Y., and Ponce de Leon, S. (1983). *Lancet* **2**, 651–652.
Sarkkinen, H. K., Tuokko, H., and Halonen, P. E. (1980). *J. Virol. Methods* **1**, 331–341.
Sarkkinen, H. K., Halonen, P. E., and Salmi, A. A. (1981a). *J. Gen. Virol.* **56**, 49–57.
Sarkkinen, H. K., Halonen, P. E., and Salmi, A. A. (1981b). *J. Med. Virol.* **7**, 213–220.
Sarkkinen, H. K., Halonen, P. E., Arstila, P. P., and Salmi, A. A. (1981c). *J. Clin. Microbiol.* **13**, 258–265.
Sathapatayavongs, B., Kohler, R. B., Wheat, L. J., White, H., Winn, W. C., Giron, J. C., and Edelstein, P. H. (1982). *Am. J. Med.* **72**, 576–582.
Schacter, J., McCormack, W. M., Smith, R. F., Parks, R. M., Bailey, R., and Ohlin, A. C. (1984). *J. Clin. Microbiol.* **19**, 399–403.
Schmitz, H. (1982). *J. Clin. Microbiol.* **16**, 361–366.
Schmitz, H., and von Deimling, U. (1980). *J. Gen. Virol.* **50**, 59–68.
Schmitz, H., and Emmerich, P. (1984). *J. Clin. Microbiol.* **19**, 664–667.
Segal, E., Berg, R., Pizzo, P., and Bennet, J. (1979). *J. Clin. Microbiol.* **10**, 116–118.
Seitz, W. R. (1984). *Clin. Biochem.* **17**, 120–125.
Sippel, J. E., and Voller, A. (1980). *Trans. R. Soc. Trop. Med. Hyg.* **74**, 644–648.
Sippel, J. E., Mamay, H. K., Weiss, E., Joseph, S. W., and Beasley, W. J. (1980). *J. Clin. Microbiol.* **7**, 372–378.
Sippel, J. E., Prato, C. M., Girgis, N. I., and Edwards, E. A. (1984). *J. Clin. Microbiol.* **20**, 259–265.
Stamm, W. E., Cole, B., Fennell, C., Bonin, P., Armstrong, A. S., Herrmann, J. E., and Holmes, K. K. (1984). *J. Clin. Microbiol.* **19**, 399–403.
Stokes, G. V., and Khan, M. W. (1984). *Microbios.* **40**, 15–23.

Streefkerk, J. G., and Deelder, A. M. (1975). *J. Immunol. Methods* **7**, 225–236.

Sugasawara, R. J., Prato, C. M., and Sippel, J. E. (1984). *J. Clin. Microbiol.* **19**, 230–234.

Svenson, S. B., and Larsen, K. (1977). *J. Immunol. Methods* **17**, 249–256.

Ukkonen, P., Vaisanen, O., and Penttinen, K. (1980). *J. Clin. Microbiol.* **11**, 319–323.

Van Weeman, B. K., and Schuurs, A. H. W. M. (1971). *FEBS Lett.* **15**, 232–236.

Veldkamp, J., and Visser, A. M. (1975). *Br. J. Vener. Dis.* **51**, 227–231.

Viljanken, M. K., Ruuskanen, D., Granberg, C., and Salmi, T. T. (1982). *Scand. J. Infect. Dis.* **14**, 117–122.

Voller, A., and Bidwell, D. E. (1976). *Br. J. Exp. Pathol.* **57**, 243–247.

Voller, A., Bidwell, D. E., and Bartlett, A. (1976). *Bull. WHO* **53**, 55–65.

Voller, A., Bartlett, A., and Bidwell, D., eds. (1981). "Immunoassays for the 80's." Univ. Park Press, Baltimore.

Warford, A. L., Levy, R. A., and Rekrut, K. A. (1984). *J. Clin. Microbiol.* **20**, 490–493.

Wiley, C. A., and Quinn, P. A. (1984). *J. Clin. Microbiol.* **19**, 421–426.

Williams, J. E., Gentry, M. K., Braden, C. A., Leister, F., and Yolken, R. H. (1984). *Bull. WHO* **62**, 463–466.

Wolters, G., Kuijpers, L., Kacaki, J., and Schuurs, A. (1976). *J. Clin. Pathol.* **29**, 873–879.

Yolken, R. H. (1982). *Rev. Infect. Dis.* **4**, 35–68.

Yolken, R. H., and Torsch, V. M. (1980). *J. Med. Virol.* **6**, 45–62.

Yolken, R. H., and Torsch, V. M. (1981). *Infect. Immun.* **31**, 742–750.

Yolken, R. H., and Wee, S. B. (1984). *J. Clin. Microbiol.* **19**, 356–360.

Yolken, R. H., Kim, W. H., Clem, T., Wyatt, R. G., Kalica, A. R., Chanock, R. M., and Kapikian, A. Z. (1977). *Lancet* **2**, 263–267.

Yolken, R. H., Wyatt, R. G., Kim, H. W., Kapikian, A. Z., and Chanock, R. M. (1978). *Infect. Immun.* **19**, 540–546.

Yolken, R. H., Stopa, P. J., and Harris, C. C. (1980). *In* "Manual of Clinical Immunology" (N. Rose and H. Friedman, eds.), 2nd Ed. Amer. Soc. Microbiol., Washington, D.C.

Yolken, R. H., Davis, D., Winkelstein, J., Russell, H., and Sippel, J. E. (1984). *J. Clin. Microbiol.* **20**, 802–805.

Ziegler, T. (1984). *J. Infect. Dis.* **150**, 149–154.

Zlotnick, H., Havas, H. F., and Buckley, H. R. (1984). *Eur. J. Clin. Microbiol.* **3**, 48–49.

The Identification of Gram-Negative, Nonfermentative Bacteria from Water: Problems and Alternative Approaches to Identification

N. ROBERT WARD,* ROY L. WOLFE,[†] CAROL A. JUSTICE,[‡] AND
BETTY H. OLSON[‡]

*BioControl Systems, Kent, Washington,
[†]The Metropolitan Water District of Southern California, La Verne,
California, and
[‡]The Program in Social Ecology,
University of California,
Irvine, California

ADVANCES IN APPLIED MICROBIOLOGY, VOLUME 31

I. Introduction

Potable water contains a wide variety of bacteria, many of which are poorly described or have not yet been studied (Reasoner and Geldreich, 1985; J. T. Staley, personal communication). A major proportion of the cultivatable heterotrophic population in drinking water falls into a broad group referred to as gram-negative, nonfermentative (GN-NF) bacteria (Reasoner and Geldreich, 1979a; LeChevallier et al., 1980). Within the GN-NF group are members that are recognized as important agents of hospital-acquired and opportunistic infections (Rubin et al., 1980; Hugh and Gilardi, 1980). Water has been suggested as a significant reservoir for these clinically significant GN-NF bacteria (Herman, 1981; Favero et al., 1971; du Moulin, 1979; Bassett et al., 1970; Berkelman et al., 1982; Smith and Massanari, 1977). The genera that are usually implicated in water-related, nosocomial outbreaks are Pseudomonas, Acinetobacter, and Flavobacterium.

GN-NF bacteria are involved in pathological situations ranging in severity from superficial, cutaneous infections to life-threatening septicemias (Rubin et al., 1980; Hugh and Gilardi, 1980; Bergan, 1981; Gilardi, 1972; Herman, 1981; Bøvre and Hagen, 1981; Glew et al., 1977). Individuals who have undergone a traumatic event, such as ocular damage, burns, or a surgical procedure, or patients who are immunocompromised are especially susceptible to these bacteria. In many instances, clinical involvement can be traced to exposure of a debilitated patient to fluids (including "sterile" water), creams, or instruments with unexpectedly high levels of GN-NF bacteria. As can be seen in Table I, GN-NF bacteria have been found in a wide variety of environments, some of which are hostile to most bacteria. The GN-NF bacteria have been detected in distilled water, antiseptic creams and solutions, chlorinated and iodinated water, and ophthalmic solutions. These bacteria have also been isolated from the surfaces of surgical instruments, respiratory therapy equipment floors, faucet taps, sinks, and sink drains (see Table I for references).

What is the health significance of GN-NF bacteria in the community water supply? To attempt to answer this question, examination from two perspectives—from the point of view of the compromised patient who is at greatest danger from these bacteria and from that of the general population—is most appropriate. Reports of outbreaks of hospital-acquired infection, traceable to GN-NF bacteria in water, are now quite numerous (Herman, 1981; Cabrera and Davis, 1961; Foley et al., 1961; Favero et al., 1975). An appreciation by

TABLE I

Isolation of GN-NF Bacteria from Various Environments

Source	Organism	Reference
Distilled water lines, dental chair spray units, humidifying units, X-ray and photo wash tanks, water baths, dead-end pipelines, sink faucets, drinking water fountains, faucet aerators	Yellow-pigmented GN-NF bacteria, *Pseudomonas*	Herman and Himmelsbach (1965), Herman (1976), Herman (1981)
Unheated room humidifiers	*Acinetobacter calcoaceticus*	Smith and Massanari (1977)
Distilled water of mist therapy units	*Pseudomonas aeruginosa*	Favero *et al.* (1971), Moffet and Williams (1967)
Peritoneal dialysis machine	*Pseudomonas cepacia*	Berkelman *et al.* (1982)
Dialysis systems	*Pseudomonas, Flavobacterium, Acinetobacter, Alcaligenes, Achromobacter, Moraxella*	Favero *et al.* (1975)
Chlorinated and iodinated pool waters	*P. aeruginosa, P. alcaligenes*	Favero and Drake (1964), Seyfried and Fraser (1980), Black *et al.* (1970)
Chlorinated and unchlorinated drinking water	*Acinetobacter, Flavobacterium, Moraxella, Pseudomonas, Alcaligenes*	LeChevallier *et al.* (1980), Reitler and Seligman (1957)
10% povidone–iodine solution	*P. cepacia*	Berkelman *et al.* (1981)
Respirator	*Pseudomonas paucimobilis*	Holmes *et al.* (1977)
Nursery resuscitation equipment	*P. aeruginosa*	Bobo *et al.* (1973)
Motel whirlpool	*P. aeruginosa*	McCausland and Cox (1975)
Urological instruments or solutions, whirlpools, lens protheses	*P. aeruginosa*	Farmer *et al.* (1982)
Peritoneal dialysis	*Achromobacter xylosoxidans*	Yabuuchi *et al.* (1974)
Baby lotion, antiseptic body rub and skin lotion, hospital body lotion, hand cream and face moisturizer, brush-on mascara	*Alcaligenes faecalis, P. aeruginosa*, other pseudomonads	Anon (1983)
Benzalkonium chloride solutions, rotting wood around hospital sink, hospital floor	*Pseudomonas multivorans*	Bassett *et al.* (1970), Adair *et al.* (1969)
Faucet aerator	*Pseudomonas*	Cross *et al.* (1966)
Ophthalmic solutions	*P. aeruginosa*	Hugh and Gilardi (1980)

health care professionals of the abilities of GN-NF bacteria to grow in "unlikely" environments, such as distilled water (Favero *et al.*, 1971) and disinfectant solutions (Berkelman *et al.*, 1981), and to colonize surfaces (Herman, 1981), has led to the development of special hospital operating procedures to minimize the potential for bacterial contamination and to the establishment of routine bacteriological surveillance programs to assure the absence of bacteria in potentially hazardous environments. It is clear that the likely presence of GN-NF bacteria must be considered with any use of water in the hospital, on every surface in contact with water, and in any moist environment. From the point of view of the hospital and the compromised patient, GN-NF bacteria in community and hospital water supplies represent an onerous problem. For the healthy individual, health assessment is much more difficult. Certainly, the presence of *Pseudomonas aeruginosa* in water signals a danger to compromised and healthy individuals alike. Ear infections, otitis externa (Sausker *et al.*, 1978; Eriksen, 1961; Hoadley and Knight, 1975; Seyfried and Fraser, 1978; McCausland and Cox, 1975), and skin rash (Vogt *et al.*, 1982; McCausland and Cox, 1975; Jacobson *et al.*, 1976; Washburn *et al.*, 1976) have been attributed to the exposure of healthy individuals to swimming and whirlpool water contaminated with high levels of *P. aeruginosa*. Aside from this organism, the pathogenicity of other members of the GN-NF group seems to be low (Rubin *et al.*, 1980; Hugh and Gilardi, 1980). It may be that, with the exception of some GN-NF bacteria, the presence of clinically significant GN-NF bacteria in the community water supply is not significant. However, caution must be interjected. At this time, little is known about the actual levels of GN-NF bacteria in water, their frequency and periodicity of occurrence, and geographical differences. Until studies are performed addressing these issues, no definite statement can be made.

Microbiologists who have investigated GN-NF bacteria have encountered numerous obstacles. At present, techniques are not available for isolating, enumerating, and identifying many of the GN-NF bacteria. Some information about the occurrences and concentrations of GN-NF bacteria in water has been obtained through the identification of bacteria recovered with plate count procedures (LeChevallier *et al.*, 1980; Armstrong *et al.*, 1981; Olivieri and Snead, 1979). However, identification of the plate count population supplies information only on those recoverable bacteria which predominate in the water; no information is obtained if the bacteria of interest exist as a minor subpopulation of the recovered group. The use of selective isolation and enumeration procedures overcomes the problem of low numbers. Unfortunately, selective techniques have only been developed for *P. aeruginosa* and *Acinetobacter* (Drake, 1966; Leven and Cabelli, 1972; Dutka and Kwan, 1978; Brodsky and Ciebin, 1978; LaCroix and Cabelli, 1982) and the utility of these procedures remains undetermined. Further work on selective procedures for the GN-NF bacteria is sorely needed.

The value of any study concerning the health significance of the GN-NF bacteria in water is contingent upon the ability of microbiologists to correctly identify the recovered bacteria. The identification of the GN-NF group, though, remains a vexing and challenging problem. One obstacle is the disarray in taxonomy. This is especially telling when reviewing the problems associated with the assignment of the yellow-pigmented bacteria to genus. Weeks (1969) described *Flavobacterium* as "more an historic concept than a taxonomic reality." Stanier (1947) labeled *Flavobacterium* "the regrettable genus." With the publication of the 9th edition of *Bergey's Manual of Systematic Bacteriology*, much improvement was made on the description of the genus (Holmes *et al.*, 1984). Other genera of the GN-NF group are also in the midst of redefinition. The 7th edition of *Bergey's* (Haynes and Burkholder, 1957) contained 160 species of *Pseudomonas*. In the 8th edition, only 29 species were recognized (Doudoroff and Palleroni, 1974). The 9th edition (Palleroni, 1984) includes 30 species assigned to four sections and 62 species assigned to a fifth, less well-described section. The genus *Achromobacter* was excluded from *Bergey's* after the 7th edition (Kendrie *et al.*, 1974; Holding and Shewan, 1974). Future studies involving numerical taxonomy, G + C ratios, transformation, and nucleic acid homology will help to delimit these and other taxa (Palleroni and Doudoroff, 1972; Palleroni *et al.*, 1973; McMeekin *et al.*, 1971, 1972; Owen and Snell, 1976; Hayes and Wilcock, 1977; Juni, 1972; Allen *et al.*, 1983). In the meantime, assignment of many of the GN-NF bacteria, particularly to the species level, must be made cautiously.

A second problem confronting the microbiologist is the general reliance of identification methodologies and schema developed for clinical laboratories. Some of the biochemical tests used in the clinical laboratory characterize the GN-NF bacteria as "inert" (MacFaddin, 1980; Blazevic, 1976; Pickett and Pedersen, 1970b). These tests belie the true metabolic versatility of these organisms. The use and development of appropriate differential tests will greatly improve our ability to correctly characterize and assign these bacteria to appropriate genera.

This paper describes methods and a phenotypic profile system that were developed to aid the microbiologist in the identification of GN-NF bacteria from water. The identification system was developed with two purposes in mind: (1) to facilitate the recognition of those GN-NF bacteria which are, at this point, considered to be of potential clinical significance and (2) to assist in the generic assignment of those GN-NF bacteria that do not match the phenotypic profiles of the organisms included in this identification scheme. Procedural steps of culture preservation, staining, and microscopic evaluation are discussed and, in some places, alternative methods are proposed. Comments are made concerning the use of schemes and commercial diagnostic kits designed for use in the clinical laboratory for identification of GN-NF bacteria from water.

II. Definition of the GN-NF Group

The GN-NF bacteria are defined operationally as those gram-negative organisms which fail to ferment carbohydrates. The GN-NF bacteria are obligate aerobes and their metabolism is strictly respiratory. Traditionally, these bacteria have been differentiated from the fermentative bacteria by their inability to grow and produce acidic by-products in a sealed OF–glucose tube. The GN-NF group comprises bacteria which are variable concerning the indophenol oxidase reaction and motility. Microscopically, these bacteria range from short, plump coccobacilli to long, thin, flexuous rods. Flagellar arrangement varies from peritrichous to monotrichous and multitrichous at the polar end. Some genera show gliding or twitching movements that are not mediated by flagella. Water-soluble and -insoluble pigments are produced by some genera and species. Optimum growth temperatures fall within the mesophilic range, and some fail to grow at temperatures above 30°C.

III. Taxonomic Uncertainty

The members of the GN-NF group that are found in water fall into two broad categories: (1) those that have been extensively studied with exhaustive, phenotypic characterizations and, possibly, genotypic characterizations and (2) those that have been fragmentarily investigated, with some information available on morphology and ecology. The recognition of this range in available information has caused bacteriologists to take one of two courses when identifying the GN-NF bacteria. The first is to assign the isolates to loosely defined groups or complexes. This approach is more flexible and, for certain members of the GN-NF group, more appropriate. The drawback is that only limited information is obtained. A second and more restrictive approach is to identify to genus or species based upon an assemblage of phenotypic characteristics. The underlying concerns with this approach are that an inadequate number of phenetic characteristics are studied to afford separation and that the tests that are performed are inappropriate for the organisms under study.

To work within the constraints of existing taxonomic uncertainty, a reasonable approach is to separate the aquatic isolates into those that can reliably be assigned to genus and those which should more appropriately be placed into a group or complex. Generally, this separation can be accomplished with the determination of cellular morphology and examination of colonial appearance.

A. Yellow-Pigmented Bacteria

The heterogeneity of the yellow-pigmented GN-NF bacteria has been frequently discussed and the authenticity of various genera has been challenged. The International Symposium on Yellow-Pigmented, Gram-Negative Bacteria of the *Flavobacterium–Cytophaga* Group (Reichenback and Weeks, 1980) was dedicated to the problems of differentiating the yellow-pigmented GN-NF bacteria. These proceedings and papers by Christensen (1977), Weeks (1969), McMeekin *et al.* (1971, 1972), Mitchell *et al.* (1969), Starr (1981), Reichenbach and Dworkin (1981), and McMeekin and Shewan (1978) provide valuable insight into the difficulties associated with the identification of these bacteria.

Table II includes predominant morphological and biochemical characteristics used to assign the yellow-pigmented bacteria to genus. Descriptions of yellow-pigmented bacteria that are not members of the GN-NF group but may be present in water are also included. These include the coryneform bacteria, *Nocardia, Mycobacterium, Micrococcus, Staphylococcus, Erwinia,* and *Enterobacter.*

The genus *Flavobacterium* has historically been a convenient repository for yellow-pigmented bacteria. As such, the treatment of the genus in *Bergey's Manual* has varied over the years. In the early editions, gram-positive and -negative, nonmotile and motile, polarly and peritrichously flagellated bacteria were included (Bergey *et al.*, 1923). The later editions of *Bergey's Manual* (Breed *et al.*, 1957) excluded gram-positive bacteria and polarly flagellated, gram-negative bacteria. The 8th edition (Weeks, 1974) included nonmotile and motile (peritrichously flagellated), oxidase-positive and -negative, aerobic and facultatively anaerobic, yellow-pigmented, gram-negative bacteria with high and low G + C ratios (30–42 and 63–70 mol%). The genus in this edition was arranged into two sections: Section 1 contained nonmotile flavobacteria with low G + C ratios and Section 2 contained motile and nonmotile flavobacteria with high G + C ratios.

Holmes *et al.* (1984) have dramatically altered the description of the genus in the 9th edition of *Bergey's Manual* by removing all species with high G + C content and three species (*Flavobacterium ferrugineum, Flavobacterium halmophilum,* and *Flavobacterium uliginosum*) previously included by Weeks (1974) in the Section 1 group. Fourteen species with legitimate standings in the *Approved Lists of Bacterial Names* (Skerman *et al.*, 1980) and three species included by Weeks (1974) in the genus, but without recog-

TABLE II

Differentiation of Yellow-Pigmented Bacteria Isolated from Fresh-Water Samples

Genus	Gram reaction	Morphology	Motility	Flagellation	Oxidase	Percentage of G + C (Tm)	Metabolism	Pigment type	Other distinguishing characteristics[a]
Flavobacterium	Negative	Short to medium rods; long, thin rods (5 μm) in liquid culture; bulbous ends	Nonmotile	− (Some strains have polar, nonfunctional flagella [Holmes et al., 1984])	+	31–42	Respiratory	Carotenoid and nonisoprenoid	Growth is not stimulated by fumarate in oxygen-limited culture (Callies and Mannheim, 1978). Starch is usually not utilized. Complex natural polymers such as chitin, agar, cellulose, and pectin are rarely hydrolyzed. Some strains have flexirubin-type pigments (Holmes et al., 1984)
Cytophaga	Negative	Medium rods to long, multicellular filaments	Gliding	−	+	30–45	Respiratory	Carotenoid and flexirubin	Growth is stimulated by fumarate in oxygen-limited culture (Callies and Mannheim, 1978). *Cytophaga* hydrolyze a variety of complex natural polymers, including chitin, agar, cellulose, pectin, and starch. Colony periphery fingerlike projections on the surface of agar media are often observed. Produce extracellular polymers, which increase the viscosity of liquid media and produce mucilaginous colonies on solid surfaces. Soil and fresh-water organisms usually contain flexirubin-type pigment (Reichenbach and Dworkin, 1981)
Flexibacter	Negative	Young culture: long (20–30 μm), agile thread cells Old culture: short rods and coccobacilli	Gliding	−	+	48	Respiratory	Flexirubin and carotenoid	The presence of flexirubin-type pigment can easily be determined with the addition of 10% KOH to a culture plate. The yellow or orange colonies will turn red or purple. The reaction is reversible with the addition of HCl (Reichenbach and Dworkin, 1981)
Lysobacter	Negative	Flexuous rods (0.2–0.5 × 2–70 μm)	Gliding	−	+	65	Respiratory	Carotenoid	Flexirubin-type pigments are never present (Reichenbach and Dworkin, 1981). Chitin is hydrolyzed, but not

Genus	Gram reaction	Cell form	Motility	Flagellation		G+C (%)	Metabolism	Pigment	Remarks
Empedobacter	Negative	Long, filamentous rods	Motile and nonmotile	Peritrichous	–	65	Facultatively anaerobic	Unknown, presumably carotenoid	agar or cellulose. Colonies are mucoid and are white, yellow, pink, or brown. Most strains produce a soluble, brown pigment (Christensen and Cook, 1978)
Pseudomonas	Negative	Straight	Motile	Polar	+(–)	73–70	Respiratory	Carotenoid and noncarotenoid	This is a new taxonomic concept. The presence of peritrichous flagellation must be shown for motile strains (McMeekin and Shewan, 1978)
Xanthomonas	Negative	Straight	Motile	Polar	–(Weak reaction)	63–70	Respiratory	Xanthomonadins (brominated, arylpolyene esters)	Observation of polar flagellation is necessary. All recognized species are plant pathogens and have only been found in association with plants and plant material. Confirmation from other habitats is lacking. Most species hydrolyze starch and Tween 80. Minimal growth requirements are complex. The absorption spectrum of *Xanthomonas* is characterized by a major absorption peak at 445 nm and secondary peaks at approximately 425 and 470 nm (Starr, 1981)
Coryneform (*Arthrobacter*, *Corynebacterium*)	Positive; may be gram variable	Rods and cocci; various angular arrangements; snapping division; strongly staining, intracellular granules	Motile and nonmotile	Polar		50–72	Respiratory	Carotenoid	Assignment of the coryneform group is based upon microscopic observation of snapping division, irregularly shaped rods and cocci, morphological change associated with growth phase, gram-variable reactions, irregularly staining cells, and cells with darkly staining granules. These bacteria are non-acid fast
Nocardia, *Mycobacterium*	Positive; poorly staining with Gram's method	Straight rods; filamentous or mycelialike	Nonmotile	–		60–72	Respiratory	Carotenoid	Most strains are acid fast. Staining of metachromatic granules and staining of cells in a band pattern is common
Micrococcus	Positive	Cocci in clusters or packets	Nonmotile	–		66–75	Respiratory	Carotenoid	Utilize sugars oxidatively
Staphylococcus	Positive	Cocci in clusters or pairs	Nonmotile	–		30–40	Facultatively anaerobic	Carotenoid	Glucose is used fermentatively with the production of lactic acid
Erwinia, *Enterobacter*	Negative	Medium straight rods	Motile	Peritrichous	–	50–58	Facultatively anaerobic	Carotenoid	Glucose is used fermentatively

301

nition on the *Approved Lists*, are now consigned to the *Species Incertae Sedis* listing.

McMeekin and Shewan (1978) suggested an expanded use of the generic epithet, *Empedobacter*. As described by these authors, this genus would contain nonmotile and motile (peritrichously flagellated), oxidase-negative, yellow-pigmented, gram-negative rods with high G + C ratios.

Bacteria of the genus *Cytophaga* are recognized by their gliding movements when in contact with a surface. Sole dependence on this characteristic has resulted, though, in the inclusion of some members of this genus into *Flavobacterium* because of their inability to demonstrate gliding using conventional microscopic and cultural methods (Christensen, 1977; Perry, 1973; McMeekin *et al.*, 1971). Many published descriptions of cytophagal movements have used wet mounts and hanging drop preparations that are considered by some investigators (Christensen, 1977; Henrichsen, 1972) to be unreliable for detecting cytophagal movements. Spreading growth on an agar surface, using a low-nutrient medium (Christensen, 1977; Henrichsen, 1972), coupled with microscopic examination of the spreading growth, is considered a better method to demonstrate gliding (Henrichsen, 1972). Although gliding remains the obligate, taxonomic characteristic (Leadbetter, 1974), evidence of degradation of complex, organic molecules such as pectin, chitin, agar, starch, insulin, cellulose, nucleic acids, and lipids (Leadbetter, 1974; Reichenbach, 1981) provides strong support for the assignment of a yellow-pigmented isolate to *Cytophaga*.

Some investigators have indicated a close, phylogenetic relationship between *Cytophaga* and *Flavobacterium*. The observation of flexirubinlike pigments in strains of *Flavobacterium* that are structurally similar to the flexirubin pigments of *Cytophaga* and *Flexibacter* support the close association between these genera (Reichenbach *et al.*, 1980). It has been suggested that the prominence that gliding now holds for defining a genus may not be justifiable, any more than is separation of a nonmotile species from an otherwise motile genus (Reichenbach *et al.*, 1980). The elimination of gliding as the predominant characteristic of these genera may be a useful step toward resolving confusion surrounding these genera. Substitution of alternative traits, such as pigment type (Reichenbach *et al.*, 1980; Reichenbach and Dworkin, 1981), and the ability to respire in the presence of fumarate (Callies and Mannheim, 1978) may prove to be useful, differential characteristics.

Two additional genera that are closely allied to *Cytophaga* and may be isolated from water (Reichenbach and Dworkin, 1981) are *Flexibacter* and *Lysobacter*. *Flexibacter* may be identified by the characteristic change in cellular shape associated with culture age (Reichenback and Dworkin, 1981). Young cultures show long, flexuous cells, whereas older cultures show short, immotile rods (Reichenbach and Dworkin, 1981). Complex polymers such as cellulose, agar, alginic acid, and chitin are not hydrolyzed (Leadbetter, 1974). *Lysobacter* is a newly described genus comprising flexuous rods and G + C ratio of approximately 65 mol% (Christensen and Cook, 1978). Most strains produce a diffusible pigment which is yellow to brown. The colonies are shiny, mucilaginous, and white, cream, yellow, pink, or brown (Christensen and Cook, 1978).

The polarly flagellated, yellow-pigmented, gram-negative rods are placed into two genera, *Pseudomonas* and *Xanthomonas*, which are not sharply demarcated and have been the center of taxonomic controversy (Starr, 1981). Historically, members of *Xanthomonas* have been known and identified exclusively from their association with and ability to cause disease in plants (Starr, 1981). Purported isolations of *Xanthomonas* from marine, aquatic, and clinical specimens have not been confirmed (Starr, 1981) and clinical isolations of "*Xanthomonas*" may actually represent isolations of *Pseudomonas paucimobilis* (Holmes *et al.*, 1977). In contrast to *Xanthomonas*, *Pseudomonas* are found in a variety of locations. They exist as free-living saprophytes in soil, fresh water, and marine environments (Doudoroff and Palleroni, 1974; Stolp and Gadkari, 1981) and in association with plants and animals (Bergan, 1981; Schroth *et al.*, 1981).

By most accounts, *Xanthomonas* and *Pseudomonas* are closely related genera (Bergan, 1981; Starr, 1981; DeLey, 1968; Murata and Starr, 1973). Bacteria included in these genera are rod shaped, motile, polarly flagellated, and oxidase positive and negative. The G + C ratios of members of these genera are indistinguishable, falling between 60 and 70 mol% (Doudoroff and Palleroni, 1974; Starr, 1981). Numerical taxonomy (Colwell *et al.*, 1968) and nucleic acid homology studies (Murata and Starr, 1977) have supported the view that some species of *Pseudomonas* (most notably *P. maltophilia*) are related to *Xanthomonas* at least on the genus level, but pigment analyses reveal distinct differences (Starr *et al.*, 1977). Xanthomonads. according to Starr *et al.* (1977), are unique in that their pigments are mixtures of brominated arylpolyene esters. This characteristic is believed to separate the *Xanthomonas* from other yellow-pigmented bacteria such as *Pseudomonas*, *Erwinia*, *Flavobacterium*, and *Corynebacterium* (Starr, 1981). If the uniqueness of the *Xanthomonas* pigments proves correct, the discrimination of *Xanthomonas* from other yellow-pigmented bacteria will be greatly improved. There is as yet no evidence that suggests that *Xanthomonas* should not be present in water, at least transiently (Starr, 1981). However, until discriminating tests are available to separate *Xanthomonas* from *Pseudomonas*, reports of aquatic isolation will be viewed with skepticism.

The taxonomic uncertainty associated with the gram-negative, yellow-pigmented bacteria sometimes necessitates the application of procedures not routinely practiced, although certainly within the scope of most laboratories.

G + C ratios, for example, may be estimated using ultraviolet light. McMeekin (1977) reported that UV sensitivity correlated with the mol% G + C of yellow-pigmented bacteria. He found that with a 90-second exposure of a 24- to 48-hour culture plated onto nutrient agar, at least a 10^5-fold decrease occurred with the low G + C bacteria and less than a 10^3-fold decrease occurred with the high G + C bacteria. Spectral absorption properties of extracted pigments can be useful for differentiating *Xanthomonas* from yellow-pigmented *Pseudomonas* (Starr, 1981). The pigments are readily extracted by methanol at 50°C (Starr, 1981). Spectral analyses of xanthomonadin pigments show a major absorption peak at 445 nm, with secondary peaks around 425 and 470 nm (Starr, 1981). The presence of flexirubin pigments of *Cytophaga, Flexibacter,* and *Flavobacterium* can be presumptively indicated with the addition of 10% KOH to a colony on a culture plate. If the flexirubin pigment is present, the yellow pigment will turn purple or red. This reaction is reversible with the addition of HCl (Reichenbach and Dworkin, 1981). The determination of fumarate-stimulated growth, which aids in the separation of *Flavobacterium* from *Cytophaga,* (Callies and Mannheim, 1978), is well within the capabilities of most laboratories. This assay only requires the addition of fumarate to a broth culture medium, incubation under static conditions, and visual inspection of growth (Callies and Mannheim, 1978).

B. NONPIGMENTED BACTERIA

Much of the taxonomic uncertainty associated with the nonpigmented GN-NF bacteria regards intrageneric taxonomy. By and large, the generic assignment of these bacteria is not as arduous as is the case with the yellow-pigmented GN-NF bacteria. Determination of cellular morphology, flagellar arrangement, oxidation of glucose, and D-C oxidase reaction is generally sufficient for identification to genus level. Table III includes prominent morphological and biochemical characteristics used to distinguish commonly occurring nonpigmented GN-NF bacteria.

Pseudomonas is a genus of related, albeit metabolically diverse, bacteria. Nucleic acid homology studies (Palleroni, 1981) have indicated that at least five clusters are sheltered within the genus. The inclusive nature of the generic description of *Pseudomonas* has contributed to uncertainty at the species level. The 7th edition of *Bergey's Manual* (Haynes and Burkholder, 1957) included some 160 species of *Pseudomonas*. The 8th edition attempted to reduce species incertitude and recognized just 29 species (Doudoroff and Palleroni, 1974). With the publication of the 9th edition of *Bergey's Manual* (Palleroni, 1984), 30 species were placed into four sections and 62 species were assigned to a fifth, less well-described section. One consequence of the

TABLE III

Microscopic and Biochemical Characteristics of the Nonpigmented GN-NF Bacteria Commonly Isolated from Water[a]

Genus	Soluble/insoluble pigment	Cellular morphology	Oxidase reaction	Motility	Flagellar arrangement	Surface translocation[b]
Acinetobacter	Some strains produce a tan, water soluble pigment	Coccobacilli (1.0–1.5 × 1.5–2.5 μm)	−	−	Not flagellated	Twitching or sliding movement in some strains
Alcaligenes	None	Rods and coccobacilli (0.5–1.0 × 0.5–2.6 μm)	+	+	Peritrichous (1–8 flagella)	Swimming
Moraxella	None	Coccobacilli (1.0–1.5 × 1.5–2.5 μm)	+	−	Not flagellated	Twitching movement in some strains
Pseudomonas	Some species produce fluorescent and nonfluorescent, soluble pigments	Rods (0.5–1.0 × 1.5–4.0 μm)	+[c]	+[d,e]	Polar; monotrichous and multitrichous	Swimming; twitching movement in some strains

[a] Microscopic and biochemical characteristics as published in the 9th edition of *Bergey's Manual of Systematic Bacteriology* (*Vol. 1*) (N. R. Krieg and J. G. Holt, eds.). Williams & Wilkins, Baltimore; and *The Prokaryotes: A Handbook on Habitats, Isolation, and Identification of Bacteria* (M. P. Starr, H. Stolp, H. G. Trüper, A. Balows, and H. G. Schlegel, eds.). Springer-Verlarg, New York.

[b] Descriptions of surface translocation are from Henrichsen (1972). Swimming, Flagella-mediated process; organisms swim at or below the medium surface depending upon their oxygen requirement. Twitching, Movement at medium surface, which may involve fimbriae; spreading usually does not cover entire medium surface and the edges are irregular; when viewed microscopically, cells move singly and motion is intermittent and does not follow the long axis of the cell. Note: Demonstration of surface translocation is dependent upon medium composition and available moisture at the medium surface.

[c] *P. maltophilia* and *P. paucimobilis* are weakly oxidase positive or oxidase negative.

[d] *P. mallei* is nonmotile.

[e] Some strains of *P. paucimobilis* are weakly motile, especially when incubated at 35°C; many yellow-pigmented pseudomonads do not move through semisolid media (Hugh and Gilardi, 1980).

intrageneric uncertainty is that species designation requires that the microbiologist perform an array of biochemical tests. Palleroni *et al.* (1970), Pickett (1970a,b), and Gilardi (1971) have studied the speciation of this genus by
using large numbers of biochemical tests.

Only one species of *Acinetobacter*, *A. calcoaceticus*, is denoted in the 9th
edition of *Bergey's Manual* (Juni, 1984). However, the study by Johnson *et al.*
(1970) using a competitive DNA–DNA hybridization assay has revealed
various subgroups within *Acinetobacter*. Gilardi (1978) has designated four
biotypes of *A. calcoaceticus* (*anitratus, haemolyticus, alcaligenes*, and *lwoffii*)
based upon biochemical studies. The wide G + C range (38–47 mol%) reflects
the intrageneric heterogeneity. It is likely that more than one species of
Acinetobacter will be recognized in the future as additional information is
available with nucleic acid homology work (Bøvre and Hagen, 1981;
Henriksen, 1973). Intergeneric studies using a transformation assay, performed with the competent recipient strain BD413 (Juni, 1973), substantiate
the legitimacy of the genus and support its delimiters, i.e., oxidase-negative,
nonmotile, glucolytic and nonglucolytic, gram-negative coccobacilli. The
transformation assay is a valuable tool for confirming *Acinetobacter* identifications (Juni, 1972).

Moraxella spp. are generally considered to be obligate parasites of humans
and animals (Bøvre and Hagen, 1981). However, reports of aquatic isolations
of "*Moraxella*" and "*Moraxella*-like" organisms have been made (LeChevallier *et al.*, 1980; Armstrong *et al.*, 1981; Olivieri and Snead, 1979). These
reports should be viewed cautiously, as the laboratory basis for *Moraxella*
identification is not well grounded. In some cases, the isolations may represent *Acinetobacter*, as these taxa have historically been confused
(Henriksen, 1973). *Moraxella* and *Acinetobacter* do share common attributes
(plump rods or coccobacilli, nonmotility, similar G + C ratios of 34–47
mol%), but can readily be distinguished with the oxidase test; *Moraxella* are
oxidase positive (Bøvre and Hagan, 1981). Additionally, some species of
Moraxella cannot be isolated unless the plating medium is supplemented
with serum (Rubin *et al.*, 1980). *Acinetobacter* are not fastidious in their
growth requirements (Juni, 1984). Nucleic acid hybridization studies will
help to clarify the relationship of aquatic "*Moraxella*" and the recognized
Moraxella spp.

The intrageneric and intergeneric taxonomy of *Alcaligenes* remains in flux.
To some extent, the genus has been a catchall for peritrichously flagellated
GN-NF bacteria (Holding and Shewan, 1974; Hendrie *et al.*, 1974). Some
species, formerly included in "*Achromobacter*," have been transferred to
Alcaligenes (Hendrie *et al.*, 1974). Kerster and DeLey (1984) reported that
"*Achromobacter xylosoxidans*" is indistinguishable from *Alcaligenes de-
nitrificans*. Hendrie *et al.* (1974) considered *A. denitrificans*, "*Alcaligenes*

odorans," and *Alcaligenes faecalis* to be synonymous. As a result, *"A. odorans"* and *A. denitrificans* were excluded from the *Approved Lists* (Skerman *et al.*, 1980). Phenotypic and genotypic studies have subsequently shown that *A. faecalis* and *"A. odorans"* are very similar (Rüger and Tan, 1983; Yamasato *et al.*, 1982), but that *A. denitrificans* and *A. faecalis* are distinct and should be regarded as separate species. Rüger and Tan (1983) proposed the revival of the name *A. denitrificans.* The 9th edition of *Bergey's Manual* (Kerster and DeLey, 1984) includes two species: *A. faecalis* and *A. denitrificans*, with subspecies for *A. denitrificans* as *A. denitrificans* subsp. *denitrificans* and *A. denitrificans* subsp. *xylosoxidans.*

IV. Problems Associated with the Identification of Bacteria Recovered with Heterotrophic (Total Plate Count) Procedures

Recent publications have indicated that new heterotrophic count procedures are superior to the standard plate count technique described in *Standard Methods for the Examination of Water and Wastewater* (APHA, AWWA, WPCF, 1980) for providing information about drinking water treatment efficiency, bacterial regrowth in distribution systems, and the presence of opportunistic pathogens in drinking water (Taylor and Geldreich, 1979; Means *et al.*, 1981; Reasoner and Geldreich, 1979b, 1985). The new heterotrophic count procedures include the use of newly developed enumeration media [m-SPC (Taylor and Geldreich, 1979) and R2A (Reasoner and Geldreich, 1979b)] and the application of low incubation temperatures (20–30°C) and longer incubation times [7 or more days (Means *et al.*, 1981; Reasoner and Geldreich, 1979a; Colwell *et al.*, 1978)]. One consequence of these new heterotrophic count methods is the increased recovery of pigmented bacteria (Reasoner and Geldreich, 1979a). This finding is not surprising, as these heterotrophic count methods are less restrictive and better suited to the recovery of yellow-pigmented bacteria, than is the *Standard Methods* procedure (APHA, AWWA, WPCF, 1980). Table IV outlines general nutritional and growth requirements of both gram-positive and gram-negative, yellow-pigmented bacteria. As can be seen in this table, many of the pigmented bacteria are slow growing and require low-temperature incubations. Further, the growth of many of these yellow-pigmented organisms is stimulated by or requires the addition of yeast extract to the recovery medium. Members of the genus *Cytophaga* prefer low-nutrient conditions and are generally recovered on media containing low concentrations of yeast extract and peptone. The total count procedure detailed in *Standard Methods* (APHA, AWWA, WPCF, 1980) uses a pour-plate technique with *Standard Methods* agar (a nutrient-rich medium containing tryptone, yeast extract, and

TABLE IV

Incubation Temperatures, Times, and Cultural Conditions
for the Recovery of Yellow-Pigmented Bacteria[a]

Genus	Optimum incubation temperature (°C)	Incubation time (days)	Culture requirements
Mycobacterium	25–35 (some species above 37)	2–14	Usually cultured on inspissated egg media or oleic acid albumin agar
Erwinia	20–30	2–4	Growth stimulated by addition of yeast extract
Arthrobacter and related coryneforms	20–30	2–7	Addition of yeast extract and glucose stimulatory
Micrococcus	20–30	2–4	Complex media usually containing yeast extract and peptone
Cytophaga, Flavobacterium, and related genera	18–30[b]	2–14	Cytophaga prefer low-nutrient conditions (low yeast extract and peptone). Sensitive to pH fluctuations but inhibited by high concentrations of many buffers. Flavobacterium requires complex media containing low concentrations of yeast extract (0.02–0.1%)
Xanthomonas	25–30	2–7	Nutrient-rich medium necessary; yeast extract fulfills nutritional requirements. Neutral pH and low ionic concentrations are preferred
Pseudomonas	30–41	2–3	Complex media usually containing yeast extract and peptone

[a] Data are from the 8th edition of Bergey's Manual of Determinative Bacteriology, 1974, Williams & Wilkins, Baltimore.

[b] Temperature ranges for Cytophaga may be lower when grown in liquid media (18–24°C) than on an agar surface (30–34°C).

glucose), with incubation at 35°C for 48 hours. These culture conditions are restrictive to the growth of many yellow-pigmented bacteria which require low-temperature incubations and long incubation times. Moreover, the use of the pour-plate technique results in the sequestration of bacteria below the surface of the medium in a region of reduced oxygen. Since these bacteria are obligate aerobes (aside from a few species of Flavobacterium and some members of the *Cytophagaceae*), growth may be prevented or reduced for bacteria present in the subsurface areas.

As discussed in Section III, the identification of pigmented GN-NF bacteria presents a formidable challenge because of the lack of strong, differential tests and the disarray in taxonomy. Until these problems are resolved, the identification of yellow-pigmented, GN-NF bacteria will remain problematic. The increased recovery of yellow-pigmented bacteria with the heterotrophic count procedures increases the difficulties confronting the microbiologist attempting to identify bacteria recovered from water.

The new heterotrophic count techniques may also increase the recovery of nonpigmented bacteria which are not familiar to many microbiologists and have not been well studied. Some of these are identified by observation of characteristic colonial and microscopic features. Bacteria that may be recovered include *Microcyclus, Spirillum, Beijerinckia*, the sheathed bacteria (*Sphaerotilus, Leptothrix,* and *Streptothrix*), the prosthecate bacteria (*Hyphomonas, Caulobacter, Asticcacaulis, Prosthecomicrobium*), the coryneforms, and acid-fast bacteria. A general description of these genera is provided in Table V.

V. Identification Schemes and Commercial Diagnostic Kits for the GN-NF Bacteria

Dichotomous keys have been utilized to identify GN-NF bacteria from clinical specimens (MacFaddin, 1980; Shayegani *et al.*, 1977) and environmental samples (Schroth *et al.*, 1981; LeChevallier *et al.*, 1980). Dichotomous keys, by definition, involve a series of differential biochemical and morphological studies that direct the user through a branching system, leading ultimately to taxonomic assignment. Technically, there are two problems inherent in most dichotomous designs: (1) Each branch or path within the key requires a unique set of biochemical tests. As such, a variety of test media and reagents must be prepared and properly stored. (2) It is time consuming. At each branch, incubation of a test medium is usually required. If three or more test results must be determined and the incubation periods for each are in excess of 2–3 days, as is often the case with the slow-growing, GN-NF bacteria, a considerable length of time is needed before identification can be completed. If a large number of isolates is being processed, data management becomes problematic, as the identifications proceed at different rates.

Aside from the technical concerns, the dichotomous keys afford a false sense that the isolate is being accurately identified since an "answer" is always obtained. If an aberrant reaction or faulty interpretation of a test result is encountered, an incorrect branch of the key is followed and, conse-

TABLE V

DISTINGUISHING MICROSCOPIC CHARACTERISTICS OF OTHER GN-NF BACTERIA
AND GN-NF-LIKE BACTERIA THAT MAY BE ISOLATED FROM WATER[a]

Genus or group	Cellular morphology	Flagellar arrangement
Acid-fast bacteria		
Nocardia	Considered gram positive but stain poorly; acid fast to partially acid fast; extensive mycelial forms to coccoid and bacillary fragments; branching often seen	Not flagellated
Mycobacterium	Considered gram positive but stain poorly; acid fast; fragmented rods or coccoid forms to myceliallike growth	Not flagellated
Coryneform bacteria		
Arthrobacter	Gram positive; however, cells may be decolorized and show gram-positive granules within gram-negative cells; rods (0.5–0.8 × 1–4 μm) in exponential phase; coccoid forms predominate in the stationary phase	Not flagellated to flagellated at the subpolar or lateral position with single flagellum
Spiral or curved bacteria		
Spirillum	Gram negative; rigid, helical cells (0.25–1.7 μm in diameter)	Bipolar flagella
Microcyclus	Gram negative; curved rods (0.5–2 × 1–10 μm); rings form when cells elongate prior to division	Not flagellated
Vibrio/Aeromonas	Gram negative; curved rods (0.5 × 1.5–3 μm); Note: the members of these genera are *fermentative* and are not recognized as GN-NF bacteria. Check fermentative properties with a sealed glucose–OF tube.	Polar flagellum
Prosthecate bacteria		
Hyphomonas	Gram negative; pear shaped (0.6–1 μm in diameter) with prosthecate extending from the narrow pole; multiplication by budding from distal portion of the hypha	Subpolar flagellum
Caulobacter	Gram negative; prosthecae extend from one pole along the long axis of the cell; cells adhere to form rosettes; asymmetrical, binary division of stalked cells	Polar flagellum
Asticcacaulis	Gram negative; rods (0.5–0.7 × 1–3	Polar flagellum

TABLE V (*Continued*)

Genus or group	Cellular morphology	Flagellar arrangement
	μm); prosthecae extend laterally along the cell; cells may form rosettes	
Prosthecomicrobium	Gram negative; prosthecae randomly distribute around the cell surface; prosthecae conical shape (1–2 μm in length); cell diameter about 1 μm	Polar–subpolar flagellum
Aerobic rods (pigmented bacteria)		
Chromobacterium (purple)	Gram negative; rods usually occurring singly (0.6–1.2 × 1.5–6 μm); bipolar staining characteristic with lipid inclusions	Polar and lateral flagella
Janthinobacterium (violet)	Gram negative; rods, usually occurring singly (0.8–1.2 × 2.5–6 μm), with rounded ends	Polar and lateral flagella
Xanthomonas (yellow)	Gram negative; rods usually occurring singly (0.2–0.8 × 0.6–2 μm)	Polar flagellum
Aerobic rods (Nonpigmented bacteria)		
Agrobacterium	Gram negative; rods (0.8 × 1.5–3 μm)	Peritrichous flagellation
Azotobacter	Gram negative or variable; large, ovoid cells at least 2 μm in diameter	Peritrichous flagellation or not flagellated
Azomonas	Gram negative; ovoid cells occurring singly, in pairs, or in clumps (at least 2 μm in diameter); pleomorphism characteristic	Polar or peritrichous flagellation
Sheathed bacteria		
Sphaerotilus	Gram negative; straight rods occurring in chains within a sheath	Single cells motile by means of a bundle of subpolar flagella
Leptothrix	Gram negative; straight rods occurring in chains within a sheath; also, free swimming cells; motile, short chains may be present	Polar flagellum
Streptothrix	Gram negative; thin rods (0.3–0.4 × 3–5 μm) occurring in chains with hyaline sheaths	Not flagellated

[a] Data are from the 8th edition of *Bergey's Manual of Determinative Bacteriology*, 1974, Williams & Wilkins, Baltimore.

quently, a misidentification is likely. The dichotomy, in addition, is discriminating only with those genera and species included in the key. Rarely encountered or poorly studied bacteria, which may not even be included in the scheme, may be processed without recognition of their distinctiveness and thus will receive a completely inappropriate taxonomic "label." The inherent inclusivity of the dichotomy is especially onerous when dealing with water isolates, as many have not been well studied (e.g., yellow-pigmented bacteria) and are not commonly included in schema. Dichotomous keys developed for clinical use should be avoided, as they may not include common water bacteria and may include some genera which are obligate animal pathogens and are unique to the medical laboratory (e.g., *Eikenella*, *Kingella*, or *Bordetella*).

Commercial kits for the identification of the GN-NF bacteria became available about a decade ago for clinical application. These systems include prepackaged biochemical tests which are housed in a variety of molded plastic configurations to facilitate inoculation. Numeric coding systems accompany the kits to aid in the matching of a set of biochemical results to an appropriate epithet. The advantages of these systems over conventional biochemical testing include the great reduction in laboratory media and reagent preparation, ease of inoculation, rapidity of identification, and a decrease in data management problems. One disadvantage of these systems is that two or more supplemental tests are often required for identification (Hofherr *et al.*, 1978; Warwood *et al.*, 1979; Shayegani *et al.*, 1978a,b). Evaluations of these test kits have shown that they are reliable for identifying two GN-NF bacteria which are most commonly encountered in the clinical laboratory, i.e., *P. aeruginosa* and *Acinetobacter*. Agreement between these kits and conventional tests for these species have been reported to be greater than 90% (Barnishan and Ayers, 1979; Shayegani *et al.*, 1978a,b; Otto and Blackman, 1979). These kits have not been shown to be reliable with other GN-NF bacteria (Otto and Blackman, 1979; Hofherr *et al.*, 1978; Warwood *et al.*, 1979; Shayegani *et al.*, 1978a,b; Barnishan and Ayers, 1979). Barnishan and Ayers (1979) reported that only 26% of weakly glucose-oxidative and -nonoxidative bacteria were identified with a Corning N/F system. Generally, these systems have had the most problems with the more "inert" GN-NF bacteria such as *P. alcaligenes*, *P. diminuta*, *P. cepacia*, fluorescent pseudomonads, *Alcaligenes spp.*, and *Flavobacterium* spp. The overall agreement with conventional tests ranges from 41% with the API 20E (Hofherr *et al.*, 1978) to 88% with the Minitek (Wellstood-Nuesse, 1979). However, these figures can be highly misleading, as they are completely dependent upon the genera and species that are tested. As mentioned above, the diagnostic systems are reliable with certain species, whereas they are unsuccessful with others. The distribution of taxa used for the evaluative studies, then, will influence the overall percentage agreement. Unless a direct comparison of the kits are made, the percentage agreement figures cannot be used to establish superiority. Further, the percentage agreement figures are only appropriate for the taxa involved in the evaluative study. In most evaluative studies, *Pseudomonas* spp. and *Acinetobacter* have been the predominant taxa tested in these studies. In the studies by Mathewson *et al.* (1983), Warwood *et al.* (1979), Hofherr *et al.* (1978), and Wellstood-Nuesse (1979), *P. aeruginosa* represented at least 25% of the pseudomonads tested.

To the authors' knowledge, no studies have been published on the usefulness of these diagnostic kits with aquatic isolates. We conducted preliminary, comparative studies with two of the commercial systems and found them to be not useful. This finding was not surprising, since these commercial kits are designed to identify commonly occurring, clinical GN-NF bacteria. Many of the bacteria, recovered with the heterotropic count procedures, may not be included in the data files accompanying the kits.

VI. Alternative Methods

A. Microscopic Evaluation

Staining and microscopic evaluation are among the more time-consuming operations in the laboratory. The determination of cell size, shape, and gram reaction, though, is indispensible to the microbiologist attempting to identify an isolate. Flagellar arrangement, which is determined following special mordant staining, and the type of cellular movement, as seen with a wet mount preparation or with microscopic examination of colonial growth on the surface of a low-nutrient agar medium, also lend strong support for the generic assignment of an isolate. It is unfortunate, but it remains, that the assignment of some isolates to genus is based solely upon microscopic evaluation. It follows, then, that microscopic study should be included into any identification scheme for these bacteria.

While microbiologists consider the gram stain a routine and invaluable tool for supplying information on cell conformation and gram reaction, many eschew flagella staining as a routine test because of its labor intensiveness and unreliability. Leifson's method (Leifson, 1951), which is a commonly used flagella staining procedure, requires scrupulous cleaning of microscopic slides, centrifugation of the bacterial culture to remove interfering materials, preparation and mixing of three separate staining solutions, and air drying of the smears. In addition, the time required for optimum flagella staining may be variable, necessitating staining for different time intervals.

We have successfully employed a modification of a procedure published by Mayfield and Innis (1977) for rapidly and simply staining flagella. This is a wet mount procedure which has the added advantage of allowing the determination of bacterial motility prior to staining. Details of the procedure are given with the directions for the identification tests used in this scheme.

Wet mount preparations are useful for distinguishing between flagella-mediated and gliding motility. Flagellated bacteria show erratic movements, with frequent runs and tumbling motion (Adler, 1966). Gliding bacteria, in contrast, often attach to the surfaces of the slides and reveal gyrating and flexing motions. Henrichsen (1972) suggests that gliding movement can best be studied with high-power microscopic examination of a colony that is growing on a thin film of low-nutrient agar in a Petri dish. Henrichsen (1972) has described gliding as coordinated movement of cells, occurring in bundles at the peripheral edges of growth, along their long axis. Gliding is influenced by the availability of moisture on the agar surface and the level of nutrients in the medium. Holmes *et al.* (1984) recommend that the medium of Anacker and Ordal (1959), which is freshly poured, be used and that a humid environment be provided.

Cultural Methods to Determine Motility

Semisolid (soft) agar media are frequently used for establishing bacterial motility. If the semisolid medium is dispensed into a Petri dish (this arrangement is referred to as a Gard plate), flagellated bacteria, inoculated at the center of the plate, will move in concentric rings to the plate periphery (Adler, 1966; Henrichsen, 1972). It has been our experience that some yellow-pigmented, isolated bacteria, which we have identified as *Flavobacterium,* show a spreading movement on the surface of the semisolid medium or within the liquid layer (syneresis fluid) between the medium and the bottom of the dish. This movement is presumably mediated by pili (Henrichsen, 1972), although some *Flavobacterium,* most notably *Flavobacterium aquatile* and *Flavobacterium meningosepticum,* have been shown to have a "nonfunctional" or "floppy" polar flagellum (Holmes *et al.,* 1984). We have also observed a polar flagellum on yellow-pigmented strains obtained from drinking water in southern California by our laboratory and strains obtained by J. T. Staley (University of Washington) from the municipal system in Seattle, Washington (unpublished data).

The composition of the semisolid medium is critical for observation of motility. The agar concentration should be as low as manageable to promote swimming and the medium should support the growth of those GN-NF bacteria, preferring low-nutrient conditions. The semisolid medium used in this laboratory contains 0.2% (w/v) casitone (Difco), 0.27% yeast extract (Difco), and 0.2–0.3% agar.

B. Culture Storage

A requisite operation following isolation and purification of an isolate is to prepare the culture for long-term storage for future availability. This is one step that should not be delayed, as the viability of some of the GN-NF bacteria is short, especially when grown on an agar surface. The experience of this laboratory is that many of the pigmented bacteria isolated from water do not remain viable for more than 1 month when stored at room temperature on the surface of an agar slant contained in a tightly sealed vial. *P. cepacia* and *P. maltophilia* are two examples of species which die quickly (3–4 days) when not adequately preserved.

Our preference is to grow the isolate on a slant of R2A or m-SPC agar, (see formulations in Section VII,B), overlay the slant with sterile 10% glycerol, and refrigerate at −80°C. In certain cases, the R2A agar may be the preferential medium, as it is buffered and will protect against detrimental levels of acidity or alkalinity. A working culture of the isolate can be maintained by growing the culture in R2A broth supplemented with 0.1% agar. These cultures may be stored at room temperature or at 4°C. Most isolates will remain viable in this semisolid medium for at least 2 months. Many of the pigmented bacteria grow extremely well in the semisolid R2A medium. This medium has been found to be useful in those cases in which the recovered bacteria are difficult to subculture.

VII. Description of the Identification System

The identification system utilizes a battery of nine biochemical tests to determine a three-digit biogrouping number. The biochemical tests include glucose oxidation, acetate alkalinization, motility, Tween 80 hydrolysis, ONPG hydrolysis, gelatin hydrolysis, oxidase, esculin, and DNA hydrolysis. Two of these tests (glucose and acetate) are performed on slanted agar media in test tubes; the remaining tests are performed in quadrant Petri plates. These media are incubated at 30°C for 3–7 days. The biogrouping code is then determined (Table VI), an identification coding list is checked and, if no auxiliary tests are required, the genus and species are recorded. In certain cases, confirmation and auxiliary tests are required for speciation. The confirmation tests involve oxidation studies with four sugars—xylose, maltose, fructose, and mannitol. The auxiliary tests include growth in the presence of 6.5% NaCl, catalase reaction, phenylalanine deaminase and urease tests, and the production of gas from nitrite and nitrate reduction. The identification coding list (Appendix A) includes instructions concerning when confirmation and auxiliary tests are required. If assignment of an isolate to genus is adequate, usually only the initial nine tests, which provide the biogrouping

TABLE VI

Identification Worksheet and Coding System

Isolate Number	Gram reaction	Morphology	Water-insoluble pigment	Nonfluorescent Water-soluble pigment	Fluorescent	Glucose (1)	Acetate (2)	Motility (4)	Tween 80 (1)	ONPG (2)	Gelatin (4)	Oxidase (1)	Esculin (2)	DNase (4)	Biogrouping Code	Xylose (1)	Maltose (2)	Fructose (1)	Mannitol (2)	Confirmation Code	Growth in 6.5% NaCl	H₂S	Catalase	Phenylalanine	Urea	Nitrite to gas	Nitrate to nitrite/gas	Flagellar arrangement	Identification
Example	Neg	Rod	Y	—	—	1	2	4 Weak	1	2	0	1	2	0	733	1	2	1	0	733-01									Pseudomonas paucimobilis

code number, are necessary. The identification coding list also contains information concerning colony and cellular morphology and pigment production.

The identification coding list was generated using information available in the literature (Gilardi, 1972; Rubin *et al.*, 1980; Hugh and Gilardi, 1980; *Bergey's Manual of Determinative Bacteriology*, 8th edition, 1974; *Bergey's Manual of Systematic Bacteriology*, 9th edition, 1984; and the *Procaryotes*, (Starr *et al.*, eds., 1981) and obtained in our laboratory. The identification coding list includes three-digit biogrouping numbers (or biogram profiles) for the organisms covered in this scheme. The biogrouping numbers corresponding to a particular species are determined by assigning "0" to all negative results and the appropriate coding number to a positive result; i.e., first digit: glucose (1), acetate (2), motility (4); second digit; Tween 80 (1), ONPG (2), gelatin (4); third digit; oxidase (1), esculin (2), DNase (4). In the generation of the biogrouping code number listing, a definitive result (positive or negative) was recorded for a particular biochemical reaction only if 90% or more of the strains are recorded as positive or negative in the literature. For reactions below 90%, positive and negative results were taken into consideration and two biogrouping numbers were generated to cover the variable reaction. Weakly positive reactions were recorded as both positive and negative results.

Biogrouping numbers for 6 genera, 33 species and biotypes, and 9 groups of bacteria with unknown affiliations (CDC group designations) are included in the identification coding list (Appendix A). These organisms are listed in Table VII, along with synonyms reported in the literature. The identification coding list was developed around those taxa that have been well studied, may be found in the aquatic environment, and are potential agents of nosocomial infection. This identification scheme facilitates recognition of those bacteria in water which is of a potential public health concern.

Certain isolates, especially those which are obtained from environmental samples (soil and water), will generate biogrouping code numbers (Appendix B) that cannot be located in the identification coding list. These atypical biogrouping code numbers are obtained because: (1) the genus or species is not included in the scheme; (2) one or more aberrant reactions are obtained; (3) one or more tests are not performing correctly; and/or (4) interpretation of one or more test results is faulty. In Table VIII, it is seen that each genus has a unique distribution within the biogrouping code numbers. Recognition of this distribution is helpful when assigning an isolate with an atypical biogrouping code number to a probable genus. To facilitate this operation, a coding list for unrecognizable strains is provided. This list gives instructions for when additional tests are needed to support the assignment to genus of an isolate with a biogram that is not included in the system.

TABLE VII

Epithets

Nomenclature used in scheme	Synonym
Pseudomonas spp.	
P. aeruginosa	*Bacterium aeruginosum, Pseudomonas pyocyaneus, Pseudomonas polycolor*
P. acidovorans	*Pseudomonas indoloxidans, Pseudomonas desmolytica, Comamonas terrigena*
P. cepacia	*Pseudomonas multivorans, Pseudomonas kingae,* CDC Group EO-1
P. diminuta	CDC Group I,A
P. fluorescens	*Pseudomonas aureofaciens, Pseudomonas chlororaphis, Pseudomonas lemonnieri, Pseudomonas marginalis, Pseudomonas boreopolis, Pseudomonas caviae, Pseudomonas effusa, Pseudomonas fairmontensis, Pseudomonas geniculata, Pseudomonas myxogenes, Pseudomonas reptilovora, Pseudomonas schuylkilliensis,* and *Pseudomonas septica*
P. maltophilia	*Pseudomonas melanogena,* CDC Group I
P. mendocina	CDC Group Vb-2
P. paucimobilis	CDC Group II,K-1
P. pickettii	CDC Group VA-2, *Pseudomonas thomasii*
P. pseudomallei	*Bacillus pseudomallei, Bacterium whitmori, Malleomyces pseudomallei, Loefflerella pseudomallei*
P. putida	*Pseudomonas convexa, Pseudomonas ovalis, Pseudomonas rugosa, Pseudomonas striata, Pseudomonas schuylkilliensis*
P. putrefaciens	*Pseudomonas rubescens, Alteromonas,* CDC Groups I,B-1, 0,B-2
P. stutzeri	*Pseudomonas stanieri,* CDC Groups Vb-1, Vb-3
P. testosteroni	*Comamonas terigena*
P. vesicularis	*Corynebacterium vesicularis*
P. pseudoalcaligenes	
P. extorquens	
P. alcaligenes	
CDC group designations	
CDC Group VA-1	Probable affiliation: *Pseudomonas* spp.
CDC Group VE-1, VE-2	Probable affiliation: *Pseudomonas* spp.
CDC Group IV,C-2	Probable affiliation: *Alcaligenes* spp.
CDC Group Vd-1, Vd-2	Probable affiliation: *Alcaligenes* spp.
Acinetobacter spp.	
A. calcoaceticus biotype *anitratus*	*Acinetobacter anitratus, Achromobacter anitratus, Moraxella glucidolytica*
A. calcoaceticus biotype *haemolyticus*	*Lingelsheimia anitrata*

TABLE VII   (*Continued*)

Nomenclature used in scheme	Synonym
A. calcoaceticus biotype *lwoffii*	*Acinetobacter lwoffii*
A. calcoaceticus biotype *alcaligenes*	*Alcaligenes, haemolysans, Moraxella lwoffii, Mima polymorpha, Achromobacter haemolytica* var. *alcaligenes*
Agrobacterium spp.	
A. radiobacter	*Achromobacter radiobacter, Alcaligenes radiobacter, Pseudomonas radiobacter*, CDC Group Vd-3
Alcaligenes spp.	
A. faecalis	CDC Group VI, *Achromobacter arsenoxydans, Alcaligenes odorans, Alcaligenes odorans* var. *viridans, Achromobacter alcaligenes*
A. dentrificans	*Alcaligenes denitrificans* subsp. *denitrificans, Alcaligenes denitrificans* susp. *xylosoxidans, Achromobacter xylosoxodans*
Flavobacterium spp.	
F. odoratum	CDC Group M-4F
F. meningosepticum	CDC Group II,A
Fl. multivorum	CDC Group II,K-2
F. balastinum	CDC Group II,B
F. breve	*Empedobacter breve*
F. spiritovorum	
Moraxella spp.	
M. osloensis	*Moraxella duplex, Mima polymorpha* var. *oxidans, M. nonliquefaciens*
M. phenylpyruvica	*Moraxella polymorpha*, CDC Group M-II
M. urethralis	*Mima polymorpha* var. *oxidans*, CDC Group M-4
M. nonliquefaciens	
CDC Group IV,E	Probable affiliation: *Alcaligenes* spp.
CDC Group II,F, II,J	Probable affiliation: *Flavobacterium* spp.

A. DIRECTIONS FOR IDENTIFICATION TESTS

1. Select well-isolated colonies from the agar or membrane surface and streak them onto m-SPC or R2A agar to check for purity. Incubate at 30°C for 2–7 days.

2. Select a well-isolated colony and transfer it onto an m-SPC agar slant. Incubate for 2–7 days at 30°C.

3. Record the isolate number onto the worksheet (Table VI).

4. (a) Perform a gram strain with growth from the m-SPC slant. The gram

TABLE VIII

Distribution of Genera within the Biogrouping Code Numbers

Genus	001–007	101–177	200–277	300–307	403–477	501–577	601–677	701–777
Moraxella spp.	X							
Flavobacterium and related genera	X	X						
Acinetobacter spp.			X	X				
Pseudomonas spp. (non-glucose oxidizers)					X		X	
Alcaligenes spp. (non-glucose oxidizers)					X		X	
Alcaligenes spp. (glucose oxidizers)						X		X
Pseudomonas spp. (glucose oxidizers)						X		X
Argobacterium spp.								X

stain should be performed as soon as growth is seen on the slant. Observe the cellular morphology and gram reaction. Record these onto the worksheet. (b) Prepare a wet mount and look for motile cells. If gliding cells are observed, record this onto the worksheet. Confirm the observation by inoculating the culture into the medium of Anacker and Ordal (1959). If flagella-mediated movement is seen, perform the modified flagella staining procedure of Mayfield and Innis (1977).

Flagella Staining

Motile cells, obtained from the m-SPC agar slant, are suspended into 0.1 ml of distilled water. Wait 5–10 minutes and transfer a small drop, approximately 5 μl, onto a clean glass slide (special cleaning of the slide is not necessary). Carefully lower a glass coverslip onto the drop, avoiding entrapped bubbles. After 10 minutes, two drops of flagella staining solution (see Section VII,B) are placed at the edge of the coverslip. The staining solution is pulled under the glass coverslip by capillary action. The slide can immediately be examined using oil immersion. Examination may be enhanced by observing the stained preparation using differential interference contrast or phase-contrast microscopy.

5. Record the presence and color of a water-insoluble (colony) pigment.

6. Record the presence and color of a water-soluble pigment (diffusing into agar medium) or discoloration of medium.

7. Inoculate the tubes and plate from the slant.

METHOD

Medium	Inoculation
Glucose–H_2S	Stab butt and streak slant
Acetate	Streak slant
Motility	Touch to surface (do not penetrate medium surface!)
Gelatin–esculin	Make a 1-in. cut in medium
DNase	Make a V-shaped streak on medium surface
Tween 80–ONPG	Make a 1-in. streak on medium surface

The glucose–H_2S tube and acetate tube are inoculated with a needle. Following inoculation of the glucose–H_2S medium, it is not necessary to obtain more inoculum for the acetate medium. The needle is not flamed until after inoculation of the acetate medium. The plate is inoculated with additional inoculum obtained from the slant. With the exception of the acetate medium, all media may be heavily inoculated with the culture.

8. Incubate the tubes and plate at 30°C for 3–7 days. Generally, test results can be recorded after 3 days. However, some strains are slow grow-

ing and may require up to 7 days of incubation. Negative acetate alkalinization tubes are incubated for 7 days.

9. Cultures on m-SPC slants are preserved by covering the slants with
sterile, 10% glycerol solution and placing them into a freezer ($-80°C$ is
preferable). Note: Some of the GN-NF bacteria die quickly following subculture (3–4 days). It is imperative that the cultures be promptly stored for
future work.

10. Following incubation, place plates in a refrigerator (4°C) for 10–15
minutes.

11. Record glucose–H_2S and acetate alkalinization results.

METHOD

Medium	Test	Positive	Negative
Glucose–H_2S	Glucose oxidation	Yellow slant	Purple slant
Glucose–H_2S	H_2S production	Brown precipitate along slab line in butt	Absence of precipitate along stab line in butt
Acetate	Acetate alkalinization	Pink–red slant	Orange color (no change in medium color)

Note: Bacteria with solely oxidative properties may be screened from
fermentative bacteria with the glucose–H_2S medium. Oxidative organisms
only grow on the slant and a color change in the medium should occur
initially at the slant. Frequently, a frank alkaline (purple color) reaction is
observed within 1–3 days of incubation, followed by a slow change to the
acidic pH (yellow). In contrast, the fermentative bacteria will grow both on
the slant and in the butt and change the slant and the butt to bright yellow
within 1–2 days of incubation. If the oxidative–fermentative properties of an
isolate are in question, the glucose–H_2S slant may be overlaid with mineral
oil following inoculation. If an organism grows in the covered medium, it is
fermentative. All fermentative bacteria should be identified via alternative
schemes, e.g., API 20E (Analysis Products, Inc., Plainview, New York) or
Enterotube (Roche Diagnostics, a division of Hoffman-LaRoche, Inc.,
Nutley, New Jersey).

An isolate is recorded as a glucose oxidizer only if yellow is observed along
the slant. With prolonged incubation, the acidic by-products of metabolism
will diffuse into the medium such that the slant and the butt are yellow.
Intermediate colors such as pink, brown, or green along the slant should be
considered negative for glucose oxidation.

12. Remove the plates from the refrigerator.

13. Record the gelatin, motility, Tween 80, and ONPG results.

METHOD

Medium	Test	Positive	Negative
Gelatin–esculin	Gelatin	Liquid along the inoculation line[a]	Solid medium along the inoculation line
Motility	Motility	Movement away from the inoculation point	No movement away from the inoculation point
Tween 80–ONPG	Tween 80 hydrolysis	Opaque zone around growth	No opaque zone around growth
Tween 80–ONPG	ONPG	Diffusible yellow color in medium	No yellow color produced

[a] Note: The observation of gelatin liquefaction can be aided by tipping the plate slightly. Any movement of the medium around the line of inoculation is positive.

If motility is equivocal or no growth occurs on the motility medium, a microscopic determination using a hanging drop or wet mount procedure with a fresh broth culture is necessary.

The opaque zone from the Tween 80 hydrolysis should extend into the surrounding medium. This haloed region, when viewed microscopically, is composed of microcrystals of calcium oleate.

The ONPG-positive reaction should not be confused with the yellow, insoluble pigment produced by some strains of bacteria, most notably, flavobacteria. The yellow o-nitrophenol, liberated by the cleavage of the ONPG reagent by the enzyme β-galactosidase, will diffuse into the surrounding medium.

14. Add a drop of oxidase reagent to the growth on the surfaces of the Tween 80–ONPG and motility media. Production of purple–black color within 15–60 seconds of reagent addition is positive. All weak or negative reactions should be verified by scraping growth from the surface of the DNase medium with an inoculating loop (a platinum loop is recommended) or an applicator stick and rubbing it onto filter paper saturated with oxidase reagent. Production of a purple–black color after 10–15 seconds indicates a positive oxidase reaction.

15. Add one drop of 10% ferric chloride solution to the line of inoculation in the gelatin–esculin medium. If the medium has not been liquefied by the isolate, it is necessary to penetrate the medium surface with an applicator stick. Production of a black to blue precipitate in 30–60 seconds is positive for esculin hydrolysis. A light-brown precipitate developing after 60 seconds is recorded as negative for esculin.

16. Add 0.5–1 ml of 1 N HCl to the DNase medium. Observe the reaction after 5 minutes. Production of clearing around the growth indicates a

positive DNase reaction. It is often helpful to view the zone of clearing by placing the quadrant plate onto a black or dark background and carefully scraping away the growth. Observation of the plate from an oblique angle is also helpful.

17. Record all reactions on the worksheet. Assign the appropriate coding number to a positive reaction. Add the coding numbers within the correct group of three tests and determine the biogrouping code. Check the coding sheets for the biogrouping code. If the code number is not located, utilize the code listing for unrecognizable strains and proceed as instructed.

18. If inoculation of confirmation sugars is necessary, the glucose–H_2S tubes are used to obtain inoculum. The sugars are inoculated by streaking the slant with a loop. The slants are incubated at 30°C for 2–7 days.

19. Record the sugar reactions.

METHOD

Sugar	Positive	Negative
Xylose	Yellow	Purple
Maltose	Yellow	Purple
Fructose	Yellow	Purple
Mannitol	Yellow	Purple

20. Record all reactions on the worksheet. Determine the confirmation code and locate the confirmation code in the code number listing (Appendixes A and B).

21. Perform all necessary auxiliary tests as instructed.

METHOD[a]

Test	Reagent addition[b]	Positive	Negative
6.5% NaCl	None	Pink–red color	Orange
Phenylalanine	10% Ferric chloride	Green color on slant	No color change
Urease	None	Pink–red color	Orange
Nitrate to gas	None	Observe gas in Durham vial	No gas in Durham vial
Nitrate to nitrite	0.5 ml of Reagent A; 0.5 ml of Reagent B	Red color in 1–2 minutes	No color change
Nitrite to gas	None	Observe gas in Durham vial	No gas in Durham vial

[a] All media are incubated at 30°C for up to 7 days.
[b] Visible growth in the broth media and on the surfaces of the agar media should be observed before the addition of reagents.

B. Reagents and Stains

1. Oxidase Reagent

Dissolve 1 g of tetramethyl-p-phenylenediamine dihydrochloride into 100 ml of distilled water. Note: (1) Discard the reagent if it turns blue; it should be colorless. (2) If a precipitate forms, discard the reagent. (3) The reagent is an irritant. Avoid contact with eyes, skin, and clothing. (4) The reagent should be protected from light. Store it in a brown bottle and refrigerate it (4°C).

2. Ferric Chloride Solution

For esculin hydrolysis reaction, prepare a 10% ferric chloride solution. Store it in a brown bottle and refrigerate (4°C).

3. Catalase Test

Use a 3% hydrogen peroxide solution.

4. Nitrate Reduction

Reagent A: Dissolve 0.6 g of dimethyl-a-naphthylamine into 100 ml of acetic acid (30% w/v) with *gentle* heating. Reagent B: Dissolve 0.8 g of sulfanilic acid into 100 ml of acetic acid (30% v/v).

Note: (1) Store Reagents A and B in brown dropper bottles and refrigerate them when not in use (4°C). (2) Prepare new reagents every three months. (3) Label bottles with date of preparation. (4) Dimethyl-1-naphthylamine may be carcinogenic. Avoid the generation of aerosols, mouth pipetting, and contact with skin.

5. Phenylalanine Deaminase Reagent

Use the ferric chloride solution prepared for the esculin hydrolysis reaction.

FLAGELLA STAIN[a]

Stock solution	g/100 ml
Potassium aluminum sulfate	15
Tannic acid	30
Mercuric chloride	6
Basic fuchsin	20 (Dissolve in ethanol)

[a] Modified from the method of Mayfield and Innis (1977).

Prepare stock solutions. On the day of use, mix $KAl(SO_4)_2$, tannic acid, $HgCl_2$, and basic fuchsin stock solution in the proportions (by volume) of 5 ml:2 ml:2 ml:0.8 ml. Remove the prepared solution to a syringe fitted with a filter holder containing a 0.2-μm polycarbonate filter. The syringe may be used to deposit stain to the slide, allowing filtration during application. Preparation of 5 ml is adequate for daily work.

C. MEDIA FORMULATIONS

GLUCOSE-H₂S

Component	g/100 ml
Glucose	1.0
Casitone	0.2
Yeast extract	0.1
Ferric ammonium Citrate	0.05
Sodium thiosulfate	0.03
Phenol red	0.004
Bromthymol blue	0.002
Agar	1.5

Dispense all components into a flask and heat to boiling to dissolve. Adjust to pH 7.1. Dispense 10 ml into 16 × 125-mm tubes. Autoclave at 121°C for 15 minutes. Allow the medium to solidify in a slanted position. Uninoculated medium should be a light-brown color.

ACETATE ASSIMILATION

Component	g/100 ml
Sodium acetate	0.2
Magnesium sulfate (anhydrous)	0.01
Potassium phosphate (monobasic)	0.05
Ammonium phosphate (dibasic)	0.05
Phenol red	0.002
Agar	1.5

Dispense all components into a flask and heat to boiling to dissolve. Adjust to pH 6.8. Dispense 10 ml into 16 × 125-mm tubes. Autoclave at 121°C for 15 minutes. Allow the medium to solidify in a slanted position. Uninoculated medium should have an orange color.

GELATIN–ESCULIN

Component	g/100 ml
Gelatin	10
Esculin	0.1
Casitone	0.2
Agar	0.15

Dispense all components into a flask and heat to boiling to dissolve. Adjust to pH 7.2. Autoclave at 121°C for 15 minutes. Cool to 45°C and aseptically dispense 5 ml into one section of the quadrant Petri dish (X plate).

MOTILITY MEDIUM

Component	g/100 ml
Casitone	0.2
Yeast extract	0.2
Agar	0.2–0.3

Dispense all components into a flask and heat to boiling to dissolve. Adjust to pH 7.2. Autoclave at 121°C for 15 minutes. Cool to 45°C and aseptically dispense 5 ml into one section of the quadrant plate. Allow to solidify.

Note: The mechanical strength of different types and lots of agar vary. Standardize the concentration of agar by preparing the motility medium with agar concentrations of 0.2, 0.3, and 0.4%. Inoculate with a motile culture of *Pseudomonas aeruginosa* and observe the movement in the semi-solid medium. Use that concentration which allows rapid motility but provides a gel with sufficient rigidity for easy manipulation of the plate. In addition, it is important to dispense a medium which has been tempered to a temperature close to the point of solidification in order to avoid condensation on the medium surface. Most agar preparations begin to solidify at around 42°C. Purified agar and agarose, which have lower methoxyl concentrations, have a lower dynamic gelling temperature and are preferable. Our laboratory employs SeaKem LE agarose (Marine Colloids Division, FMC Corporation, Rockland, Maine) at a concentration of 0.175%.

DNASE MEDIUM

Component	g/100 ml
Casitone	0.2
Yeast extract	0.2
DNA	0.2
Agar	1.5

Dispense all components into a flask and heat to boiling to dissolve. Adjust to pH 7.2. Autoclave at 121°C for 15 minutes. Cool to 45°C and aseptically dispense 5 ml into one section of the quadrant plate. An alternative method would be to use DNase Test Agar (Difco).

TWEEN 80–ONPG

Component	g/100 ml
Lactose	0.02
Yeast extract	0.2
Casitone	0.2
Calcium chloride	0.02
o-Nitrophenyl-β-D-galactopyranoside	0.2
Tween 80 (polyethylene sorbitan monooleate)	1% (v/v)
Agar	1.5

Dispense all components (minus ONPG and Tween 80) into a flask and heat to boiling to dissolve. Adjust to pH 7.2. Autoclave at 121°C for 15 minutes. Cool to 45°C.

Tween 80: Autoclave Tween 80 at 121°C for 30 minutes. Aseptically add to a final concentration of 1% (v/v). ONPG: Filter sterilize a 1% solution. Store at 4°C. Aseptically add to a final concentration of 0.2%.

D. CONFIRMATION TESTS

CARBOHYDRATE BASAL MEDIUM[a]

Component	g/100 ml
Yeast extract	0.1
Casitone	0.2
Phenol red	0.004
Bromthymol blue	0.002
Agar	1.5

[a] For xylose, maltose, fructose, and mannitol oxidation studies.

Dispense all components into a flask and heat to boiling to dissolve. Adjust to pH 7.1. Autoclave at 121°C for 15 minutes. Cool to 45°C. Aseptically add sterile sugar solutions to a final concentration of 1% (w/v). Dispense 5 ml of each sugar into sterile 13 × 100-mm tubes. Sugars: Prepare 10% (w/v) solutions and filter sterilize through a 0.2-μm membrane.

E. Auxiliary Tests

6.5% NaCl Broth

Component	g/100 ml
Casitone	0.2
Yeast extract	0.2
Sodium chloride	6.5
Phenol red	0.004

Dispense all components into a flask and heat to dissolve. Adjust to pH 6.8. Dispense 5 ml into 16 × 125-mm tubes. Autoclave at 121°C for 15 minutes.

1. Phenylalanine Deaminase Agar

Use phenylalanine agar (BBL). Prepare as specified by the manufacturer.

2. Urea Test Agar

Use Christensen's urea agar (BBL). Prepare as specified by the manufacturer.

Nitrate Reduction Broth

Component	g/100 ml
Casitone	0.2
Yeast extract	0.2
Potassium nitrate	0.2

Dispense all components into a flask and heat to dissolve. Adjust to pH 7.0. Dispense 7 ml into 16 × 125-mm tubes with Durham vials. Autoclave at 121°C for 15 minutes.

Nitrite Reduction Broth

Component	g/100 ml
Casitone	0.2
Yeast extract	0.2
Potassium nitrite (chemically pure)	0.2

Dispense all components into a flask and heat to dissolve. Adjust to pH 7.0. Dispense 7 ml into 16 × 125-mm tubes with Durham vials. Autoclave at 121°C for 15 minutes.

R2A[a]

Ingredient	g/100 ml
Yeast extract	0.05
Difco Proteose Peptone #3	0.05
Difco Casamino Acids	
(Difco, 0230-02-0)	0.05
Glucose	0.05
Soluble starch	0.05
K_2HPO_4	0.03
$MgSO_4 \cdot 7H_2O$	0.005
Sodium pyruvate	0.03
Agar	1.5

[a] From Reasoner and Geldreich (1985).

Heat to dissolve components. Adjust to pH 7.0 with crystalline K_2HPO_4 or KH_2PO_4. Autoclave at 121°C for 15 minutes.

m-SPC[a]

Ingredient	g/100 ml
Bactopeptone	2.0
Gelatin	2.5
Glycerol	1 ml
Agar	1.5

[a] From Taylor and Geldreich (1979).

Dispense peptone, gelatin, and agar into a flask and heat to dissolve. Adjust to pH 7.1. Autoclave for 15 minutes. Autoclave glycerol at 121°C for 30 minutes. Add sterile glycerol aseptically.

COMMERCIAL PRODUCT AVAILABILITY

o-Nitrophenyl-β-D-galactopyranoside (ONPG)
   a. Sigma Chemical
   b. Cal-biochem
Agarose
   a. Marine Colloids, Division of FMC Corporation
N,N,N',N'-Tetramethyl-p-phenylenediamine dihydrochloride
   a. Sigma Chemical
   b. Fisher
   c. Eastman
Esculin [6-(β-glucosyloxy)-7-hydroxycoumarin]
   a. Aldrich
   b. Sigma Chemical

Yeast extract
   a. Difco
Casitone
   a. Difco
Tween 80 (polyoxyethylene sorbitan monooleate)
   a. Difco
   b. Fisher
   c. Baker
   d. Sigma
DNA (deoxyribonucleic acid sodium salt)
   a. Eastman
   b. Sigma
DNase Test Agar
   a. Difco

VIII. Biochemical Basis for Tests Employed in the Identification Scheme

A. GLUCOSE OXIDATION

Bacteria that oxidize glucose and other carbohydrates differ from fermentative bacteria by their requirement for oxygen (or an inorganic compound such as nitrate) as a terminal electron acceptor and by the lack of initial phosphorylation of glucose before degradation.

1. Fermentation

Fermentation is an anaerobic process requiring initial phosphorylation of glucose to glucose 6-phosphate prior to its degradation. The main pathway of glucose fermentation is the Embden–Meyerhof pathway, although the pentose phosphate shunt or Entner–Doudoroff pathway may be utilized. Fermentation requires an organic compound as the terminal electron acceptor (substrate level phosphorylation). The end products of fermentation vary, depending upon the species and the conditions of culture, but pyruvic acid is always an intermediate breakdown product. Fermentation produces more acidic by-products than oxidation (MacFaddin, 1980).

2. Oxidation

Oxidation is an aerobic process which requires no initial phosphorylation of glucose before degradation; rather, the glucose is oxidized to gluconic acid, then to 2-ketogluconic acid (Norris and Campbell, 1949; Stokes and Campbell, 1951; Juni, 1978). On media rich in glucose and other carbohydrates, gluconic acid and other oxidized derivatives of glucose accumulate and are excreted (Juni, 1978). Intermediates of the tricarboxylic acid cycle, such as α-ketoglutaric acid, may be produced. Energy is generated through a series of oxidation–reduction reactions, involving intermediate carriers that readily undergo reversible redox reactions (electron transport). The last carrier in this series catalyzes the reduction of oxygen to water. The energy liberated in these redox reactions is conserved in the phosphorylation of ADP to ATP (oxidative phosphorylation) (Juni, 1978).

3. Oxidation–Fermentation Test

The medium of Hugh and Leifson (1953) has traditionally been utilized to differentiate fermentative and oxidative organisms. This medium contains a high level of carbohydrate (at least 1% w/v) with a low concentration of peptone. The ratio of carbohydrate to peptone is important because of the tendency of many of the oxidative bacteria to produce little acidity from glucose degradation but high concentrations of alkaline products from the metabolism of amino acids in the peptone (Snell and Lapage, 1971). The oxidation of glucose, then, may be masked if the production of acidic by-products is too low to overcome the alkaline degradation products of amino acid metabolism (Otto and Pickett, 1976). Media such as triple-sugar iron agar (TSI) or Kligler's iron agar (KIA) are inappropriate for the observation of the oxidation of glucose because of the high amino acid concentration in these media.

The test is conducted by inoculating two tubes; one of which is sealed with petrolatum and the other of which remains open. Nonfermentative bacteria are capable of growing only in the open tube, whereas the fermentative bacteria can grow in both the sealed and open tubes. The medium contains 0.2–0.3% agar (w/v), which inhibits the diffusion of oxygen into the medium. As such, the nonfermentative bacteria grow only at the medium surface in the open tube.

A glucose oxidizer will change the color of the glucose medium from the uninoculated green (pH 7.1) to yellow (near pH 6.0), whereas a glucose nonoxidizer will change the color of the medium from green to blue (above pH 7.6). Frequently, a frank alkaline reaction will be seen within 1–3 days of incubation, followed by an acidic reaction in 3–7 days. This occurs because many of the glucose oxidizers initially metabolize the amino acids in the medium, then the glucose.

The disadvantages of the test are: (1) Two tubes are required to screen fermentative bacteria from nonfermentative bacteria; (2) some of the nonfermentative bacteria do not grow in the medium and supplementation with 0.1% yeast extract or 2.0% serum is required (MacFaddin, 1980); (3) some nonfermentative bacteria require 1–3 weeks of incubation before a positive oxidation reaction is observed (MacFaddin, 1980); (4) the yellow insoluble pigment produced by some of the nonfermentative bacteria, most notably the flavobacteria, make interpretation of the glucose reaction difficult, since a positive glucose test is indicated by the production of a yellow color at the medium surface; and (5) the nonfermentative bacteria must decrease the medium pH to near 6 before complete development of the yellow color. Some weak acid producers are unable to overcome the alkaline products of amino acid metabolism to sufficiently decrease the pH for easy interpretation of oxidation.

The oxidation–fermentation medium recommended by Tatum *et al.* (1974) employs a low peptone concentration (0.2% w/v of a pancreatic digest of casein) and phenol red indicator. The uninoculated medium (pH 7.3) has an orange color. Above pH 7.8, the medium develops a pink–red color and below pH, 6.8 a yellow color. The advantage of this formulation over the medium of Hugh and Leifson (1953) is that the pH of medium has to be decreased to just below pH 7 for a complete change in the indicator color.

4. Glucose Oxidation–H_2S Medium

The medium employed in this scheme for establishing glucose-oxidative capability includes a low peptone concentration [0.2% casitone (a pancreatic digest of casein)], yeast extract (0.1%), an indicator system for H_2S production (0.03% sodium thiosulfate and 0.05% ferric ammonium citrate), a mixture of pH indicators (0.004% phenol red and 0.002% bromthymol blue), glucose (1%), and agar (1.5%). Following sterilization, the medium is allowed to solidify in a slanted position. Both the slant and the butt of the medium are inoculated.

The advantages of this medium include the following: (1) Only one tube is required to separate fermentation and oxidative bacteria. Oxidative bacteria grow only along the slant and the pH change is noted at the surface of slant, not in the butt of the tube. Fermentative bacteria grow in the butt of the tube and a yellow color is developed throughout the medium in 1–2 days. (2) Incorporation of yeast extract into the medium enhances the growth of those bacteria, which normally do not grow well in the Hugh and Leifson OF medium (Tatum *et al.*, 1974). (3) The mixture of the phenol red and bromthymol blue pH indicators improves the interpretation of oxidative activity. Minor changes in pH are noted with these indicators. The colors which are observed over a range of pH are as follows: below pH 6.8, bright yellow; pH 6.8–6.9, light green; pH 7.0–7.2, brown; pH 7.3–7.6, red; and above pH 7.6, purple. Glucose nonoxidizers produce a strong purple color along the slant of this medium. Weak acid producers are readily detected because only a decrease below pH 7 is necessary for an unambiguous change in the medium color. (4) Incorporation of the H_2S detection system aids in the identification of *Pseudomonas putrefaciens*. This organism is the only known nonfermentative species capable of producing H_2S (Rubin *et al.*, 1980).

B. ACETATE ALKALINIZATION

Certain aerobic bacteria are able to utilize acetate as a sole source of carbon. The oxidation of acetate proceeds via the glyoxylate pathway (Moat, 1979). In this pathway, isocitric acid (a normal intermediate in the tricarboxylic acid cycle) is cleaved, yielding succinic acid and glyoxylic acid. Acetyl–coenzyme A is condensed with glyoxylate to form malate, and acetate is condensed with succinate to form isocitrate. Energy and assimilatory products are produced through the functional TCA and electron transport systems.

The acetate medium is a minimal-basal-salts medium with acetate and phenol red. The ability of an isolate to utilize acetate as a sole source of carbon is indicated by an alkaline reaction (orange to red change in the pH indicator) along the slant, due to the production of ammonia and alkaline by-products of metabolism.

C. TWEEN 80 HYDROLYSIS

Some aerobic bacteria elaborate a lipase which is able to hydrolyze long-chain fatty acid esters such as C_{12}, C_{16}, C_{18} (Sierra, 1957). This is an intracellular enzyme which is released with lysis of the cell. Polyoxyethylene sorbitan monooleate (Tween 80) is hydrolyzed by the lipase, releasing oleic acid.

Oleic acid released from the hydrolysis of Tween 80 binds with the calcium in the medium, forming microcrystals of calcium oleate. Lipase activity of aerobic bacteria is easily observed with the presence of a haloed region of microcrystals around the bacterial growth (Sierra, 1957).

D. ONPG TEST

Galactoside compounds are cleaved by the enzyme β-galactosidase, which is an intracellular, inducible enzyme (Cohen and Monod, 1957). Production of this enzyme is induced by the presence in the medium of lactose, melibiose, methyl-α-D-galactoside, or the gratuitous inducer, isopropyl-β-D-thiogalactoside (IPTG) (MacFaddin, 1980). The ONPG test is utilized to assay for the presence of the enzyme β-galactosidase (MacFaddin, 1980).

The compound *o*-nitrophenyl-β-D-galactopyranoside (ONPG) is cleaved by the enzyme β-galactosidase. ONPG is a colorless compound which, when cleaved by the enzyme, liberates a yellow, chromogenic *o*-nitrophenol (Mac-

Faddin, 1980). A positive ONPG test is indicated by the production of a yellow color in the medium. Note: The liberated o-nitrophenol may impart a yellow color the media in the surrounding wells of the quadrant plate.

E. Gelatin Liquefaction Test

The gelatin liquefaction test assays for the presence of extracellular gelatinases, which break down proteins into smaller peptides and amino acids for cellular transport and metabolism (MacFaddin, 1980).

The gelatin medium is solid below 28°C and a loose semisolid (due to the presence of a 0–15% concentration of agar) above 28°C. If an organism can proteolyze gelatin, the medium remains a semisolid with refrigeration.

F. Oxidase Test

The oxidase test determines the presence of cytochrome c in the electron transport system of aerobic bacteria (Jurtshuk et al., 1975). This test is performed with a p-phenylenediamine dye which is an artificial electron acceptor (Steel, 1961). Cytochrome c oxidizes the colorless p-phenylenediamine compound to a purple–black indophenol compound in 10–60 seconds (Steel, 1961).

One drop of a 1% tetramethyl-p-pheylenediamine solution is added to the bacterial growth on the surface of the Tween 80–ONPG medium. Oxidase-positive colonies become purple to black in 10–60 seconds. A discoloration of the surrounding medium should be ignored.

G. Esculin Hydrolysis Test

Esculin [6-(β-glucosyloxy)-7-hydroxycoumarin] is hydrolyzed to β-D-glucose and esculetin (6,7-dihydroxycoumarin) by certain oxidative bacteria (Edberg et al., 1976). Esculin hydrolysis is indicated by the formation of a brown–black precipitate (MacFaddin, 1980), formed by the reaction of esculetin with iron in the medium.

H. DNase Test

DNases are a class of extracellular endonucleases which are produced by some GN-NF bacteria and are able to hydrolyze polymerized DNA into smaller oligonucleotide pieces (MacFaddin, 1980; Black et al., 1971). The depolymerization of DNA results in changes in the physical properties of the nucleic acid molecule, such as a decrease in the viscosity of a DNA solution and an increase in absorption with ultraviolet light (MacFaddin, 1980).

Polymerized DNA is precipitated with an adjustment of pH to below 4, whereas depolymerized DNA is not precipitated (Jeffries *et al.*, 1957). Elaboration of deoxyribonuclease is indicated by a clearing around the growth when the plates are flooded with 0.1 *N* HCl.

I. XYLOSE, MALTOSE, FRUCTOSE, MANNITOL OXIDATION

See Section VIII,A.

J. GROWTH IN 6.5% NaCl BROTH

The ability of aerobic bacteria to tolerate high salt concentrations is determined with this test. Growth is indicated by a change in the phenol red indicator from orange (pH 6.8) to pink–red (above pH 7.4) with the accumulation of alkaline products from amino acid metabolism.

K. PHENYLALANINE DEAMINASE AGAR

The phenylalanine deaminase test assays for the oxidative deamination of the amino acid phenylalanine to the keto acid phenylpyruvic acid (Singer and Volcani, 1955). The oxidative deamination of phenylalanine is indicated with the addition of 10% ferric chloride to an incubated, slanted medium. The production of a green color within 1 minute of reagent addition is a positive test. The positive $FeCl_3$ reaction is due to the formation of a hydrazone.

L. UREASE TEST

Urea is hydrolyzed by the constitutive enzyme urease to ammonia and carbon dioxide (MacFaddin, 1980). The hydrolysis of urea is indicated by a change in the phenol red indicator from orange (pH 6.8) to pink–red (above pH 7.4), due to the release of the ammonia.

M. NITRATE/NITRITE REDUCTION TESTS

The ability of an organism to reduce nitrate to nitrite or free nitrogen gas or nitrite to gas is determined in these tests (MacFaddin, 1980). The end product possibilities of nitrate reduction include the production of nitrite (NO_2), ammonia (NH_3), molecular nitrogen (N_2), nitrous oxide (N_2O), or hydroxylamine-type compounds (R-NHOH). The reduction of nitrate to nitrogen gas or nitrous oxide is termed "denitrification." Nitrate reduction is evidence by either the absence of nitrate in the medium or the presence of nitrite or nitrogen gas. Nitrite reduction is indicated by the presence of nitrogen gas.

The nitrite reduction test is a two-part test. The reduction of nitrate to nitrite is indicated by the production of a red color with the addition of sulfanilic acid and α-naphthylamine to the broth. Denitrification is evidenced by the sequestration of nitrogen gas in the Durham vial. If no gas or production of nitrite is detected, zinc dust is added to the broth to confirm the presence of nitrate. Zinc reduces nitrate to nitrite, which then reacts with the sulfanilic acid and α-naphthylamine to form the red color. If no nitrate is observed with addition of zinc, then nitrate has been reduced to an end product other than nitrogen gas or nitrite.

APPENDIX A

IDENTIFICATION CODING LIST

Biocode	Characteristics
001	**Genus:** *Moraxella* (Note: the members of this genus are nonpigmented; also, aquatic isolations of these bacteria have not been confirmed.) **Probable species:** *M. nonliquefaciens* **Distinguishing characteristics of *Moraxella* spp.:** *Microscopic*: Small bacilli (1 × 2 μm), often appearing as coccobacilli in short chains. *Macroscopic*: Generally slow-growing bacteria. Some species (e.g., *M. lacunata* and *M. atlantae*) do not grow without serum enrichment. Colonies are nonpigmented. Occasional strains are mucoid.
005	**Genus:** *Flavobacterium* (Note: The bacteria included in this genus are yellow pigmented.) **Probable species:** *F. aquatile* (Note: This species grows poorly on nutrient agar containing peptone and meat extract (Holmes *et al.*, 1984). *F. aquatile* grows well on media containing yeast extract, enzymatic digest of casein, and glucose (Weeks, 1974). **Distinguishing characteristics of *Flavobacterium* spp.:** *Microscopic*: Rods with rounded ends, typically 0.5–0.7 μm in width, 1–3 μm in length. Some species (e.g., *F. aquatile* and *F. meningosepticum*) approach coccobacillary forms in young cultures. Others produce longer rods (*F. breve*), filaments (*F. meningosepticum*), and chains of rods (*F. odoratum*). No intracellular granules of poly-β-hydroxybutyrate or endospores formed. *Macroscopic*: Colonies are yellow pigmented, although nonpigmented strains occur (e.g., *F. meningosepticum*). Some strains are slow growing and require 7 or more days of incubation. Colonies are entire and convex; occasionally mucoid (*F. aquatile*), with butyrous consistency [*F. balustinum* (CDC Group II,B), *F. meningosepticum*] or with spreading edges (*F. odoratum*).
011	**Genus:** *Moraxella* **Probable species:** *M. phenylpyruvica* **Distinguishing characteristics of *Moraxella* spp.:** See Biocode 001

(continued)

APPENDIX A (*Continued*)

Biocode	Characteristics

041	**Genus:** *Flavobacterium*
	Probable species: CDC Group II,F, CDC Group II,J
	Improbable species: *F. odoratum*[a]
	Tests required for speciation: Urea
	  CDC Group II,F: urea, negative; soluble tan pigment, positive.
	  CDC Group II,J: urea, positive; soluble tan pigment, positive.
	  F. odoratum: urea, positive; soluble tan pigment, negative.
	Distinguishing characteristics of *Flavobacterium* **spp.:** See Biocode 001
	  F. odoratum (macroscopic): Colonies have a pale yellow pigment. A "fruity" odor is characteristic. *F. odoratum* is usually Tween 80 positive.

045	**Genus:** *Flavobacterium*
	Probable species: *F. odoratum, F. breve*
	Tests required for speciation: Run confirmation tests
	  Confirmation codes
	    F. odoratum: 045-00
	    F. breve: 045-20
	Distinguishing characteristics of *Flavobacterium* **spp.:** See Biocode 001.
	  F. breve (macroscopic): Colonies have a pale yellow pigmentation. Species is slow growing, producing colonies of 0.5 mm after 7 days of incubation at 30°C.
	  F. odoratum: See Biocode 041

051	**Genus:** *Flavobacterium* **Probable species:** *F. odoratum* **Distinguishing characteristics of** *Flavobacterium* **spp.:** See Biocode 001 *F. odoratum*: See Biocode 041

055	**Genus:** *Flavobacterium*
	Probable species: *F. odoratum, F. breve*
	Tests required for speciation: Run confirmation tests
	  Confirmation codes
	    F. odoratum: 055-00
	    F. breve: 055-20
	Distinguishing characteristics of *Flavobacterium* **spp.:** See Biocode 001
	  F. odoratum: See biocode 041
	  F. breve: See Biocode 045

123 133 137	**Genus:** *Flavobacterium* **Probable species:** *F. multivorum* **Distinguishing characteristics of** *Flavobacterium* **spp.:** See Biocode 001 *F. multivorum* (macroscopic): Colonies are pale yellow until 48 hours of incubation and sometimes surrounded by a tan, soluble pigment.
145	**Genus:** *Flavobacterium* **Probable species:** *F. breve* **Distinguishing characteristics of** *Flavobacterium* **spp.:** See Biocode 001 *F. breve*: See Biocode 045

APPENDIX A (*Continued*)

Biocode	Characteristics

153   **Genus:** *Flavobacterium*
Probable species: *F. balustinum* (CDC Group II,B)
Distinguishing characteristics of *Flavobacterium* spp.: See Biocode 001
    F. balustinum (CDC Group II,B) (macroscopic): Colonies are brightly yellow
       pigmented. Colonies become mucoid with prolonged incubation. Gelatin
       liquefaction may be weak or delayed.

155   **Genus:** *Flavobacterium*
Probable species: F. breve
Distinguishing characteristics of *Flavobacterium* spp.: See Biocode 001
    F. breve: See Biocode 045

157   **Genus:** *Flavobacterium*
Probable species: *F. balustinum* (CDC Group II,B)
Distinguishing characteristics of *Flavobacterium* spp.: See Biocode 001
    F. balustinum: See Biocode 153

167   **Genus:** *Flavobacterium*
Probable species: *F. meningosepticum, F. spiritovorum*
Tests required for speciation: Run confirmation tests
    F. meningosepticum: 167-20, 167-21, 167-22, 167-23
    F. spiritovorum: 167-33
Distinguishing characteristics of *Flavobacterium* spp.: See Biocode 001
    F. spiritovorum (macroscopic): Some strains are nonpigmented; others are a
       pale yellow until 48 hours of incubation. This species is not actively proteo-
       lytic.
    F. meningosepticum (macroscopic): Colonies are 1–1.5 mm with 24- to 48-hour
       incubation at 30°C. The colonial appearance ranges from smooth and convex
       with an entire edge to slightly mottled and butyrous. A slight yellow pigment
       is typical.

173   **Genus:** *Flavobacterium*
Probable species: *F. balustinum, F. multivorum*
Tests required for speciation: Run confirmation tests
    Confirmation codes
       F. balustinum: 173-01
       F. multivorum: 173-31
Distinguishing characteristics of *Flavobacterium* spp.: See Biocode 001
    F. balustinum: See Biocode 153
    F. multivorum: See Biocode 123

177   **Genus:** *Flavobacterium*
Probable species: *F. meningosepticum, F. multivorum, F. spiritovorum*
Tests required for speciation: Run confirmation tests
    Confirmation codes
       F. meningosepticum: 177-20, 177-21, 177-22, 177-23
       F. spiritovorum: 177-33
       F. multivorum: 177-31

(*continued*)

APPENDIX A (*Continued*)

Biocode	Characteristics
177 (*cont.*)	**Distinguishing characteristics of** *Flavobacterium* **spp.:** See Biocode 001 *F. meningosepticum*: See Biocode 167 *F. spiritovorum*: See Biocode 167 *F. multivorum*: See Biocode 123
201	**Genus:** *Moraxella* **Probable species:** *M. osloensis, M. urethralis* **Tests required for speciation:** Phenylalanine deaminease *M osloensis*: phenylalanine negative *M. urethralis*: phenylalanine negative **Distinguishing characteristics of** *Moraxella* **spp.:** See Biocode 001
210	**Genus:** *Acinetobacter* **Probable species:** *A. calcoaceticus* biotype *lwoffii* **Distinguishing characteristics of** *Acinetobacter sp.*: *Microscopic*: Cultures obtained from an agar slant will show diplococcal forms predominantly, approximately 0.7×1 μm. Bacillary forms, 1.2×2 μm, are typical from a broth culture. *Macroscopic*: 1- to 2-mm colonies, gray–white to cream colored, which are convex, circular and entire with a mucoid or butyrous consistency, are produced with 24- to 48-hour incubation at 30°C. Some strains produce a tan, water-soluble pigment.
211	**Genus:** *Moraxella* **Probable species:** *M. osloensis* **Distinguishing characteristics of** *Moraxella* **spp.:** See Biocode 001
250	**Genus:** *Acinetobacter* **Probable species:** *A. calcoaceticus* biotype *alcaligenes* **Distinguishing characteristics of** *Acinetobacter* sp.: See Biocode 210
310	**Genus:** Acinetobacter **Probable species:** *A. calcoaceticus* biotype *anitratus* **Distinguishing characteristics of** *Acinetobacter sp.*: See Biocode 210
322[a] 323[a] 332[a] 333[a]	**Genus:** *Pseudomonas* **Probable species:** *P. paucimobilis* **Distinguishing characteristics of** *Pseudomonas* **spp.:** *Microscopic*: Cells are polar flagellated, either monotrichous or lophotrichous. Rods are small (0.5×1–2 μm) and generally straight. *Macroscopic*: Colonies vary from smooth, mucoid, and spreading to dry, wrinkled, and refractile. Bacteria may produce soluble (fluorescent or non-fluorescent) or insoluble (yellow to orange–red) pigments. *P. paucimobilis* (macroscopic): Colonies are yellow and occasionally surrounded by a tan–brown discoloration of the agar medium. Some strains are nonmotile and give a weak or negative oxidase reaction. Acid production from carbohydrates is often delayed and weak.

APPENDIX A (*Continued*)

Biocode	Characteristics
350	**Genus:** *Acinetobacter* **Probable species:** *A. calcoaceticus* biotype *haemolyticus* **Distinguishing characteristics of** *Acinetobacter* **sp.:** See Biocode 210
401	**Genera:** *Pseudomonas, Alcaligenes* **Probable species:** *P. acidovorans, P. alcaligenes, P. diminuta, P. extorquens, P. testosteroni, A. denitrificans, A. faecalis,* CDC Group IV,E **Tests required for differentiation of genera:** Flagellar staining *P. acidovorans*: polar, multitrichous *P. alcaligenes*: polar, monotrichous (wavelength about 1.6 μm) *P. diminuta*: polar, monotrichous (wavelength about 0.6–1.0 μm) *P. extorquens*: polar, monotrichous *P. testosteroni*: polar, multitrichous *A. denitrificans*: peritrichous *A. faecalis*: peritrichous CDC Group IV,E: peritrichous **Tests required for specification:** *Pseudomonas*: Run confirmation tests; compare flagellar characteristics *Confirmation codes* *P. acidovorans*: 401-03 *P. alcaligenes*: 401-00 *P. diminuta*: 401-00 *P. extorquens*: 401-11 *P. testosteroni*: 401-00 *Alcaligenes*: Nitrate reduction, phenylalanine *A. denitrificans*: nitrate to nitrite, +; nitrate to gas, +; phenylalanine, − *A. faecalis*: nitrate to nitrite, +/−; nitrate to gas, −; phenylalanine, − CDC Group IV,E: nitrate to nitrite, +; nitrate to gas, +; phenylalanine, + **Distinguishing characteristics of** *Pseudomonas* **spp.:** See Biocode 322 *P. acidovorans* (macroscopic): Colonies are nonpigmented but may be surrounded by a tan to brown zone of discoloration. *P. alcaligenes* (macroscopic): Colonies are nonpigmented, 1–2 mm after 24–48 hours of incubation at 30°C, and convex with an entire edge. *P. diminuta* (microscopic): Rods, 0.5 × 1.04 μm, occurring singly. Monotrichous flagellation is characteristic. Short wavelength of flagellum is the distinctive differential property. (Macroscopic): Colonies are nonpigmented. *P. testosteroni* (microscopic): 0.7 × 2–3 μm, occurring singly or in pairs, with polar tuft of flagella. (Macroscopic): Colonies are nonpigmented and occasionally surrounded by a tan–brown discoloration of the agar medium. **Distinguishing characteristics of** *Alcaligenes* **spp.:** *microscopic*: Cocci or coccobacilli, 0.5 × 0.5–1 μm, which occur singly. *macroscopic*: Colonies are nonpigmented. Colony shape varies between species.

(*continued*)

APPENDIX A (*Continued*)

Biocode	Characteristics
401 (*cont.*)	*A. faecalis* (macroscopic): Biotype I produces colonies of about 0.5 mm with 24-hour incubation at 30°C. These colonies are convex, low, and glistening with an entire edge. Biotype II produces colonies of 1–1.5 mm which are umbonate with a spreading edge. Some strains (previously named "*A. odorans*") produce a fruity odor. *A. denitrificans* (macroscopic): Colonies are about 0.5 mm, convex, entire, and glistening. CDC Group IV,E (microscopic): Morphology may be coccoid, rod-shaped, or filamentous. (Macroscopic): Colonies are convex, entire and nonpigmented; 1- to 2-mm colonies are produced with 24- to 48-hour incubation at 30°C.
403	**Genus:** *Pseudomonas* **Probable species:** *P. vesicularis* **Distinguishing characteristics of** *Pseudomonas* **spp.:** See Biocode 322 *P. vesicularis* (macroscopic): Some strains produce an orange–red carotenoid pigment and cause a brown discoloration in the surrounding medium.
405[a]	**Genus:** *Pseudomonas* **Probable species:** *P. diminuta* **Distinguishing characteristics of** *Pseudomonas* **spp.:** See Biocode 322 *P. diminuta*: See Biocode 401
411	**Genera:** *Pseudomonas, Alcaligenes* **Probable species:** *P. acidovorans, P. alcaligenes, P. diminuta, P. testosteroni, A. denitrificans, A. faecalis* **See Biocode 401 for description of organisms and additional tests**
413	**Genus:** *Pseudomonas* **Probable species:** *P. vesicularis* **Distinguishing characteristics of** *Pseudomonas* **spp.:** See Biocode 322 *P. vesicularis*: See Biocode 403
415[a]	**Genus:** *Pseudomonas* **Probable species:** *P. diminuta* **Distinguishing characteristics of** *Pseudomonas* **spp.:** See Biocode 322 *P. diminuta*: See Biocode 401
423	**Genus:** *Pseudomonas* **Probable species:** *P. vesicularis* **Distinguishing characteristics of** *Pseudomonas* **spp.:** See Biocode 322 *P. vesicularis*: See Biocode 403
441	**Genus:** *Pseudomonas* **Probable species:** *P. diminuta* **Distinguishing characteristics of** *Pseudomonas* **spp.:** See Biocode 322 *P. diminuta*: See Biocode 401

APPENDIX A (*Continued*)

Biocode	Characteristics

443
Genus: *Pseudomonas*
Probable species: *P. vesicularis*
Distinguishing characteristics of *Pseudomonas* **spp.:** See Biocode 322
  P. vesicularis: See Biocode 403

445[a]
Genus: *Pseudomonas*
Probable species: *P. diminuta*
Distinguishing characteristics of *Pseudomonas* **spp.:** See Biocode 322
  P. diminuta: See Biocode 401

453
Genus: *Pseudomonas*
Probable species: *P. vesicularis*
Distinguishing characteristics of *Pseudomonas* **spp.:** See Biocode 322
  P. vesicularis: See Biocode 403

455[a]
Genus: *Pseudomonas*
Probable species: *P. diminuta*
Distinguishing characteristics of *Pseudomonas* **spp.:** See Biocode 322
  P. diminuta: See Biocode 401

463
Genus: *Pseudomonas*
Probable species: *P. vesicularis*
Distinguishing characteristics of *Pseudomonas* **spp.:** See Biocode 322
  P. vesicularis: See Biocode 403

501
Genus: *Alcaligenes*-like
Probable species: CDC Group Vd-1, CDC Group Vd-2
Tests required for speciation: Run confirmation tests
  Confirmation codes
    CDC Group Vd-1: 501-11
    CDC Group Vd-2: 501-33
Distinguishing characteristics of *Alcaligenes* **spp.:** See Biocode 401

503
Genera: *Alcaligenes*-like, *Pseudomonas*
Probable species: CDC designation, CDC Group Vd-2; *P. vesicularis*
Tests required for differentiation: Run confirmation tests
  Confirmation codes
    CDC Group Vd-2: 503-33
    P. vesicularis: 503-10, 503-20, 503-30
Distinguishing characteristics of *Alcaligenes*: See Biocode 401
Distinguishing characteristics of *Pseudomonas* **spp.:** See Biocode 322
  P. vesicularis: See Biocode 403

513
Genus: *Pseudomonas*
Probable species: *P. vesicularis*
Distinguishing characteristics of *Pseudomonas* **spp.:** See Biocode 322
  P. vesicularis: See Biocode 403

(*continued*)

APPENDIX A (*Continued*)

Biocode	Characteristics
515[b]	**Genus:** *Pseudomonas*
	Probable species: *P. putrefaciens*
517[b]	**Distinguishing characteristics of** *Pseudomonas* **spp.:** See Biocode 322
	P. putrefaciens (macroscopic): Mucoid colonies which are pink–red. A yellow, fluorescent pigment diffuses into agar medium. Most strains produce brown–black precipitate in the butt of the glucose–H_2S agar medium.
523	**Genus:** *Pseudomonas*
533	**Probable species:** *P. vesicularis*
543	**Distinguishing characteristics of** *Pseudomonas* **spp.:** See Biocode 322
553	*P. vesicularis*: See Biocode 403
555[a]	**Genus:** *Pseudomonas*
557	**Probable species:** *P. putrefaciens*
	Distinguishing characteristics of *Pseudomonas* **spp.:** See Biocode 322
	P. putrefaciens: See Biocode 515
563	**Genus:** *Pseudomonas*
573	**Probable species:** *P. vesicularis*
	Distinguishing characteristics of *Pseudomonas* **spp.:** See Biocode 322
	P. vesicularis: See Biocode 403
576	**Genus:** *Pseudomonas*
577[a]	**Probable species:** *P. maltophilia*
	Distinguishing characteristics of *Pseudomonas* **spp.:** See Biocode 322
	P. maltophilia (macroscopic): Colonies often produce a faint, yellow pigment which does not diffuse into the agar medium. A brown discoloration of the agar medium may accompany growth. Some strains produce a weak oxidase reaction. An ammoniacal odor is frequently apparent. Weak oxidase and acetate alkalinization reactions are typical.
601	**Genera:** *Alcaligenes, Pseudomonas*
	Probable species: *A. faecalis, A. denitrificans,* CDC Group IVC-2, *P. alcaligenes, P. testosteroni, P. pseudoalcaligenes*
	Test required for differentiation of genera: Flagellar staining
	P. alcaligenes: polar, monotrichous
	P. pseudoalcaligenes: polar, monotrichous
	P. testosteroni: polar, multitrichous
	A. faecalis: peritrichous
	A. denitrificans: peritrichous
	CDC Group IVC-2: peritrichous
	Tests required for speciation:
	Pseudomonas: Run confirmation tests; compare flagellar characteristics
	Confirmation codes and flagellar arrangement:
	P. alcaligenes: 601-00; polar, monotrichous
	P. pseudoalcaligenes: 601-01; polar, monotrichous
	P. testosteroni: 601-00; polar, multitrichous

APPENDIX A (*Continued*)

Biocode	Characteristics

601 (*cont.*)   **Tests required for speciation:**
    Alcaligenes: Nitrate broth, nitrite broth, urea
      A. *faecalis*: nitrate to nitrite, $+/-$; nitrate to gas, $-$; nitrite to gas, $-$;
      urea, $-$.
      A. *denitrificans*: nitrite to nitrite, $+$; nitrate to gas, $+$; nitrite to gas, $+$;
      urea, $+/-$.
      CDC Group IVC-2: nitrate to nitrite, $-$; nitrate to gas, $-$; nitrite to gas, $-$;
      urea, $+$.
  Distinguishing characteristics of *Pseudomonas* **spp.:** See Biocode 322
    P. alcaligenes: See Biocode 401
    P. pseudoalcaligenes (macroscopic): Colonies are nonpigmented, 1–2 mm after
      24–48 hours of incubation at 30°C, and convex with an entire edge.
    P. testosteroni: See Biocode 401
  Distinguishing characteristics of *Alcaligenes* **spp.:** See Biocode 401
    A. faecalis: See Biocode 401
    A. denitrificans: See Biocode 401
    CDC Group IVC-2 (macroscopic): Colonies are nonpigmented, about 1 mm
      with 24–48 hour incubation at 30°C, and convex with an entire edge.

611   **Genera:** *Alcaligenes, Pseudomonas*
  Probable species: *A. faecalis, A. denitrificans*, CDC Group IVC-2, *P. al-caligenes, P. testosteroni*
  Improbable species; *P. pseudoalcaligenes*[a]
  See Biocode 601 for a description of organisms and additional tests

641   **Genus:** *Alcaligenes*
  Probable species: *A. faecalis*
  Distinguishing characteristics of *Alcaligenes* **spp.:** See Biocode 401
    A. faecalis; See Biocode 401

676[a]   **Genus:** *Pseudomonas*
677[a]   **Probable species:** *P. maltophilia*
  Distinguishing characteristics of *Pseudomonas* **spp.:** See Biocode 322
    P. maltophilia: See Biocode 576

700   **Genus:** *Pseudomonas*-like
  Probable species: CDC Group Ve-2
  Distinguishing characteristics of *Pseudomonas* **spp.:** See Biocode 322
    CDC Group Ve (microscopic): Cells stain irregularly and appear to contain
    vacuoles. They are short to medium in length (1–2 μm) with tapered ends,
    giving the appearance of being spindle shaped. (Macroscopic): 1- to 2-mm
    colonies are produced in 24–48 hours of incubation at 30°c. Colony mor-
    phology ranges from smooth and mucoid to rough, wrinkled, and adherent.
    These organisms produce a yellow, intracellular pigment.

701   **Genera:** *Alcaligenes, Pseudomonas*
  Probable species: CDC Groups Vd-1 or Vd-2, *A. denitrificans* subsp. *xylosoxy-dans, P. putida, P. aeruginosa*

(*continued*)

APPENDIX A (*Continued*)

Biocode	Characteristics

701 (*cont.*)   **Improbable species:** *P. pseudoalcaligenes*,[a] *P. mendocina*[a]

Test required for differentiation of genera: Flagellar staining. (Note: If isolate has a soluble, nonfluorescent or fluorescent pigment, flagellar staining is not necessary. Isolate is a member of the genus *Pseudomonas*.)

CDC Group Vd-1: peritrichous
CDC Group Vd-2: peritrichous
A. denitrificans subsp. *xylosoxydans*: peritrichous
P. putida: polar, multitrichous
P. aeruginosa: polar, monotrichous
P. pseudoalcaligenes: polar, monotrichous
P. mendocina: polar, monotrichous

Tests required for speciation:

Alcaligenes: Run confirmation tests
  Confirmation codes
    CDC Group Vd-1: 701-11
    CDC Group Vd-2: 701-33
    A. denitrificans subsp. *xylosoxidans*: 701-10

Tests required for speciation:

Pseudomonas: Run confirmation tests. (Note: *P. aeruginosa* is the only known gram-negative rod to produce pyocyanin. Production of pyocyanin is diagnostic for *P. aeruginosa* and auxiliary tests are not necessary.)

Species	Confirmation code	Flagellar arrangement	Pyocyanin	Pyoverdin
P. putida	701-11, 701-13 701-31, 701-33	Polar, multi-trichous	−	Usually +
P. aeruginosa	701-01, 701-03 701-21, 701-23	Polar, mono-trichous	+	Usually +
P. pseudoalcaligenes	701-01, 701-21	Polar, mono-trichous	−	−
P. mendocina	701-11	Polar, mono-trichous	−	−

Distinguishing characteristics of *Alcaligenes*: See Biocode 401
Distinguishing characteristics of *Pseudomonas* spp.: See Biocode 322

  P. putida (macroscopic): 1- to 2-mm, nonpigmented colonies produced in 24–48 hours of incubation at 30°C. Most strains produce fluorescent pyoverdin pigments which diffuse into the agar medium.

  P. aeruginosa (macroscopic): Most strains produce pyocyanin, a water-soluble, blue, nonfluorescent phenazine pigment. Many strains also produce a yellow, water-soluble pigment (pyoverdin) which, in combination with pyocyanin, imparts a characteristic blue–green color to culture media. This species is the only one known to produce pyocyanin, and the presence of a bluish water soluble pigment is diagnostic for *P. aeruginosa*. However, strains may produce combinations of pycyanin, pyoverdin, pyorubin (red,

APPENDIX A (*Continued*)

Biocode	Characteristics
701 (*cont.*)	water-soluble), and pyomelanin (brown to black, water-soluble), which mask the presence of pyocyanin. Highly mucoid strains may lack motility. A "grapelike" odor as aminoacetophenone is characteristic. *P. pseudoalcaligenes*: See Biocode 601 *P. mendocina* (macroscopic): Strains produce a brown–yellow, intracellular, carotenoid pigment. Colonies are flat, smooth, and butyrous.
703	**Genus:** *Alcaligenes*-like **Probable species:** CDC Group Vd-2 **Distinguishing characteristics of** *Alcaligenes*: See Biocode 401
705[a]	**Genus:** *Pseudomonas* **Probable species:** *P. aeruginosa* **Distinguishing characteristics of** *Pseudomonas* **spp.:** See Biocode 322 *P. aeruginosa*: See Biocode 701
710	**Genus:** *Pseudomonas* and *Pseudomonas*-like **Probable species:** CDC Group Ve-2 **Improbable species:** *P. cepacia*[a] **Tests required for speciation:** Observe fluorescent pigment CDC Group Ve-2: fluorescent pigment, − *P. cepacia*: fluorescent pigment, + **Distinguishing characteristics of** *Pseudomonas* **spp.:** See Biocode 322 CDC Group Ve-2: See Biocode 700 *P. cepacia* (macroscopic): Most strains produce a fluorescent, diffusible pigment which is violet under ultraviolet light. Nonfluorescent, water-soluble pigments, which may be sulfur-yellow, brown, red, or purple in color, are frequently produced. A "sweet" odor is produced by some strains. Note: Fresh isolates should be subcultured frequently, as they may become nonviable in 2–3 days on the surface of an agar medium.
711	**Genus:** *Pseudomonas* **Probable species:** *P. cepacia, P. aeruginosa, P. mendocina, P. pickettii, P. stutzeri* **Improbable species:** *P. pseudoalcaligenes*[b] **Tests required for speciation:** Run confirmation tests. Growth in 6.5% NaCl, pyocyanin production

Species	Confirmation code	6.5% NaCl	Pyocyanin
P. cepacia	711-33	−	−
P. stutzeri	711-31, 711-33	+	−
P. mendocina	711-11	+	−
P. pickettii	711-11	−	−
P. aeruginosa	711-01, 711-03, 711-21, 711-23	−	+
P. pseudoalcaligenes	711-01	−	−

(*continued*)

APPENDIX A *(Continued)*

Biocode	Characteristics
711 *(cont.)*	**Distinguishing characteristics of** *Pseudomonas* **spp.:** See Biocode 322

711 *(cont.)* **Distinguishing characteristics of** *Pseudomonas* **spp.:** See Biocode 322
  P. cepacia: See Biocode 710
  P. stutzeri (macroscopic): Colony morphology ranges from dry, adherent, and wrinkled to smooth and mucoid. Most stains produce a buff-colored colony, although some strains are nonpigmented.
  P. mendocina: See Biocode 701
  P. pickettii (macroscopic): Colonies are nonpigmented and no soluble pigments are produced. Convex colonies with an entire edge are typical. Acid production from carbohydrate utilization is slow.
  P. aeruginosa: See Biocode 701
  P. pseudoalcaligenes: See Biocode 601

712[a]
713 **Genus:** *Pseudomonas*
  Probable species: *P. cepacia*
  Distinguishing characteristics of *Pseudomonas* **spp.:** See Biocode 322
    P. cepacia: See Biocode 710

715[a] **Genus:** *Pseudomonas*
  Probable species: *P. aeruginosa, P. putrefaciens*
  Distinguishing characteristics of *Pseudomonas* **spp.:** See Biocode 322
    P. aeruginosa: See Biocode 701
    P. putrefaciens: See Biocode 515

717[a] **Genus:** *Pseudomonas*
  Probable species: *P. putrefaciens*
  Distinguishing characteristics of *Pseudomonas* **spp.:** See Biocode 322
    P. putrefaciens: See Biocode 515

720[a] **Genus:** *Pseudomonas*-like
  Probable species: CDC Group Ve-2
  Distinguishing characteristics of *Pseudomonas* **spp.:** See Biocode 322
    CDC Group Ve-2: See Biocode 700

722 **Genus:** *Pseudomonas* and *Pseudomonas*-like
  Probable species: CDC Ve-1
  Improbable species: *P. paucimobilis*[a]
  Tests required for speciation: Run confirmation tests
    Confirmation codes
      CDC Ve-1: 722-33
      P. paucimobilis: 722-31
  Distinguishing characteristics of *Pseudomonas* **spp.:** See Biocode 322
    CDC Ve-1: See Biocode 700
    P. paucimobilis: See Biocode 322

723 **Genus:** *Agrobacterium, Pseudomonas*
  Probable species: *A. radiobacter, P. paucimobilis*
  Tests required for differentiation of genera: Run confirmation tests

APPENDIX A (*Continued*)

Biocode	Characteristics
723 (*cont.*)	*Confirmation codes* *A. radiobacter*: 723-33 *P. paucimobilis*: 723-31 **Distinguishing characteristics of** *Agrobacterium* **spp.:** *Microscopic*: Medium-sized (0.8 × 1.5–3 μm), straight rods which are peritrichously flagellated. *Macroscopic*: Colonies are nonpigmented, with growth on carbohydrate-containing media usually accompanied by copious amounts of extracellular polysaccharide material. **Distinguishing characteristics of** *Pseudomonas* **spp.:** See Biocode 322 *P. paucimobilis*: See Biocode 322
730[a]	**Genus:** *Pseudomonas* and *Pseudomonas*-like **Probable species:** *P. cepacia*, CDC Group Ve-2 **Tests required for speciation:** Observe fluorescent pigment *P. cepacia*: fluorescent pigment, + CDC Group Ve-2: fluorescent pigment, − **Distinguishing characteristics of** *Pseudomonas* **spp.:** See Biocode 322 *P. cepacia*: See Biocode 710 CDC Group Ve-2: See Biocode 700
731	**Genus:** *Pseudomonas* **Probable species:** *P. cepacia* **Distinguishing characteristics of** *Pseudomonas* **spp.:** See Biocode 322 *P. cepacia*: See Biocode 710
732	**Genus:** *Pseudomonas* **Probable species:** *P. paucimobilis* **Distinguishing characteristics of** *Pseudomonas* **spp.:** See Biocode 322 *P. paucimobilis*: See Biocode 322
733	**Genus:** *Pseudomonas* **Probable species:** *P. cepacia*, *P. paucimobilis* **Tests required for speciation:** Run confirmation tests *Confirmation codes* *P. cepacia*: 733-33 *P. paucimobilis*: 733-31 **Distinguishing characteristics of** *Pseudomonas* **spp.:** See Biocode 322 *P. cepacia*: See Biocode 710 *P. paucimobilis*: See Biocode 322
741	**Genus:** *Pseudomonas* **Probable species:** *P. fluorescens*, *P. aeruginosa* **Improbable species:** *P. pseudomallei*[a] **Tests required for speciation:** Observe the presence of pyocyanin, fluorescent pigment *P. fluorescens*: pyocyanin, −; fluorescent pigment, +

(*continued*)

Biocode	Characteristics

741 *(cont.)*   *P. aeruginosa*: pyocyanin, +; fluorescent pigment, +
P. pseudomallei: pyocyanin, −; fluorescent pigment, −
Distinguishing characteristics of *Pseudomonas* spp.: See Biocode 322
P. fluorescens (microscopic): Straight rod with a polar tuft of flagella.   (Macroscopic): Cultures produce a diffusible, fluorescent pigment. Some biotypes produce soluble and insoluble, phenazine pigments: green (*P. chlororaphis*), orange (*P. chlororaphis*, *P. aureofaciens*), and blue (*P. lemonnieri*). Some strains produce extracellular polysaccharide material on carbohydrate-containing media.
P. aeruginosa: See Biocode 701
P. pseudomallei (macroscopic): Colonies vary from mucoid and smooth to rough and wrinkled. Colonies are pigmented, ranging from bright orange to buff colored. A pungent and putrefactive-type odor accompanies growth. Identification should be confirmed serologically (not common in the Western Hemisphere).

743[a]   **Genus:** *Pseudomonas*
Probable species: *P. pseudomallei*
Distinguishing characteristics of *Pseudomonas* spp.: See Biocode 322
P. pseudomallei: See Biocode 741

745[a]   **Genus:** *Pseudomonas*
Probable species: *P. aeruginosa*
Distinguishing characteristics of *Pseudomonas* spp.: See Biocode 322
P. aeruginosa: See Biocode 701

750   **Genus:** *Pseudomonas*
Probable species: *P. cepacia*
Distinguishing characteristics of *Pseudomonas* spp.: See Biocode 322
P. cepacia: See Biocode 710

751   **Genus:** *Pseudomonas*
Probable species: *P. aeruginosa*, *P. cepacia*, *P. fluorescens*, *P. mendocina*, *P. pickettii*, *P. pseudomallei*
Tests required for speciation: Run confirmation tests; observe pyocyanin and fluorescent pigment production; growth in 6.5% NaCl broth

Species	Confirmation code	Pyocyanin	Fluorescent pigment	6.5% NaCl broth
P. aeruginosa	751-01, 751-03, 751-21, 751-23	+	Usually +	−
P. cepacia	751-33	−	+ (Violet)	−
P. fluorescens	751-13, 751-33	−	+ (Blue–green)	−
P. mendocina	751-11	−	−	+
P. pickettii	751-11	−	−	−
P. pseudomallei	751-33	−	−	−

Distinguishing characteristics of *Pseudomonas* spp.: See Biocode 322
P. aeruginosa: See Biocode 701

Biocode	Characteristics
751 (*cont.*)	*P. cepacia*: See Biocode 710 *P. fluorescens*: See Biocode 741 *P. mendocina*: See Biocode 701 *P. pickettii*: See Biocode 711 *P. pseudomallei*: See Biocode 741
752[a]	**Genus:** *Pseudomonas* **Probable species:** *P. cepacia* **Distinguishing characteristics of** *Pseudomonas* **spp.:** See Biocode 322 *P. cepacia*: See Biocode 710
753	**Genus:** *Pseudomonas* **Probable species:** *P. cepacia, P. pseudomallei* **Tests required for speciation:** Observe fluorescent pigment production *P. cepacia*: fluorescent pigment, + (violet) *P. pseudomallei*: fluorescent pigment, − **Distinguishing characteristics of** *Pseudomonas* **spp.:** See Biocode 322 *P. cepacia*: See Biocode 710 *P. pseudomallei*: See Biocode 741
755	**Genus:** *Pseudomonas* **Probable species:** *P. putrefaciens* **Improbable species:** *P. aeruginosa*[a] **Tests required for speciation:** Observe pyocyanin production; H_2S production (glucose tube) *P. putrefaciens*: pyocyanin, −; H_2S, + *P. aeruginosa*: pyocyanin, +; H_2S, − **Distinguishing characteristics of** *Pseudomonas* **spp.:** See Biocode 322 *P. putrefaciens*: See Biocode 515 *P. aeruginosa*: See Biocode 701
757	**Genus:** *Pseudomonas* **Probable species:** *P. putrefaciens* **Distinguishing characteristics of** *Pseudomonas* **spp.:** See Biocode 322 *P. putrefaciens*: See Biocode 515
762	**Genus:** *Pseudomonas*-like **Probable species:** CDC Group Ve-1 **Distinguishing characteristics of** *Pseudomonas* **spp.:** See Biocode 322 CDC Group Ve-1: See Biocode 700
770[a] 771 772[a] 773	**Genus:** *Pseudomonas* **Probable species:** *P. cepacia* **Distinguishing characteristics of** *Pseudomonas* **spp.:** See Biocode 322 *P. cepacia*: See Biocode 710
776[a] 777[a]	**Genus:** *Pseudomonas* **Probable species:** *P. maltophilia* **Distinguishing characteristics of** *Pseudomonas* **spp.:** See Biocode 322 *P. maltophilia*: See Biocode 576

[a] Unusual biogrouping number for species.
[b] Unlikely biogrouping number for species.

CODING LIST FOR UNRECOGNIZABLE STRAINS[a]

Code and pigmentation	Oxidase positive	Oxidase negative
000–077		
Nonpigmented	Recheck motility microscopically using a hanging drop or wet mount procedure with a fresh broth culture.	Recheck motility microscopically using a hanging drop or wet mount procedure with a fresh broth culture.
	If motile, determine new biogrouping number and check coding list.	If motile and polarly flagellated, determine new biogrouping number and check coding list. If new biogrouping number is not located, record as *Pseudomonas*-like (possibly *P. malto-philia*). Inoculate confirmation tests. Look for the rapid acidification of maltose.
	If peritrichously flagellated, record it as *Alcaligenes*-like. See Biocode 401 for a description of the genus.	
	If polarly flagellated, record it as *Pseudomonas*-like. See Biocode 322 for a description of the genus.	If positive for maltose, fructose, DNase, Tween 80, and esculin, record it as *P. maltophilia*. See Biocode 322 for a description of the genus and Biocode 576 for a description of the species.
	If nonmotile, record it as *Moraxella*-like. See Biocode 001 for a description of the genus. These bacteria are generally ONPG, DNase, and esculin negative. Hydrolysis of Tween 80 and gelatin is variable. Catalase is usually positive.	If nonmotile, record as *Acinetobacter*-like. See Biocode 210 for a description of the genus. Note: *Acinetobacter* spp. generally utilize acetate as a sole carbon source.
Pigmented	Recheck motility microscopically using a hanging drop or wet mount procedure with a fresh broth culture.	Recheck motility microscopically using a hanging drop or wet mount procedure with a fresh broth culture.
	If motile and polarly flagellated, determine new biogrouping number and check coding list. If new biogrouping number is not located, record it as *Pseudomonas* (*Xanthomonas*)-like. See Biocode 322 for a description of the genus *Pseudomonas*. *Xanthomonas* spp. are very weak acid producers from carbohydrate oxidation. The oxidase reaction is weak or negative. These bacteria are polarly flagellated, catalase positive, and usually hydrolyze Tween 80 and esculin. All species recognized in this genus are plant pathogens. As is known so far, these bacteria are only in association with plants and plant material.	If motile and polarly flagellated, determine new biogrouping number and check coding list. If new biogrouping number is not located, record as *Pseudomonas* (*Xanthomonas*)-like. See Oxidase-positive entry.
		If motile and peritrichously flagellated, or nonmotile, record it as *Empedobacter*-like. This is a new taxonomic concept. See Section IV, on pigmented bacteria.

X. campestris can be differentiated from P. paucimobilis by determining the absorbance spectrum of the yellow carotenoid pigment solubilized in petroleum ether. X. campestris has peak maxima at 418, 437, and 463 nm.

If gliding movement is seen microscopically or swarming growth is observed on the surfaces of the agar media, assignment to the family Cytophagaceae is warranted. This family includes the genera Cytophaga and Flexibacter.

If nonmotile, record as Flavobacterium (Cytophaga)-like. See Biocode 041 for a description of the genus Flavobacterium. These bacteria are variable concerning the hydrolysis of esculin, Tween 80, and gelatin. ONPG reaction may be positive.

100–177
Nonpigmented

Recheck motility microscopically using a hanging drop or wet mount procedure with a fresh broth culture.

If motile, determine new biogrouping number and check coding list.

If peritrichously flagellated, record it as Alcaligenes-like. See Biocode 501 for a description of the genus.

If polarly flagellated, record it as Pseudomonas-like. See Biocode 322 for a description of the genus.

If nonmotile, it is a possible member of CDC Group II,K. CDC Group II,K bacteria are ONPG and esculin positive, gelatin and DNase negative, and variable for Tween 80 hydrolysis. Most strains of this group are motile, with a yellow pigment. CDC Group II,K, Biotype I is currently designated P. paucimobilis. CDC Group II,K, Biotype 2 is currently designated F. multivorum.

P. mallei is the only nonmotile species of Pseudomonas and is primarily a disease of horses, reported to be eradicated from the United States and Canada.

Recheck motility microscopically using a hanging drop or wet mount procedure with a fresh broth culture.

If motile and polarly flagellated, determine new biogrouping number and check coding list. If new biogrouping code number is not located, record it as Pseudomonas-like (possibly P. maltophilia). Inoculate a confirmation plate. Look for rapid acidification of maltose.

If positive for maltose, fructose, DNase, Tween 80, and esculin, record it as P. maltophilia. See Biocode 322 for a description of the genus and Biocode 575 for a description of the species.

If nonmotile, record as Acinetobacter-like. See Biocode 210 for a description of the genus. Note: Acinetobacter spp. generally utilize acetate as a sole carbon source.

(continued)

Code and pigmentation	Oxidase positive	Oxidase negative
Pigmented[b]	Recheck motility microscopically using a hanging drop or wet mount procedure with a fresh broth culture. If motile, determine new biogrouping number and check coding list. If new biogrouping number is not located, record it as *Pseudomonas* (*Xanthomonas*)-like. See Biocode 322 for a description of the genus *Pseudomonas*. *Xanthomonas* spp. are very weak acid producers from carbohydrate oxidation. The oxidase reaction is weak or negative. These bacteria are poorly flagellated, catalase positive, and usually hydrolyze Tween 80 and esculin. All species recognized in this genus are plant pathogens. As is known so far, these bacteria are only in association with plants and plant material. *X. campestris* can be differentiated from *P. paucimobilis* by determining the absorbance spectrum of the yellow carotenoid pigment solubilized in petroleum ether. *X. campestris* has peak maxima at 418, 437, and 463 nm. If gliding movement is seen microscopically or swarming growth is observed on the surfaces of the agar media, assignment to the family *Cytophagaceae* is warranted. This family includes the genera *Cytophaga* and *Flexibacter*. Note: Most members of this group will not grow on the motility medium. If nonmotile, record it as *Flavobacterium* (*Cytophaga*)-like. See Biocode 041 for a description of the genus *Flavobacterium*. These bacteria are variable concerning the hydrolysis of esculin, Tween 80, and gelatin. ONPG reaction is variable.	Check motility microscopically using a hanging drop or wet mount procedure with a fresh broth culture. If motile and polarly flagellated, determine new biogrouping number and check coding list. If new biogrouping number is not located, record it as *Pseudomonas* (*Xanthomonas*)-like. See Oxidase-positive entry. If motile and peritrichously flagellated, or nonmotile, record it as *Empedobacter*-like. This is a new taxonomic concept. See Section IV, on pigmented bacteria.
200–277 Nonpigmented	Check motility microscopically using a hanging drop or wet mount procedure with a fresh broth culture.	Check motility microscopically using a hanging drop or wet mount procedure with a fresh broth culture.

If motile, determine new biogrouping number and check coding list. If new biogrouping number is not located, perform a flagellar stain.

If peritrichously flagellated, record as *Alcaligenes*-like. See Biocode 401 for a description of the genus.

If polarly flagellated, record it as *Pseudomonas*-like. See Biocode 322 for a description of the genus.

If nonmotile, record it as *Moraxella*-like. See Biocode 001 for a description of the genus. These bacteria are ONPG, DNase, and esculin negative. Hydrolysis of Tween 80 and gelatin is variable. Catalase is usually positive.

Pigmented[b]

Check motility microscopically using a hanging drop or wet mount procedure with a fresh broth culture.

If motile and polarly flagellated, determine new biogrouping number and check coding list. If new biogrouping number is not located, record it as *Pseudomonas*-like. See Biocode 322 for a description of the genus.

If nonmotile, record it as "identity unknown."

300–377 Nonpigmented

Check motility microscopically using a hanging drop or wet mount procedure with a fresh broth culture.

If motile, determine new biogroup in number and check coding list.

If peritrichously flagellated, record it as *Alcaligenes*-like. See Biocode 501 for a description of the genus.

If polarly flagellated, record as *Pseudomonas*-like. See Biocode 322 for a description of the genus.

If nonmotile, record it as "identity unknown." This is a possible, immotile variant of the genus *Pseudomonas*. *P. mallei* is the only nonmotile species of this genus. This species, though, is generally limited to Asia, the Middle East, and Africa.

If motile, determine new biogrouping number and check coding list. If new biogrouping number is not located, record it as *Pseudomonas*-like. See Biocode 322 for a description of the genus.

If nonmotile, record it as *Acinetobacter*-like. See Biocode 210 for a description of the genus.

Check motility microscopically using a hanging drop or wet mount procedure with a fresh broth culture.

If motile and polarly flagellated, determine new biogrouping code number and check coding list. If new biogrouping number is not located, record it as *Pseudomonas*-like. See Biocode 322 for a description of the genus.

If nonmotile, record it as "identity unknown."

Check motility microscopically using a hanging drop or wet mount procedure with a fresh broth culture.

If motile, determine new biogroup in number and check coding list. If new biogrouping number is not located, record it as *Pseudomonas*-like. See Biocode 322 for a description of the genus.

If nonmotile, record it as *Acinetobacter*-like. See Biocode 210 for a description of the genus.

(continued)

APPENDIX B (Continued)

Code and pigmentation	Oxidase positive	Oxidase negative
Pigmented[b]	Check motility microscopically using a hanging drop or wet mount procedure with a fresh broth culture. If motile and polarly flagellated, determine new biogrouping number and check coding list. If new biogrouping number is not located, record it as *Pseudomonas*-like. See Biocode 322 for a description of the genus. If nonmotile, possibly *P. paucimobilis*. If positive for ONPG and esculin and negative for gelatin and DNase, record it as *P. paucimobilis*. See Biocode 322 for a description of the genus and species.	Check motility microscopically using a hanging drop or wet mount procedure with a fresh broth culture. If motile and polarly flagellated, determine new biogrouping number and check coding list. If new biogrouping number is not located, record it as *Pseudomonas*-like (possibly *P. pauci-mobilis*). If positive for ONPG and esculin and negative for gelatin and DNase, record it as *P. paucimobilis*. See Biocode 322 for a description of the genus and species.
400–477 Nonpigmented	If peritrichously flagellated, record it as *Alcaligenes*-like. See Biocode 401 for a description of the genus. If polarly flagellated, record as *Pseudomonas*-like. See Biocode 322 for a description of the genus.	If polarly flagellated, record it as *Pseudomonas*-like. See Biocode 322 for a description of the genus.
Pigmented[b]	If polarly flagellated, record it as *Pseudomonas (Xanthomonas)*-like. See Biocode 322 for a description of the genus. *Xantho-monas* spp. are very weak acid producers from carbohydrate oxidation. The oxidase reaction is weak or negative. These bacteria are polarly flagellated, catalase positive, and usually hydrolyze Tween 80 and esculin. All species recognized in this genus are plant pathogens. As is known so far, these bacteria are only in association with plants and plant material. *X. cam-pestris* can be differentiated from *P. paucimobilis* by determining the absorbance spectrum of the yellow carotenoid pigment solubilized in petroleum ether. *X. campestris* has peak maxima at 428, 437, and 463 nm.	If polarly flagellated, record it as *Pseudomonas (Xanthomonas)*-like. See Oxidase-positive entry.

500–577 Nonpigmented	If peritrichously flagellated, record it as *Alcaligenes*-like. See Biocode 501 for a description of the genus. If polarly flagellated, record as *Pseudomonas*-like. See Biocode 322 for a description of the genus.	If polarly flagellated, record it as *Pseudomonas*-like (possibly *P. maltophilia*). Inoculate a confirmation plate. Look for rapid acidification of maltose. If positive for maltose, fructose, DNase, Tween 80, and esculin, record it as *P. maltophilia*. See Biocode 322 for a description of the genus and 575 for a description of the species.
Pigmented[b]	If polarly flagellated, record it as *Pseudomonas* (*Xanthomonas*)-like. See Biocode 322 for a description of the genus *Pseudomonas*. *Xanthomonas* spp. are very weak acid producers from carbohydrate oxidation. The oxidase reaction is weak or negative. These bacteria are polarly flagellated, catalase positive, and usually hydrolyze Tween 80 and esculin. All species recognized in this genus are plant pathogens. As is known so far, these bacteria are only in association with plants and plant material. *X. campestris* can be differentiated from *P. pauci-mobilis* by determining the absorbance spectrum of the yellow carotenoid pigment solubilized in petroleum ether. *X. campestris* has peak maxima at 418, 437, and 463 nm.	Record it as *Pseudomonas* (*Xanthomonas*)-like. See Oxidase-positive entry.
600–677 Nonpigmented	If peritrichously flagellated, record it as *Alcaligenes*-like. See Biocode 401 for a description of the genus. If polarly flagellated, record it as *Pseudomonas*-like. See Biocode 322 for a description of the genus.	If polarly flagellated, record it as *Pseudomonas*-like. See Biocode 322 for a description of the genus.
Pigmented[b]	If polarly flagellated, record it as *Pseudomonas*-like. See Biocode 322 for a description of the genus.	If polarly flagellated, record it as *Pseudomonas*-like. See Biocode 322 for a description of the genus.

(continued)

357

Code and pigmentation	Oxidase positive	Oxidase negative
700–777		
Nonpigmented	Perform a flagellar stain. If peritrichously flagellated and ONPG positive, record it as *Alcaligenes*-like. See Biocode 723 for a description of the genus. If peritrichously flagellated and ONPG negative, record it as *Achromobacter*-like. See Biocode 501 for a description of the genus. If polarly flagellated, record it as *Pseudomonas*-like. See Biocode 322 for a description of the genus.	Record it as *Pseudomonas*-like. See Biocode 322 for a description of the genus.
Pigmented[b]	Record it as *Pseudomonas*-like. See Biocode 322 for a description of the genus.	Record it as *Pseudomonas*-like. See Biocode 322 for a description of the genus.

[a] These biocodes may be obtained because (1) the genus or species is not included in the identification scheme; (2) aberrant reactions are obtained; or (3) tests were not performed correctly or interpretation is faulty.

[b] The taxonomy of peritrichously flagellated, yellow-pigmented bacteria is uncertain. These bacteria would have previously been considered candidates for Section 2, *Flavobacterium* (with high mol% G + C) (Weeks, 1974). See Section III, on the taxonomic uncertainty of pigmented bacteria.

REFERENCES

Adair, R. W., S. G. Geftic, and J. Gelzer (1969). Resistance of *Pseudomonas* to quarternary ammonium compounds. I. Growth in benzalkonium chloride solution. *Appl. Microbiol.* **18**, 299–302.

Adler, J. (1966). Chemotaxis in bacteria. *Science* **153**, 708–715.

Airoldi, T., and W. Litsky (1972). Factors contributing to the microbial contamination of coldwater humidifiers. *Am. J. Med. Technol.* **38**, 491–495.

Allen, D. A., B. Austin, and R. R. Colwell (1983). Numerical taxonomy of bacterial isolates associated with freshwater fishery. *J. Gen. Microbiol.* **129**, 2043–2062.

Anacker, R. L., and E. J. Ordal (1959). Studies on the myxobacterium *Chondrococcus columnaris* I. Serological typing. *J. Bacteriol.* **78**, 25–32.

Anonymous (1983). Microbiology of cosmetics. *In* "The Microbiological Update" (M. S. Cooper, ed.), Vol. 1, #3. Microbiological Applications, Dumont, New Jersey.

APHA, AWWA, and WPCF (1980). "Standard Methods for the Examination of Water and Wastewater," 15th Ed. Washington, D.C.

Armstrong, J. L., D. S. Shigeno, J. J. Calomiris, and R. J. Seidler (1981). Antibiotic-resistant bacteria in drinking water. *Appl. Environ. Microbiol.* **42**, 277–283.

Barnishan, J., and L. W. Ayers (1979). Rapid identification of nonfermentative, Gram negative rods by the Corning N/F system. *J. Clin. Microbiol.* **9**, 239–243.

Bassett, D. C. J., K. J. Stokes, and W. R. G. Thomas (1970). Wound infection with *Pseudomonas multivorans:* A water-borne contaminant of disinfectant solutions. *Lancet* **1**, 1188–1191.

Bennett, J. V. (1974). Nosocomial infections due to *Pseudomonas. J. Infect. Dis.* **130**, 54–57.

Bergan, T. (1981). Human- and animal-pathogenic members of the genus *Pseudomonas. In* "The Procaryotes: A Handbook on Habitats, Isolation, and Identification of Bacteria" (M. P. Starr, H. Stolp, H. G. Trüper, A. Balows, and H. G. Schlegel, eds.), pp. 666–700. Springer-Verlag, New York.

Bergey, D. H., F. C. Harrison, R. S. Breed, B. W. Hammer, and F. M. Huntoon (1923). "Bergey's Manual of Determinative Bacteriology," 1st Ed., pp. 1–442. Williams & Wilkins, Baltimore.

Berkelman, R. L., S. Lewin, J. R. Allen, R. L. Anderson, L. D. Budnick, S. Shapiro, S. M. Friedman, P. 'Nicholas, R. S. Holzman, and R. W. Haley (1981). Pseudobacteremia attributed to contamination of providone-iodine with *Pseudomonas. Ann. Intern. Med.* **95**, 32–36.

Berkelman, R. L., J. Godley, J. A. Weber, R. L. Anderson, A. M. Lerner, N. J. Petersen, and J. R. Allen (1982). *Pseudomona cepacia* peritonitis associated with contamination of automatic peritoneal dialysis machines. *Ann. Intern. Med.* **96**, 456–458.

Black W. A., R. Hodgson, and A. McKechie (1971). Evaluation of three methods using deoxyribonuclease production as a screening test for *Serratia marcesans. J. Clin. Pathol.* **24**, 313.

Blazevic, D. J. (1976). Current taxonomy and identification of nonfermentative Gram negative bacilli. *Hum. Pathol.* **7**, 265–275.

Bobo, R. A., E. J. Newton, L. F. Jones, L. H. Farmer, and J. J. Farmer III (1973). Nursery outbreak of *Pseudomonas aeriginosa:* Epidemiological conclusions from five different typing methods. *Appl. Microbiol.* **25**, 414–420.

Bøvre, K., and N. Hagen (1981). The family Neisseriaceae: Rod-shaped species of the genera *Moraxella, Acinetobacter, Kingella,* and *Neisseria,* and the *Branhanella* group of cocci. *In* "The Procaryotes: A Handbook on Habitats, Isolation, and Identification of Bacteria" (M.

P. Starr, H. Stolp, H. G. Trüper, A. Balows, and H. G. Schlegel, eds.), pp. 1056–1529. Springer-Verlag, New York.

Breed, R. S., E. G. D. Murray, and N. R. Smith (1957). "Bergey's Manual of Determinative Bacteriology," 7th Ed., pp. 1–498. Williams & Wilkins, Baltimore.

Brodsky, M. H., and B. W. Ciebin (1978). Improved medium for recovery and enumeration of *Pseudomonas aeruginosa* from water using membrane filters. *Appl. Environ. Microbiol.* 36, 36–42.

Cabrera, H. A., and G. H. Davis (1961). Epidemic meningitis of the newborn caused by flavobacteria. *Am. J. Dis. Child.* 101, 289–295.

Callies, E., and W. Mannheim (1978). Classification of the *Flavobacterium–Cytophaga* complex on the basis of respiratory quinones and fumarate respiration. *Int. J. Syst. Bacteriol.* 28, 14–19.

Christensen, P. J. (1977). The history, biology, and taxonomy of the *Cytophaga* group. *Can. J. Microbiol.* 23, 1601–1653.

Christensen, P. J., and F. D. Cook (1978). *Lysobacter*, a new genus of non-fruiting, gliding bacteria with a high base ratio. *Int. J. Syst. Bacteriol.* 28, 367–373.

Cohen, G. N., and J. Monod (1957). Bacterial permeases. *Bacteriol. Rev.* 21, 169–194.

Colwell, R. R., M. L. Moffett, and M. D. Sutton (1968). Computer analysis of relationships among phytopathogenic bacteria. *Phytopathology* 58, 1207–1217.

Colwell, R. R., B. Austin, and L. Wan (1978). Public health considerations of the microbiology of "potable" water. *In* "Proceedings of the Conference on the Evaluation of the Microbiology Standards for Drinking Water." EPA-570/9-78-OOC. USEPA, Washington, D.C.

Cross, D. F., A. Benchimol, and E. G. Dimond (1966). The faucet aerator—a source of *Pseudomonas* infection. N. Engl. J. Med. 274, 1430–1431.

DeLey, J. (1968). DNA base composition and hybridization in the taxonomy of phytopathogenic bacteria. *Annu. Rev. Phytopathol.* 6, 63–90.

Doudoroff, M., and N. J. Palleroni (1974). *Pseudomonas* Migula 1894. *In* Bergey's Manual of Determinative Bacteriology" (R. E. Buchanan and N. E. Gibbons, eds.), 8th Ed., pp. 217–243. Williams & Wilkins, Baltimore.

Drake, C. H. (1966). Evaluation of culture media for the isolation and enumeration of *Pseudomonas aeruginosa. Health Lab. Sci.* 3, 10–19.

du Moulin, G. C. (1979). Airway colonization by *Flavobacterium* in an intensive care unit. *J. Clin. Microbiol.* 10, 155–160.

Dutka, B. J., and K. K. Kwan (1978). Confirmation of the single-step membrane filtration procedure for estimating *Pseudomonas aeruginosa* densities in water. *Appl. Environ. Microbiol.* 33, 240–245.

Edberg, S. C., K. Gam, C. J. Bottenbley, and J. M. Singer (1976). Rapid spot test for the determination of esculin hydrolysis. *J. Clin. Microbiol.* 4, 180–184.

Eriksen, K. R. (1961). Nosocomial *Pseudomonas* infections: Water as source of infection. *Nord. Med.* 66, 1386–1388.

Farmer, J. J., III, R. A. Weinstein, C. H. Zierdt, and C. D. Brokopp (1982). Hospital outbreaks caused by *Pseudomonas aeruginosa*: Importance of Serogroup 011. *J. Clin. Microbiol.* 16, 266–270.

Favero, M. S., and C. H., Drake (1964). Comparative study of microbial flora of iodinated and chlorinated pools. *Publ. Health Rep.* 79, 251–257.

Favero, M. S., L. A. Carson, W. W. Bond, and N. J. Petersen (1971). *Pseudomonas aeruginosa*: Growth in distilled water from hospitals. *Science* 173, 836–838.

Favero, M. S., N. J. Petersen, L. A. Carson, W. W. Bond, and S. H. Hindman (1975). Gram-negative water bacteria in hemodialysis systems. *Health Lab. Sci.* 12, 321–334.

Foley, J. R., C. R. Gravelle, W. E. Englehard, and T. D. Y. Chin (1961). Achromobacter septicemia—fatalities in prematures. *Am. J. Dis. Child.* 101, 279–288.

Gilardi, G. L. (1971). Characterization of *Pseudomonas* species isolated from clinical specimens. *Appl. Microbiol.* **21**, 414–419.

Gilardi, G. L. (1972). Infrequently encountered *Pseudomonas* species causing infection in humans. *Ann. Intern. Med.* **77**, 211–215.

Gilardi, G. L. (1978). Identification of miscellaneous glucose nonfermenting gram-negative bacteria. *In* "Glucose Nonfermenting Gram-Negative Bacteria in Clinical Microbiology" (G. L. Gilardi, ed.), pp. 45–65. CRC Press, Cleveland, Ohio.

Glew, R. H., R. C. Moellering, Jr., and L. J. Kunz (1977). Infections with *Acinetobacter calcoaceticus* (*Herellea vaginicola*): Clinical and laboratory studies. *Medicine* **56**, 79–97.

Hayes, P. R., and A. P. D. Wilcock (1977). Deoxyribonucleic acid base composition of flavobacteria and related Gram negative yellow pigmented rods. *J. Appl. Bacteriol.* **43**, 111–115.

Haynes, W. C., and W. H. Burkholder (1957). *Pseudomonas* Migula 1894. *In* "Bergey's Manual of Determinative Bacteriology" (R. S. Breed, E. G. D. Murray, and N. R. Smith, eds.), 7th Ed., pp. 89–152. Williams & Wilkins, Baltimore.

Hendrie, M. S., A. J. Holding, and J. M. Shewan (1974). Amended descriptions of the genus *Alcaligenes* and of *Alcaligenes faecalis* and proposal that the generic name *Achromobacter* be rejected: Status of the named species of *Alcaligenes* and *Achromobacter:* Request for an opinion. *Int. J. Syst. Bacteriol.* **24**, 534–550.

Henrichsen, J. (1972). Bacterial surface translocation: A study and a classification. *Bacteriol. Rev.* **36**, 478–503.

Henriksen, S. D. (1973). *Moraxella, Acinetobacter* and the *Mimeae. Bacteriol. Rev.* **37**, 522–561.

Herman, L. G. (1976). Sources of the slow-growing pigmented water bacteria. *Health Lab. Sci.* **13**, 5–10.

Herman, L. G. (1981). The slow growing gram negative pigmented water bacteria. *In* "The *Flavobacterium-Cytophaga* Group" (H. Reichenbach and O. B. Weeks, eds.), pp. 169–178. Gesellschaft für Biotechnologische Forschung, Braunschweig.

Herman, L. G. and C. K. Himmelsbach (1965). Detection and control of hospital sources of flavobacteria. *Hospitals* **39**, 72–76.

Hoadley, A. W., and D. E. Knight (1975). External otitis among swimmers and nonswimmers. *Arch. Environ. Health* **30**, 445–448.

Hofherr, L., H. Votava, and D. J. Blazevic (1978). Comparison of three methods for identifying nonfermenting gram-negative rods. *Can. J. Microbiol.* **24**, 1140–1144.

Holding, A. J., and J. M. Shewan (1974). Genus *Alcaligenes* Castellani and Chalmers 1919. *In* "Bergey's Manual of Determinative Bacteriology" (R. E. Buchanan and N. E. Gibbons, eds.), 8th Ed., pp. 273–275. Williams & Wilkins, Baltimore.

Holmes, B., R. J. Owen, A. Evans, H. Malnick, and W. R. Wilcox (1977). *Pseudomonas paucimobilis*, a new species isolated from human clinical specimens, the hospital environment, and other sources. *Int. J. Syst. Bacteriol.* **27**, 133–146.

Holmes, B., R. J. Owen, and T. A. McMeekin (1984). Genus *Flavobacterium* Bergey et al., 1923. *In* "Bergey's Manual of Determinative Bacteriology" (N. E. Kreig and J. G. Holt (eds.), 9th Ed., pp. 353–360. Williams & Wilkins, Baltimore.

Hugh, R., and G. L. Gilardi (1980). *Pseudomonas. In* "Manual of Clinical Microbiology" (E. H. Lennette, A. Balows, W. J. Hansler, and J. P. Truant, eds.), 3rd Ed. ASM, Washington, D.C.

Hugh, R., and E. Leifson (1953). The taxonomic significance of fermentative versus oxidative metabolism of carbohydrates by various gram negative bacteria. *J. Bacteriol.* **66**, 24–26.

Jacobson, J. A., A. W. Hoadley, and J. J. Farmer, III (1976). *Pseudomonas aeruginosa* serogroup II and pool-associated skin rash. *Am. J. Publ. Health* **66**, 1092–1093.

Jeffries, C. D., D. F. Holtman, and D. G. Guse (1957). Rapid method for determining the activity of microorganisms on nucleic acids. *J. Bacteriol.* **73**, 590–591.

Johnson, J. L., R. S. Anderson, and E. J. Ordal (1970). Nucleic acid homologies among oxidase-negative *Moraxella* species. *J. Bacteriol.* **101**, 568–573.

Juni, E. (1972). Interspecies transformation of *Acinetobacter:* Genetic evidence for a ubiquitous genus. *J. Bacteriol.* **112**, 917–931.

Juni, E. (1978). Genetics and physiology of *Acinetobacter*. *Annu. Rev. Microbiol.* **32**, 349–371.

Juni, E. (1984). Genus III *Acinetobacter* Brison and Prevot 1954. *In* "Bergey's Manual of Systematic Bacteriology" (N. E. Kreig and J. G. Holt, eds.), 9th Ed., pp. 303–307. Williams & Wilkins, Baltimore.

Jurtshuk, P., T. J. Mueller, and W. C. Acord (1975). Bacterial terminal oxidases. *Crit. Rev. Microbiol.* **3**, 399–468.

Kerster, K., and J. DeLey (1984). Genus *Alcaligenes* Castellani and Chalmers 1919. *In* "Bergey's Manual of Systematic Bacteriology" (N. E. Kreig and J. G. Holt, eds.), 9th Ed., pp. 361–373. Williams & Wilkins, Baltimore.

LaCroix, S. J., and V. J. Cabelli (1982). Membrane filter method for enumeration of *Acinetobacter calcoaceticus* from environmental waters. *Appl. Environ. Microbiol.* **43**, 90–96.

Leadbetter, E. R. (1974). Genus I. *Cytophaga* Winogradsky 1929; Lewin, 1969. *In* "Bergey's Manual of Determination Bacteriology" (R. E. Buchanan and N. E. Gibbons, eds.), 8th Ed., pp. 101–105. Williams & Wilkins, Baltimore.

LeChevallier, M. W., R. J. Seidler, and T. M. Evans (1980). Enumeration and characterization of standard plate count bacteria in chlorinated and raw water supplies. *Appl. Environ. Microbiol.* **40**, 922–930.

Leifson, E. (1951). Staining, shape, and arrangement of bacterial flagella. *J. Bacteriol.* **62**, 377–389.

Levin, M. A., and V. J. Cabelli (1972). Membrane filter technique for enumeration of *Pseudomonas aeruginosa*. *Appl. Microbiol.* **24**, 864–870.

McCausland, R. S., and P. J. Cox (1975). *Pseudomonas* infection traced to motel whirlpool. *J. Environ. Health* **37**, 455–459.

MacFaddin, J. F. (1980). "Biochemical Tests for Identification of Medical Bacteria," 2nd Ed. Williams & Wilkins, Baltimore.

McMeekin, T. A. (1977). Ultraviolet light sensitivity as an aid for the identification of Gram-negative, yellow-pigmented rods. *J. Gen. Microbiol.* **103**, 149–151.

McMeekin, T. A., and J. M. Shewan (1978). A review: Taxonomic strategies for *Flavobacterium* and related genera. *J. Appl. Bacteriol.* **45**, 321–332.

McMeekin, T. A., J. T. Patterson, and J. G. Murray (1971). An initial approach to the taxonomy of some Gram-negative, yellow pigmented rods. *J. Appl. Bacteriol.* **34**, 699–716.

McMeekin, T. A., D. B. Stewart, and J. G. Murray (1972). The adansonian taxonomy and the deoxyribonucleic acid base composition of some Gram negative, yellow pigmented rods. *J. Appl. Bacteriol.* **35**, 129–137.

Mathewson, J. J., R. B. Simpson, and F. L. Brooks (1983). Evaluation of the Microscan Urinary Combo Panel and API 20E system for identification of glucose-nonfermenting gram-negative bacilli isolated from clinical veterinary materials. *J. Clin. Microbiol.* **17**, 129–142.

Mayfield, C. I., and W. E. Innis (1977). A rapid, simple method for staining bacterial flagella. *Can. J. Microbiol.* **23**, 1311–1313.

Means, E. G., L. Hanami, H. F., Ridgway, and B. H. Olson (1981). Evaluating mediums and plating techniques for enumerating bacteria in water distribution systems. *AWWA* **73**, 585–590.

Mitchell, T. G., M. S. Hendrie, and J. M. Shewan (1969). The taxonomy, differentiation and identification of *Cytophaga* species. *J. Appl. Bacteriol.* **32**, 40–50.

Moat, A. G. (1978). "Microbial Physiology," pp. 159–161. Wiley (Interscience), New York.

Moffet, H. L., and T. Williams (1967). Bacteria recovered from distilled water and inhalation therapy equipment. *Am. J. Dis. Child.* **114**, 7–12.

Murata, N., and M. P. Starr (1973). A concept of the genus *Xanthomonas* and its species in light of segmental homology of deoxyribonucleic acids. *Phytopathol. Z.* **77**, 285–323.

Norris, F. C., and J. J. R. Campbell (1949). The intermediate metabolism of *Pseudomonas aeruginosa* III. The application of paper chromotography to the identification of gluconic and 2-ketogluconic acids, intermediates in glucose oxidation. *Can. J. Res.* **27**, 253–264.

Olivieri, V. P., and M. C. Snead (1979). Plate count microorganisms from the water distribution system. *Proc. Water Quality Technol. Conf. Am. Waterworks Assoc.* **7**, 167–184.

Otto, L. A., and U. Blachman (1979). Nonfermentative bacilli: Evaluation of three systems for identification. *J. Clin. Microbiol.* **10**, 147–154.

Otto, L. A., and M. J. Pickett (1976). Rapid method for identification of Gram-negative, nonfermentative bacilli. *J. Clin. Microbiol.* **3**, 566–575.

Owen, R. J., and J. J. S. Snell (1976). Deoxyribonucleic acid reassociation in the classification of flavobacteria. *J. Gen. Microbiol.* **93**, 89–102.

Palleroni, N. J. (1981). Introduction to the family *Pseudomonadaceae*. In "The Procaryotes: A Handbook on Habitats, Isolation, and Identification of Bacteria" (M. P. Starr, H. Stolp, H. G. Trüper, A. Balows, and H. G. Schlegel, eds.), pp. 655–665. Springer-Verlag, New York.

Palleroni, N. J. (1984). Genus I. *Pseudomonas* Migula 1894. In "Bergey's Manual of Systematic Bacteriology" (N. E. Kreig and J. G. Holt, eds.), 9th Ed., pp. 141–199. Williams & Wilkins, Baltimore.

Palleroni, N. J., and M. Doudoroff (1972). Some properties and taxonomic subdivisions of the genus *Pseudomonas. Annu. Rev. Phytopathol.* **10**, 73–100.

Palleroni, N., J., M. Doudoroff, R. V. Stainer, R. E. Solanes, and M. Mandel (1970). Taxonomy of the aerobic pseudomonads: The properties of the *Pseudomonas stutzeri* group. *J. Gen. Microbiol.* **60**, 219–231.

Palleroni, N. J., R. W. Ballard, E. Ralston, and M. Doudoroff (1973). Nucleic acid homologies among some *Pseudomonas* species. *Int. J. Syst. Bacteriol.* **23**, 333–339.

Palmquist, A. F., and D. Jarkow (1973). Evaluation of *Pseudomonas* and *Staphylococcus aureus* as indicators of bacterial quality of swimming pools. *J. Environ. Health* **36**, 230–232.

Perry, L. B. (1973). Gliding motility in some non-spreading flexibacteria. *J. Appl. Bacteriol.* **36**, 227–232.

Pickett, M. J., and M. M. Pedersen (1970a). Characterization of saccharolytic nonfermentative bacteria associated with man. *Can. J. Microbiol.* **16**, 351–362.

Pickett, M. J., and M. M. Pedersen (1970b). Salient features of nonsaccharolytic and weakly saccharolytic nonfermentative rods. *Can. J. Microbiol.* **16**, 401–409.

Reasoner, D. J., and E. E. Geldreich (1979a). Significance of pigmented bacteria in water supplies. *Proc. 7th Annu. AWWA Water Quality Technol. Conf.* **7**, 187–196.

Reasoner, D. J., and E. E. Geldreich (1979b). A new medium for the enumeration and subculture of bacteria from potable water. *Annu. Meet. Am. Soc. Microbiol.*, Los Angeles Abstr. N7.

Reasoner, D. J., and E. E. Geldreich (1985). A new medium for the enumeration and subculture of bacteria from potable water. *Appl. Environ. Microbiol.* **49**, 1–7.

Reichenbach, H., and M. Dworkin (1981). The order Cytophagales (with addenda on the genra *Herpetosiphon, Saprospira,* and *Flexithrix*). In "The Procaryotes: A Handbook on Habitats, Isolation, and Identification of Bacteria" (M. P. Starr, H. Stolp, H. G. Trüper, A. Balows, and H. G. Schlegel, eds.), pp. 356–379. Springer-Verlag, New York.

Reichenbach, H., W. Kohl, A. Bottger-Vetter, and H. Achenbach (1980). Flexirubin-type pigments in *Flavobacterium. Arch. Microbiol.* **126**, 291–293.

Reitler, R., and R. Seligmann (1957). *Pseudomonas aeruginosa* in drinking water. *J. Appl. Bacteriol.* **20**, 145–150.

Rubin, S. J., P. A. Granato, and B. L. Wasilauskas (1980). Glucose-nonfermenting gram-negative bacteria. *In* "Manual of Clinical Microbiology" (E. H. Lennette, A. Balows, W. J. Hausler, and J. P. Truant, eds.), 3rd Ed. ASM, Washington, D.C.

Rüger, H. J., and T. L. Tar (1983). Separation of *Alcaligenes denitrificans* sp. nov., nom. rev. from *Alcaligenes faecalis* on the basis of DNA base composition, DNA homology, and nitrate reduction. *Int. J. Syst. Bacteriol.* 33, 85–89.

Sausker, W. F., J. L. Aeling, J. E. Fitzpatrick, and F. N. Judson (1978). *Pseudomonas* folliculitis acquired from a health spa whirlpool. *J. Am. Med. Assoc.* 239, 2362–2365.

Schroth, M. N., D. C., Hilderbrand, and M. P. Starr (1981). Phytopathogenic members of the genus *Pseudomonas*. *In* "The Procaryotes: A Handbook on Habitats, Isolation, and Identification of Bacteria" (M. P. Starr, H. Stolp, H. G. Trüper, A. Balows, and H. G. Schlegel, eds.), pp. 701–718. Springer-Verlag, New York.

Seyfried, P. L., and D. J. Fraser (1978). *Pseudomonas aeruginosa* in swimming pools related to the incidence of otitis externa infection. *Health Lab. Sci.* 15, 50–57.

Seyfried, P. L., and D. J. Fraser (1980). Persistence of *Pseudomonas aeruginosa* in chlorinated swimming pools. *Can. J. Microbiol.* 26, 350–355.

Shayegani, M. A., A. M. Lee, and L. M. Parsons (1977). A scheme for identification of non-fermentative bacteria. *Health Lab. Sci.* 14, 83–94.

Shayegani, M., A. M. Lee, and D. D. McGlynn (1978a). Evaluation of the Oxi/Ferm tube system for identification of nonfermentative gram-negative bacilli. *J. Clin. Microbiol.* 7, 533–538.

Shayegani, M., D. S. Maupin, and D. M. McGlynn (1978). Evaluation of the API 20E system for identification of nonfermentative gram-negative bacteria. *J. Clin. Microbiol.* 7, 539–545.

Sierra, G. (1957). A simple method for detection of lipolytic activity of micro-organisms and some observations on the influence of the contact between cells and fatty substrates. *Antonie van Leeuwenhoek J. Microbiol. Serol.* 23, 15–22.

Singer, J., and B. E., Volcani (1955). An improved ferric chloride test for differentiating *Proteus-Providence* group from Enterobacteriaceae. *J. Bacteriol.* 69, 303–306.

Skerman, V. B. D., V. McGowan, and P. H. A. Sneath (1980). Approved lists of bacterial names. *Int. J. Syst. Bacteriol.* 30, 225–420.

Smith, P. W., and R. M. Massanari (1977). Room humidifiers as the source of *Acinetobacter* infections. *JAMA* 237, 795–797.

Snell, J. J. S., and S. P. Lapage (1971). Comparison of four methods for demonstrating glucose breakdown by bacteria. *J. Gen. Microbiol.* 68, 221–225.

Stanier, R. Y. (1947). Studies on non-fruiting myxobacteria. I. *Cytophaga johnsonae* n. sp., a chitin-decomposing myxobacterium. *J. Bacteriol.* 53, 297–315.

Starr, M. (1981). The genus *Xanthomonas*. *In* "The Procaryotes: A Handbook on Habitats, Isolation and Identification of Bacteria" (M. P. Starr, H. Stolp, H. G. Trüper, A. Balows, and H. G. Schlegel, eds.), pp. 742–763. Springer-Verlag, New York.

Starr, M. P., C. L. Jenkins, L. B. Bussey, and A. G. Andrews (1977). Chemotaxonomic significance of the xanthomonadins, novel brominated arylpolyene pigments produced by bacteria of the genus *Xanthomonas*. *Arch. Microbiol.* 113, 1–9.

Starr, M. P., H. Stolp, H. G. Trüper, A. Balows, and H. G. Schlegel (1981). "The Prokaryotes: A Handbook on Habitats, Isolation, and Identification of Bacteria," pp. 1–2284. Springer-Verlag, New York.

Steel, K. J. (1961). The oxidase reagent as a taxonomic tool. *J. Gen. Microbiol.* 25, 297–306.

Stokes, F. C., and J. J. R. Campbell (1951). The oxidation of glucose and gluconic acid by dried cells of *Pseudomonas aeruginosa*. *Arch. Biochem.* 30, 121–130.

Stolp, H., and D. Gadkari (1981). Nonpathogenic members of the genus *Pseudomonas*. *In* "The

Procaryotes: A Handbook on Habitatas, Isolation, and Identification of Bacteria" (M. P. Starr, H. Stolp, H. G. Trüper, A. Balow, and H. G. Schlegel, eds.), pp. 719–741. Springer-Verlag, New York.

Tatum, H. W., W. H. Ewing, and R. E. Weaver (1974). Miscellaneous Gram-negative bacteria. *In* "Manual of Clinical Microbiology," (E. H. Lennette, E. H. Spauldig, and J. P. Truant, eds.), 2nd Ed., pp. 270–294. American Society for Microbiology, Washington, D.C.

Taylor, R. H., and E. E. Geldreich (1979). A new membrane filter procedure for bacterial counts in potable water and swimming pool samples. *J. Am. Water Works Assoc.* **71**, 402–405.

Vogt, R., D. LaRue, M. F. Parry, C. D. Brokopp, D. Klaucke, and J. Allen (1982). *Pseudomonas aeruginosa* skin infections in persons using a whirlpool in Vermont. *J. Clin. Microbiol.* **15**, 571–574.

Warwood, N. M., D. J. Blazevic, and L. Hofherr (1979). Comparison of the API 20E and Corning N/F systems for identification of nonfermentative Gram-negative rods. *J. Clin. Microbiol.* **10**, 175–179.

Washburn, J., J. A. Jacobson, E. Marston, and B. Thorsen (1976). *Pseudomonas aeruginosa* rash associated with a whirlpool. *J. Am. Med. Assoc.* **235**, 2205–2207.

Weeks, O. B. (1969). Problems concerning the relationships of cytophages and flavobacteria. *J. Appl. Bacteriol.* **32**, 13–18.

Weeks, O. B. (1974). Genus *Flavobacterium* Bergey *et al.*, 1923. *In* "Bergey's Manual of Determinative Bacteriology" (R. E. Buchanan and N. E. Gibbons, eds.), 8th Ed., pp. 357–364. Williams & Wilkins, Baltimore.

Weeks, O. B. (1981). The genus *Flavoacterium. In* "The Procaryotes: A Handbook on Habitats, Isolation, and Identification of Bacteria" (M. P. Starr, H. Stolp, H. G. Trüper, A. Balow, and H. G. Schlegal, eds.), pp. 1365–1370. Springer-Verlag, New York.

Wellstood-Nuesse, S. (1979). Comparison of the minitek system with conventional methods for identification of nonfermentative and oxidase-positive fermentative gram negative bacilli. *J. Clin. Microbiol.* **9**, 511–516.

Yabuuchi, E., I. Yano, S. Goto, E. Tanimura, T. Ito, and A. Ohyana (1974). Description of *Achromobacter xylosoxidans* Yabuuchi and Ohyana 1971. *Int. J. Syst. Bacteriol.* **24**, 470–477.

Yamasato, K., M. Akagawa, N. Oishi, and H. Kuraishi (1982). *J. Gen. Appl. Microbiol.* **28**, 195–213.

INDEX